STOCHASTIC PROCESSES

Basic Theory and Its Applications

STOCHASTIC PROCESSES

Basic Theory and Its Applications

Narahari U Prabhu

Cornell University, USA

NEW JERSEY • LONDON • SINGAPORE • BEIJING • SHANGHAI • HONG KONG • TAIPEI • CHENNAI

Published by

World Scientific Publishing Co. Pte. Ltd.

5 Toh Tuck Link, Singapore 596224

USA office: 27 Warren Street, Suite 401-402, Hackensack, NJ 07601

UK office: 57 Shelton Street, Covent Garden, London WC2H 9HE

British Library Cataloguing-in-Publication Data
A catalogue record for this book is available from the British Library.

STOCHASTIC PROCESSES
Basic Theory and Its Applications

ISBN-13 978-981-270-626-3
ISBN-10 981-270-626-7

Typeset by Stallion Press
Email: enquiries@stallionpress.com

Printed in Singapore by World Scientific Printers (S) Pte Ltd

To my friend Joseph M. Gani
You were good to me, Joe!

Contents

PREFACE

This book is based on a beginning graduate level course I have taught at Cornell University since 1965. Research developments over the last several years made it necessary to include in the course topics such as stopping times, strong Markov property, martingales in discrete time, regenerative phenomena, and Tauberian theorems. These topics are not only important theoretically, but also provide essential tools to deal with applications. Presentation of this material includes a certain amount of measure theory, but only at an elementary level, not likely to intimidate the serious student.

Also, the presentation of older topics such as stationary processes, Markov chains, Brownian Motion, Poisson process, branching processes, renewal theory, and random walks is updated, using insights gained from research.

Apart from the Brownian Motion and Poisson process, other Lévy processes that occur in applications are also treated. These include the Gaussian, inverse Gaussian, and the randomized Bernoulli random walk. The characteristics of a general Lévy process are also described.

For each class of stochastic processes treated here I have only attempted to describe the conceptual background and derive the basic results. Also, my treatment of applications is for the purpose of illustrating the theoretical results. Several monographs are now available dealing with special classes of stochastic processes and their important applications.

N. U. Prabhu
Ithaca, New York
January, 2007

ABBREVIATIONS AND NOTATIONS

Abbreviations

Term	Abbreviation
Characteristic function	c.f.
Distribution function	d.f.
If and only if	iff
Laplace transform	L.T
Probability generating function	p.g.f

The notation F is used both for a d.f. and a distribution measure, leading to the use of $F(x)$ and $F\{dx\}$, respectively.

Notation

1. The normal density and its d.f. are

$$n(x;\mu,\sigma^2) = \frac{1}{\sqrt{2\pi}\sigma} e^{-\frac{1}{2\sigma^2}(x-\mu)^2}$$
$$N(x;\mu,\sigma^2) = \int_{-\infty}^{x} n(y;\mu,\sigma^2)dy \quad (-\infty < \alpha < \infty).$$

For convenience write

$$n(x) = n(x;0,1), \quad N(x) = N(x;0,1).$$

2. One-sided normal d.f.s

(i)

$$N_+(x) = 0 \quad \text{for } x < 0, \text{ and } = 2N(x) - 1 \quad \text{for } x \geq 0.$$

This distribution has density $n+(x)$, where

$$n+(x) = 0 \quad \text{for } x < 0, \text{ and } = 2n(x) \quad \text{for } x \geq 0.$$

Its mean and variance are, respectively, $\sqrt{2/\pi}$ and $1 - 2/\pi$.

(ii)

$$N_-(x) = 2N(x) \quad \text{for } x < 0, \text{ and} = 1 \quad \text{for } x \geq 0.$$

This density in this case is

$$n_-(x) = 2n(x) \quad \text{for } x < 0, \text{ and} = 0 \quad \text{for } x \geq 0.$$

3. The stable d.f. with exponent 1/2:

$$G_{1/2}(x) = 0 \quad \text{for } x < 0, \text{ and} = 2\left[1 - N\left(\frac{1}{\sqrt{x}}\right)\right] \quad \text{for } x \geq 0.$$

The corresponding density is

$$g_{1/2}(x) = 0 \quad \text{for } x < 0, \text{ and} = \frac{1}{\sqrt{2\pi x^3}} e^{-1/2x} \quad \text{for } x \geq 0.$$

The Laplace transform of this distribution is given by $e^{-\sqrt{2\theta}}$ $(\theta > 0)$. Its mean is $+\infty$.

CHAPTER 1

A Review of Probability Distributions and Their Properties

1.1 Introduction

In this chapter we present a review of some standard continuous probability distributions. We derive the moments (when they exist) of these distributions, but the emphasis is on structural properties of the distributions such as convolutions and relationships with each other. For many of these distributions there are discrete counterparts, but these are not discussed here. We include infinitely divisible distributions (and the subclass of stable distributions) which have become increasingly important in applications.

1.2 The Exponential Density

This is defined by the function

$$\begin{aligned} f(x) &= 0 && \text{for } x < 0 \\ &= \lambda e^{-\lambda x} && \text{for } x \geq 0, \end{aligned} \tag{2.1}$$

where $0 < \lambda < \infty$. The corresponding d.f. $F(x)$ is given by

$$\begin{aligned} F(x) &= \int_{-\infty}^{x} f(y)dy = 0 && \text{for } x < 0 \\ &= \int_{0}^{x} \lambda e^{-\lambda y} dy = 1 - e^{-\lambda x} && \text{for } x \geq 0. \end{aligned} \tag{2.2}$$

We have

$$E(X) = \int_0^\infty \lambda e^{-\lambda x} x dx = 1/\lambda, \quad E(X^2) = \int_0^\infty \lambda e^{-\lambda x} x^2 dx = 2/\lambda^2,$$

which give

$$\operatorname{Var}(X) = 2/\lambda^2 - (1/\lambda)^2 = 1/\lambda^2. \tag{2.3}$$

The lack of memory property. Suppose that a random variable X has the exponential density (2.2) and let $a > 0$. We then have

$$\begin{aligned}\Pr\{X - a \le x | X > a\} &= \frac{\Pr\{a < X \le a + x\}}{\Pr\{X > a\}} = \frac{F(a+x) - F(a)}{1 - F(a)} \\ &= 1 - e^{-\lambda x} = F(x).\end{aligned} \tag{2.4}$$

Thus the "residual" random variable $X - a$ has the same distribution as X. For a physical interpretation, let X denote the lifetime of an article (such as electric bulbs) and a its present age; then the residual lifetime of the bulb will have the same distribution as that of a new bulb. This property is usually called the lack of memory (or Markov) property of the exponential density.

1.3 The Gamma Density

This is defined by

$$\begin{aligned}g(x; \lambda, \alpha) &= 0 && \text{for } x < 0 \\ &= e^{-\lambda x} \lambda^\alpha \frac{x^{\alpha-1}}{\Gamma(\alpha)} && \text{for } x \ge 0\end{aligned} \tag{3.1}$$

where $0 < \lambda < \infty$, $0 < \alpha < \infty$. For $\alpha = 1$ this reduces to the exponential density. We have

$$E(X^k) = \int_0^\infty e^{-\lambda x} \lambda^\alpha \frac{x^{\alpha+k-1}}{\Gamma(\alpha)} dx = \frac{\lambda^\alpha}{\Gamma(\alpha)} \frac{\Gamma(\alpha+k)}{\lambda^{\alpha+k}} = \frac{1}{\lambda^k} \frac{\Gamma(\alpha+k)}{\Gamma(\alpha)}$$

and therefore

$$E(X) = \frac{\alpha}{\lambda}, \quad \operatorname{Var}(X) = \frac{\alpha}{\lambda^2}. \tag{3.2}$$

Suppose that X, Y are independent random variables with densities $g(x; \lambda, \alpha)$ and $g(y; \lambda, \beta)$, respectively. Then the density $h(z)$ of their sum $X + Y$ is given by the convolution formula

$$\begin{aligned}h(z) &= \int_0^z g(x; \lambda, \alpha) g(z - x; \lambda, \beta) dx \\ &= \int_0^z e^{-\lambda x} \lambda^\alpha \frac{x^{\alpha-1}}{\Gamma(\alpha)} e^{-\lambda(z-x)} \lambda^\beta \frac{(z-x)^{\beta-1}}{\Gamma(\beta)} dx\end{aligned}$$

$$= e^{-\lambda z}\frac{\lambda^{\alpha+\beta}}{\Gamma(\alpha)\Gamma(\beta)}\int_0^z x^{\alpha-1}(z-x)^{\beta-1}dx$$

$$= e^{-\lambda z}\lambda^{\alpha+\beta}\frac{z^{\alpha+\beta-1}}{\Gamma(\alpha+\beta)} = g(z;\lambda,\alpha+\beta).$$

Thus $X+Y$ has the gamma density with the parameters $\lambda, \alpha+\beta$. We write

$$g(x;\lambda,\propto) * g(x;\lambda,\beta) = g(x;\lambda,\alpha+\beta). \tag{3.3}$$

This is also called the *additive property* of the gamma density. In particular we find that if the random variables $X_1, X_2, \ldots, X_n$ are independent and have the same exponential density $\lambda e^{-\lambda x}$, then their sum $X_1+X_2+\cdots+X_n$ has the gamma density $g(x;\lambda,n)$. For a fixed λ we shall write this density as $g_n(x)$ and the corresponding d.f. as $G_n(x)$. Thus

$$G_n(x) = \int_0^x e^{-\lambda y}\lambda^n \frac{y^{n-1}}{(n-1)!}dy \tag{3.4}$$

for $x \geq 0$ and $G_n(x) = 0$ for $x < 0$. An integration by parts yields the result

$$G_n(x) = G_{n-1}(x) - e^{-\lambda x}\frac{(\lambda x)^{n-1}}{(n-1)!}(n \geq 2). \tag{3.5}$$

For $n = 1$, we have $G_1(x) = 1 - e^{-\lambda x}$ for $x \geq 0$. Putting $n = 2$ in (3.5) we find that

$$G_2(x)1 - e^{-\lambda x} - e^{-\lambda x}\frac{\lambda x}{1!},$$

and by induction

$$G_n(x) = 1 - \sum_{r=0}^{n-1} e^{-\lambda x}\frac{(\lambda x)^r}{r!}(x \geq 0). \tag{3.6}$$

1.4 The Beta Density

This is defined by the function

$$\begin{aligned} b(x;\alpha,\beta) &= \frac{1}{B(\alpha,\beta)}x^{\alpha-1}(1-x)^{\beta-1} \quad \text{for } 0 < x < 1 \\ &= 0 \quad \text{elsewhere,} \end{aligned} \tag{4.1}$$

where $\alpha > 0, \beta > 0$. For a positive integer r we have

$$\begin{aligned} E(X^r) &= \int_0^1 x^r b(x;\alpha,\beta)dx = \frac{1}{B(\alpha,\beta)}\int_0^1 x^{\alpha+r-1}(1-x)^{\beta-1}dx \\ &= \frac{B(\alpha+r,\beta)}{B(\alpha,\beta)} = \frac{\Gamma(\alpha+r)}{\Gamma(\alpha)}\frac{\Gamma(\alpha+\beta)}{\Gamma(\alpha+\beta+r)}. \end{aligned}$$

This gives

$$E(X) = \frac{\alpha}{\alpha+\beta}, \quad E(X^2) = \frac{\alpha(\alpha+1)}{(\alpha+\beta)(\alpha+\beta+1)}, \tag{4.2}$$

so that

$$\mathrm{Var}(X) = \frac{\alpha\beta}{(\alpha+\beta)^2(\alpha+\beta+1)}. \tag{4.3}$$

The beta density (4.1) can be derived from the gamma density (3.1) as follows. Let X, Y be independent random variables with the densities $g(X;\lambda,\alpha)$ and $g(Y;\lambda,\beta)$, respectively, and let

$$Z = \frac{X}{X+Y} \quad (0 \le Z \le 1). \tag{4.4}$$

The d.f. of Z is given by

$$\begin{aligned} F(z) &= \Pr\{Z \le z\} = \mathrm{P}\left\{X \le \frac{z}{1-z}Y\right\} \\ &= \iint_{x \le \frac{zy}{1-z}} g(x;\lambda,\alpha)g(y;\lambda,\beta)dxdy \\ &= \int_0^\infty g(y;\lambda,\beta)G\left(\frac{zy}{1-z};\lambda,\alpha\right)dy. \end{aligned}$$

Therefore the density of Z is given by

$$\begin{aligned} f(z) &= \frac{1}{(1-z)^2}\int_0^\infty yg(y;\lambda,\beta)g\left(\frac{zy}{1-z};\lambda,\alpha\right)dy \\ &= \frac{z^{\alpha-1}}{(1-z)^{\alpha+1}} \cdot \frac{\lambda^{\alpha+\beta}}{\Gamma(\alpha)\Gamma(\beta)}\int_0^\infty e^{-(\lambda y)/(1-z)}y^{\alpha+\beta+1}dy \\ &= \frac{1}{B(\alpha,\beta)}z^{\alpha-1}(1-z)^{\beta-1} \qquad (0 < z < 1), \end{aligned} \tag{4.5}$$

which is the beta density (4.1).

1.5 The Uniform Density

The uniform (or rectangular) density in the interval $[a, b]$ is defined as follows:

$$f(x) = \frac{1}{b-a} \quad \text{for a} \leq x \leq b$$
$$= 0 \quad \text{elsewhere.} \tag{5.1}$$

The d.f. is given by

$$F(x) = \int_{-\infty}^{x} f(u)du = 0 \quad \text{for } x \leq a$$
$$= \int_{a}^{b} \frac{du}{b-a} = 1 \quad \text{for } x \geq b$$
$$= \int_{a}^{x} \frac{du}{b-a} = \frac{x-a}{b-a} \quad \text{for } a < x < b. \tag{5.2}$$

We have

$$E(X^r) = \int_{a}^{b} \frac{x^r}{b-a} dx = \frac{1}{r+1} \cdot \frac{b^{r+1} - a^{r+1}}{b-a}$$

so that

$$E(X) = \frac{a+b}{2}, \quad \text{Var}(X) = (1/3)\frac{b^3 - a^3}{b-a} - (1/4)(a+b)^2 = \frac{(b-a)^2}{12}. \tag{5.3}$$

1.6 The Cauchy Density

This is defined by

$$c(x; a) = \frac{1}{\pi} \cdot \frac{a}{a^2 + x^2} \quad (-\infty < x < \infty), \tag{6.1}$$

where $a > 0$. The d.f. is given by

$$C(x; a) = \int_{-\infty}^{x} \frac{1}{\pi} \cdot \frac{a}{a^2 + u^2} du = \frac{1}{2} + \frac{1}{\pi} \tan^{-1}(x/a) \tag{6.2}$$

for $-\infty < x < \infty$. We have

$$\int_{-\infty}^{\infty} |x| c(x; a) dx = \frac{2}{\pi} \int_{0}^{\infty} \frac{ax}{a^2 + x^2} dx = \frac{a}{\pi} [\log(a^2 + x^2)]_0^{\infty} = \infty \tag{6.3}$$

so that the mean does not exist.

Suppose that the random variable θ has a uniform density in $\left[-\frac{1}{2}\pi, \frac{1}{2}\pi\right]$ and let $X = a\tan\theta$. The d.f. of X is given by

$$\begin{aligned} F(x) &= \Pr\{a\tan\theta \le x\} \\ &= \Pr\{\theta \le \tan^{-1}(x/a)\} = \int_{-(\pi/2)}^{\tan^{-1}(x/a)} \frac{d\theta}{\pi} \\ &= \frac{1}{2} + \frac{1}{\pi}\tan^{-1}\left(\frac{x}{a}\right). \end{aligned} \tag{6.4}$$

This proves that X has the Cauchy density (6.1).

1.7 The Normal Density in One Dimension

The normal (or Gaussian) density in one dimension is defined as follows:

$$n(x;\mu,\sigma) = \frac{1}{\sqrt{2\pi}\sigma} e^{-(1/2\sigma^2)(x-\mu)^2} \quad (-\infty < x < \infty). \tag{7.1}$$

We shall denote the corresponding d.f. by $N(x;\mu,\sigma)$. We have

$$E(X-\mu) = \int_{-\infty}^{\infty} (x-\mu)n(x;\mu,\sigma)dx = \int_{-\infty}^{\infty} t(1/\sqrt{2\pi})e^{-(1/2)t^2 dt} = 0$$

since the integrand is an odd function of t, so we obtain $E(X) = \mu$. Moreover

$$E(X-\mu)^2 = \int_{-\infty}^{\infty} (x-\mu)^2 n(x;\mu,\sigma)dx = \sigma^2 \int_{-\infty}^{\infty} \frac{t^2}{\sqrt{2\pi}} e^{-(1/2)t^2} dt = \sigma^2.$$

We thus find that the mean and the variance of the normal density (7.1) are given by

$$E(X) = \mu, \quad \operatorname{Var}(X) = \sigma^2. \tag{7.2}$$

When $\mu = 0, \sigma = 1$ we shall call (7.1) the standardized normal density $n(x) = n(x;0,1)$.

1.7.1 *Convolution Property*

Suppose X, Y are independent random variables with the normal densities $n(x;\mu_1,\sigma_1)$ and $n(y;\mu_2,\sigma_2)$. Then the sum $X+Y$ has the density function given by

$$\int_{-\infty}^{\infty} n(y;\mu_2,\sigma_2)n(z-y;\mu_1,\sigma_1)dy$$

$$= \frac{1}{2\pi\sigma_1\sigma_2}\int_{-\infty}^{\infty} e^{-(1/2\sigma_2^2)(y-\mu_2)^2-(1/2\sigma_1^2)(z-y-\mu_1)^2 dy}. \tag{7.3}$$

Now we find that the expression in the exponent is $(-1/2)$ times

$$(1/\sigma_2^2)(y-\mu_2)^2 + (1/\sigma_1^2)[(z-\mu_1-\mu_2)-(y-\mu_2)]^2$$

$$= \frac{\sigma^2}{\sigma_1^2\sigma_2^2}(y-\mu_2)^2 - \frac{2}{\sigma_1^2}(y-\mu_2)(z-\mu_1-\mu_2)^2 + \frac{1}{\sigma_1^2}(z-\mu_1-\mu_2)^2,$$

where $\sigma^2 = \sigma_1^2+\sigma_2^2$. The substitution

$$u = \frac{\sigma}{\sigma_1\sigma_2}(y-\mu_2) - \frac{\sigma_2}{\sigma\sigma_1}(z-\mu_1-\mu_2)$$

reduces the second integral in (7.3) to

$$= (1/\sqrt{2\pi}\sigma)e^{-(1/2\sigma^2)(z-\mu_1-\mu_2)^2}\int_{-\infty}^{\infty} n(u;0,1)du$$

$$= (1/\sqrt{2\pi}\sigma)e^{-(1/2\sigma^2)(z-\mu_1-\mu_2)^2},$$

which is the normal density with mean $\mu_1+\mu_2$ and variance $\sigma_1^2+\sigma_2^2$. Thus

$$n(x;\mu_1,\sigma_1) * n(x;\mu_2,\sigma_2) = n\left(x;\mu_1+\mu_2,\sqrt{\sigma_1^2+\sigma_2^2}\right). \tag{7.4}$$

1.8 The Normal Density in n Dimensions

The two-dimensional normal (Gaussian) density is

$$f(x_1,x_2) = ce^{-\frac{1}{2(1-\rho^2)}}\left[\frac{(x-\mu_1)^2}{\sigma_1^2} - 2\rho\frac{(x_1-\mu_1)(x_2-\mu_2)}{\sigma_1\sigma_2} + \frac{(x_2-\mu_2)^2}{\sigma_2^2}\right], \tag{8.1}$$

where $c = (2\pi\sigma_1\sigma_2\sqrt{1-\rho^2})^{-1}$, μ_1,μ_2 are the means, σ_1^2,σ_2^2 the variances, and ρ is the correlation coefficient.

In higher dimensions it is convenient to write the normal density in a compact form using matrix notation. Let us consider the random vector $X' = (X_1, X_2, \ldots, X_n)$ and denote by C the matrix which has $\mathrm{Var}(X_1), \mathrm{Var}(X_2), \ldots, \mathrm{Var}(X_n)$ for its diagonal elements and Covar

(X_j, X_k) for the element in the jth row and kth column $(j \neq k)$. Then C is the covariance matrix of the distribution of $(X_1, X_2, \ldots, X_n)$. Assuming C to be nonsingular, denote $C^{-1} = A = (a_{jk})$. Also, let $E(X_j) = \mu_j$ $(j = 1, 2, \ldots, n)$ and $\mu' = (\mu_1, \mu_2, \ldots, \mu_n)$. Then the n-dimensional normal density is defined as

$$f(x_1, x_2, \ldots, x_n) = \frac{|A|^{1/2}}{(2\pi)^{n/2}} e^{-\frac{1}{2}\sum_{j=1}^{n}\sum_{k=1}^{n} a_{jk}(x_j-\mu_j)(x_k-\mu_k)}, \tag{8.2}$$

where the quadratic form in the exponent can be written as $(X - \mu)'$ $A(X - \mu)$. It should be noted that this density is uniquely determined by the mean vector μ and the covariance matrix C. For $n = 2$ we have

$$C = \begin{bmatrix} \sigma_1^2 & \rho\sigma_1\sigma_2 \\ \rho\sigma_2\sigma_1 & \sigma_2^2 \end{bmatrix}$$

and

$$A = C^{-1} = \begin{bmatrix} \frac{1}{\sigma_1^2(1-\rho^2)} & \frac{-\rho}{\sigma_1\sigma_2(1-\rho^2)} \\ \frac{-\rho}{\sigma_2\sigma_1(1-\rho^2)} & \frac{1}{\sigma_2^2(1-\rho^2)} \end{bmatrix}.$$

Thus (8.2) reduces to (8.1).

We summarize below some of the important properties of density (8.2):

(a) If the components $X_1, X_2, \ldots, X_n$ are uncorrelated, then C and its inverse A are both diagonal matrices and (8.2) reduces to

$$\prod_{i=1}^{n}\left(\frac{1}{\sqrt{2\pi}\sigma i}\right)e^{-\frac{1}{2\sigma_i^2}(x_i-\mu_i)^2}, \tag{8.3}$$

which shows that $X_1, X_2, \ldots, X_n$ are independent. More generally, if $\mathrm{Covar}(X_j, X_k) = 0$ for $1 \leq j \leq r, r < k \leq n$, then the vectors $(X_1, X_2, \ldots, X_r)$ and $(X_{r+1}, X_{r+2}, \ldots, X_n)$ are independent.

(b) All marginal densities are normal.

(c) The conditional density of X_n, given $X_1 = x_1, X_2 = x_2, \ldots, X_{n-1} = x_{n-1}$, is normal with mean

$$-\frac{a_{1n}}{a_{nn}}x_1 - \frac{a_{2n}}{a_{nn}}x_2 - \cdots - \frac{a_{n-1,n}}{a_{nn}}x_{n-1}$$

and variance $(a_{nn})^{-1}$. These results mean that the regression is linear and the residual variance is a constant.

(d) If $Y = LX$ is a nonsingular linear transformation (that is, with $|L| \neq 0$), then the vector $(Y_1, Y_2, \ldots, Y_n)$ has a normal density with mean vector L_μ and covariance matrix LCL'. In particular, there is always a linear transformation for which $Y_1, Y_2, \ldots, Y_n$ are mutually uncorrelated (and therefore independent by (a) above).

1.9 Infinitely Divisible Distributions

Definition 9.1 A random variable X (or equivalently, its distribution) is infinitely divisible if for each $n \geq 1$ it has the representation

$$X \stackrel{d}{=} X_{1n} + X_{2n} + \cdots + X_{nn}, \tag{9.1}$$

where $X_{kn} (k = 1, 2, \ldots, n)$ are IID random variables.

1.9.1 *Examples*

(9.1) If $X = c$ (a constant), then representation (9.1) holds with the distribution of X_{kn} concentrated at c/n. Therefore a degenerate distribution is infinitely divisible.

(9.2) If X has a normal density with mean μ and variance σ^2, then representation (9.1) holds with X_{kn} having a normal density with mean μ/n and variance σ^2/n. Thus the normal distribution is infinitely divisible.

(9.3) If X has the Cauchy density with scale parameter a, namely,

$$f(x) = \frac{a}{\pi} \cdot \frac{1}{a^2 + x^2} \quad (a > 0),$$

then representation (9.1) holds with X_{kn} having a Cauchy density with parameter a/n. Thus the Cauchy distribution is infinitely divisible.

(9.4) Suppose X has the gamma density with parameters (λ, α), namely,

$$f(x) = e^{-\lambda x} \lambda^\alpha \frac{x^{\alpha-1}}{\Gamma(\alpha)} \quad (\lambda > 0, \alpha > 0).$$

Then representation (9.1) holds with X_{kn} having a gamma density with parameters $(\lambda, \alpha/n)$. Thus the gamma distribution is infinitely divisible. In particular, the exponential distribution is infinitely divisible. Their discrete counterparts, the negative binomial and the geometric distributions, are also infinitely divisible.

(9.5) A simple Poisson random variable with parameter λ is infinitely divisible, since for each $n \geq 1$ it can be represented as the sum of n independent Poisson random variables with parameter λ/n. It follows that a compound Poisson random variable X is also infinitely divisible, since $X = Y_1 + Y_2 + \cdots + Y_N$, where the Y_k are IID and N is a simple Poisson random variable with parameter λ, that is independent of the Y_k.

(9.6) If $|X| < a$ with probability 1, then for each $n \geq 1$ we must have $|X_{1n} + X_{2n} + \cdots + X_{nn}| < a$ with probability 1. This implies that for $k = 1, 2, \ldots, n$, $|X_{kn}| < an^{-1}$ with probability 1, so $\text{Var}(X_{kn}) < a^2 n^{-2}$. Therefore

$$\text{Var}(X) \leq n \cdot a^2 n^{-2} = a^2 n^{-1} \to 0 \quad \text{as } n \to \infty,$$

which means that X is a constant. Thus a distribution concentrated on a finite interval is not infinitely divisible unless it is a constant (see Example 9.1 above). In particular, the binomial and the uniform distributions are not infinitely divisible.

To represent the family of infinitely divisible distributions F we use their characteristic function (c.f.)

$$\phi(\omega) = E(e^{i\omega X}) = \int_{-\infty}^{\infty} e^{i\omega x} F\{dx\} \quad (\omega, \text{ real}, \ i = \sqrt{-1}). \tag{9.2}$$

It is known that ϕ is of the form $\phi = e^{\psi}$, where

$$\psi(\omega) = i\omega b + \int_{-\infty}^{\infty} \frac{e^{i\omega x} - 1 - i\omega \sin x}{x^2} M\{dx\}, \tag{9.3}$$

b being a real constant and M a measure such that $M\{I\} < \infty$ for every bounded interval I, and

$$M^+(x) = \int_{x-}^{\infty} \frac{1}{y^2} M\{dy\}, \quad M^-(-x) = \int_{-\infty}^{-x} \frac{1}{y^2} M\{dy\},$$

$M^+(x) < \infty$, $M^-(-x) < \infty$ for $\alpha > 0$. Special choices of this measure lead to the distributions described above. Two particular cases are the following:

(i) If M is concentrated at the origin with weight $\sigma^2 > 0$, then $\phi(\omega) = e^{i\omega b - \frac{1}{2}\sigma^2\omega^2}$, which is the c.f. of a normal distribution.

(ii) Suppose M is concentrated on $(0, \infty)$ with density $\alpha e^{-\lambda x} x (x > 1)$. It is easily verified that M satisfies the above conditions. We have

$$\int_0^\infty \frac{e^{i\omega x} - 1}{x^2} M\{dx\} = \log \left(1 - \frac{i\omega}{\lambda}\right)^{-\alpha}.$$

Choosing

$$b = \alpha \int_0^\infty \frac{\sin x}{x} e^{-\lambda x} dx < \infty$$

we find that

$$\phi(\omega) = \left(1 - \frac{i\omega}{\lambda}\right)^{-\alpha}.$$

This is the c.f. of the gamma density of $e^{-\lambda x} \lambda^\alpha x^{\alpha - 1} / \Gamma(\alpha)$.

Representation (9.2) is due to W. Feller. The measure M can be expressed in the classical form due to P. Lévy (see Chap. 5).

1.10 Stable Distributions

Definition 10.1 A random variable X (or, equivalently, its distribution is called stable if for each $n \geq 1$, there exist real numbers $c_n > 0, d_n$ such that

$$X_1 + X_2 + \cdots + X_n \stackrel{d}{=} c_n X + d_n, \tag{10.1}$$

where $X_1, X_2, \ldots, X_n$ are IID random variables with the same distribution as X.

1.10.1 *Examples*

(10.1) If X has a distribution concentrated at a single point, then representation (10.1) is satisfied with $c_n = n, d_n = 0$. Thus a degenerate distribution is (trivially) stable. We shall ignore this from our consideration.

(10.2) If X has a normal density with mean M and variance σ^2, then since

$$\frac{X_1 + X_2 + \cdots + X_n - nm}{\sigma\sqrt{n}} \stackrel{d}{=} \frac{X - m}{\sigma}$$

representation (10.1) holds with $c_n = \sqrt{n}$ and $d_n = m(n - \sqrt{n})$. Accordingly, the normal density is stable.

(10.3) If X has a Cauchy density, then since

$$\frac{X_1 + X_2 + \cdots + X_n}{n} \stackrel{d}{=} X$$

representation (10.1) holds with $c_n = n, d_n = 0$, and X is therefore stable.

Stable distributions are infinitely divisible since representation (9.1) holds with

$$X_{kn} = \left(X_k - \frac{d_n}{n}\right)c_n^{-1} \quad (k = 1, 2, \ldots, n).$$

The c.f. of a stable distribution is of the form $\phi = e^{\psi}$, where

$$\psi(\omega) = \lambda w\gamma - e|\omega|^{\alpha}[1 + i\beta\frac{\omega}{|\omega|}\Omega(|\omega|, \alpha)], \tag{10.2}$$

where $c > 0, |\beta| \leq 1, 0 < \alpha \leq 2$, and

$$\begin{aligned} \Omega(|\omega|, \alpha) &= \tan\frac{\pi\alpha}{2} \quad \text{if } \alpha \neq 1 \\ &= \frac{2}{\pi}\log|\omega| \quad \text{if } \alpha = 1. \end{aligned} \tag{10.3}$$

It is easily seen that the case $\alpha = 2$ corresponds to the normal density and the case $\beta = 0, \alpha = 1$ corresponds to the Cauchy density.

1.11 Problems for Solution

1. *Laplace's second law.* Suppose that the random variables X_1, X_2 are independent and have the same density $\lambda e^{-\lambda x}$. Let $Y = X_1 - X_2$. Prove the following:

   (a) Y has the density $(\lambda/2)e^{-\lambda|x|}(-\infty < x < \infty)$.
   (b) $E(Y^r) = 0$ for odd r, while $E(Y^r) = r!\ \lambda^{-r}$ for even r.

2. *The arc sine law* arises as a special case of the beta density $b(x; \alpha, \beta)$ with $\alpha = \beta = 1/2$. Prove the following:

   (i) The distribution function is given by $F(x) = 0$ for $x < 0, = 2\pi^{-1}$; arc sin $\sqrt{x}$ for $0 \leq x \leq 1$; and =1 for $x > 1$.
   (ii) The density is a minimum at $x = 1/2$, which is the mean of the distribution.

3. *A variant of the beta density.* Let X be a random variable with the beta density $b(x;\alpha,\beta)$ and $Y = X^{-1} - 1$. Show that Y has the density

$$\frac{1}{B(\alpha,\beta)}\frac{x^{\beta-1}}{(1+x)^{\alpha+\beta}}(x \geq 0).$$

4. The random variable X_1, X_2 are independent and have gamma densities $g(x;\lambda,\alpha), g(x;\lambda,\beta)$, respectively. Show that the ratio X_2/X_1 has the density given in Problem 3.
5. *The Polya distribution.* This is defined as the discrete distribution $\{p_k\}$, where

$$p_k = \frac{\binom{-\alpha}{k}\binom{-\beta}{n-k}}{\binom{-\alpha-\beta}{n}} \quad (k = 0, 1, \ldots, n),$$

with $\alpha > 0$, $\beta > 0$, and n is a positive integer.

(i) Show that p_k can be expressed as

$$p_k = \int_0^1 \binom{n}{k} p^k(1-p)^{n-k} b(p;\alpha,\beta)dp,$$

where $b(p;\alpha,\beta)$ is the beta density. This shows that $\{p_k\}$ is the mixture of the binomial with the beta distribution.

(ii) If the random variable X_n has the Polya distribution, show that as $n \to \infty, X_n/n$ has the limit density $b(x;\alpha\beta)$.

6. *A density related to the normal.* Let $n(x) = n(x;0,1)$ be the standard normal density and $N(x)$ the corresponding distribution function. For the density defined by

$$\begin{aligned} n_+(x) &= 2n(x) \quad \text{for } x \geq 0 \\ &= 0 \quad\quad\quad \text{for } x < 0, \end{aligned}$$

prove the following:

(a) The distribution function corresponding to $n_+(x)$ is given by

$$\begin{aligned} N_+(x) &= 2N(x) - 1 \quad \text{for } x \geq 0 \\ &= 0 \quad\quad\quad\quad\quad \text{for } x < 0. \end{aligned}$$

(b) The mean and the variance are, respectively, given by

$$\sqrt{\frac{2}{\pi}} \text{ and } 1 - \frac{2}{\pi}.$$

7. *Stable density with exponent* 1/2. Suppose that the random variable X has a standard normal density.

   (i) Show that $a^2 x^{-2}$ has the density

$$g_{1/2}(x;a) = \frac{a}{\sqrt{2\pi x^3}} e^{-a^2/2x} \quad \text{for } x > 0$$
$$= 0 \quad \text{for } x \leq 0.$$

   (ii) Show that the corresponding distribution function is given by

$$G_{1/2}(x;a) = 0 \quad \text{for } x \leq 0$$
$$= 2\left[1 - N\left(\frac{a}{\sqrt{x}}\right)\right] \quad \text{for } x \geq 0.$$

   (iii) For this density show that the moment of order β exists if and only if $\beta < 1/2$.

   (iv) Establish the convolution property

$$g_{1/2}(x;a) * g_{1/2}(x;b) = g_{1/2}(x;a+b).$$

8. Characterize the class of densities such that if the random variable X has this density, so does X^{-1}. Show that the Cauchy density belongs to this class.
9. The random variables X_1, X_2 have the joint density of the form $g(x_1^2 + x_2^2)$ for all x_1, x_2. Show that their ratio has a Cauchy density.
10. For $t > 0$ denote by $c(x;t)$ the Cauchy density

$$c(x;t) = \frac{1}{\pi} \cdot \frac{t}{t^2 + x^2} \quad (-\infty < x < \infty).$$

   (i) Prove the convolution property

$$c(x;t) * c(x;s) = c(x;t+s).$$

   (ii) Let U, V be two independent random variables with the Cauchy density $c(x;t)$, and let $X_1 = \alpha U + \beta V, X_2 = \gamma U + \delta V$ $(\alpha, \beta, \gamma, \delta > 0)$. Show that the density of $X_1 + X_2$ is the convolution of the densities of X_1 and X_2, where X_1, X_2 are clearly dependent.

11. Let $X_1, X_2, \ldots, X_n$ be mutually independent random variables with a uniform distribution in $(0, a)$, and $U_n(x)$ be the distribution function

of their sum $X_1 + X_2 + \cdots + X_n$. Show that

$$U_{n+1}(x) = \frac{1}{a} \int_{x-a}^{x} U_n(y)dy \quad (n \geq 1).$$

Verify that $U_1(x) = a^{-1}[x_+ - (x-a)_+]$ and by induction prove that

$$U_n(x) = \frac{1}{a^n n!} \sum_{r=0}^{n} (-1)^n \binom{n}{r} (x - ra)_+^n.$$

(Here $x_+ = \max(0, x)$ for any real number x.)

12. Let $(X_1, X_2, \ldots, X_n)$ be a random sample with replacement from the uniform distribution on $(1, 2, \ldots, N)$. If X is the largest number appearing in the sample, find the mean and the variance of X. Show that for large N

$$E(X) \sim \frac{nN}{n+1}, \quad \text{Var}(X) \sim \frac{nN^2}{(n+1)^2(n+2)}.$$

13. Let $(X_1, X_2, \ldots, X_n)$ be a random sample from the uniform distribution over (0,1), and $R_n = \max(X_1, X_2, \ldots, X_n) - \min(X_1, X_2, \ldots, X_n)$. Show that

$$E(R_n) = \frac{n-1}{n+1}, \quad \text{Var}(R_n) = \frac{2(n-1)}{(n+1)^2(n+2)}.$$

14. Let $X_1, X_2, \ldots, X_n$ be mutually independent random variables with distribution F. Let these be re-ordered in increasing order of magnitude to obtain $X_{(1)}, X_{(2)}, \ldots, X_{(n)}$, where $X_{(1)} \leq X_{(2)} \leq \cdots \leq X_{(n)}$. Show that

$$P\{X_{(k)} \leq x\} = n \binom{n-1}{k-1} \int_0^{F(x)} y^{k-1}(1-y)^{n-k} dy.$$

Show also that for fixed k, the distribution of $nX_{(k)}$ converges to G_k as $n \to \infty$ in the two cases where (i) $F(x) = 1 - e^{-\lambda x}$, and (ii) $F(x) = x$. Here G_k is the gamma distribution with density $e^{-\lambda x} \lambda^k x^{k-1}/(k-1)!$ with an appropriate parameter λ.

15. Let the notation be the same as in Problem 14 and $F(x) = 1 - e^{-\lambda x}$. Let $Y_1 = X_{(1)}, Y_2 = X_{(2)} - X_{(1)}, \ldots, Y_n = X_{(n)} - X_{(n-1)}$.

    (a) Show that the random variables $Y_1, Y_2, \ldots, Y_n$ are independent with Y_k having density $\lambda(n-k+1)e^{-\lambda(n-k+1)x} (k = 1, 2, \ldots, n)$.

(b) For $X_{(n)} = \max(X_1, X_2, \ldots, X_n)$ prove that

$$EX_{(n)} \sim \frac{1}{\lambda} \log n, \quad \mathrm{Var} X_{(n)} \sim \frac{1}{\lambda^2} \cdot \frac{\pi^2}{6}$$

and

$$P\{\lambda X_{(n)} - \log n \le x\} \to e^{-e^{-x}} \quad \text{as } n \to \infty.$$

16. With notation the same as in Problem 14, let F be a uniform distribution over $(0, t)$ and $Y_1 = X_{(1)}, Y_2 = X_{(2)} - X_{(1)}, \ldots, Y_n = X_{(n)} - X_{(n-1)}, Y_{n+1} = t - X_{(n)}$. Prove the following:

(a) $P\{Y_1 > y_1, Y_2 > y_2, \ldots, Y_{n+1} > y_{n+1}\} = t^{-n}(t - y_1 - y_2 - \cdots - y_{n+1})_+^n$.

(b) $P\{\max(Y_1, Y_2, \ldots, Y_{n+1}) \le y\} = \sum_{r=0}^{n+1} \binom{n+1}{r} (1 - r\frac{y}{t})_+^n$.

(Here $x_+ = \max(0, x)$ for any real number x.)

17. *Extremal distributions.* Let $X_1, X_2, \ldots, X_n$ be mutually independent random variables with distribution F, and $X_{(n)} = \max(X_1, X_2, \ldots, X_n)$. Prove the following:

(a) If F has the Cauchy density $\pi^{-1}(1 + x^2)^{-1}$, then

$$P\left\{\frac{\pi X_{(n)}}{n} \le x\right\} \to e^{-x^{-1}} \quad \text{as } n \to \infty.$$

(b) If F has the stable density $(2\pi x^3)^{-1/2}\, e^{-1/2x}$, then

$$P\left\{\frac{\pi X_{(n)}}{2n^2} \le x\right\} \to e^{-x^{-1/2}} \quad \text{as } n \to \infty.$$

[Hint: In each case prove that $1 - F(x) = bx^{-a} + o(x^{-a})$ as $x \to \infty$ with $b > 0, a > 0$.]

18. If F and G are infinitely divisible distributions, show that so is their convolution $F * G$.

19. Let $X = \sum_{k=1}^{\infty} X_k/k$, where the random variables X_k are IID with density $\frac{1}{2}e^{-|x|}$. Show that X is infinitely divisible and find the associated measure M.

20. *Mixtures of exponential (geometric) distributions.* Let

$$f(x) = \sum_{k=1}^{n} p_k \lambda_k e^{-\lambda_k x},$$

where $p_k > 0, \sum p_k = 1$ and for definitions $0 < \lambda_1 < \lambda_2 < \cdots < \lambda_n$. Show that $f(x)$ is infinitely divisible. (Similarly a mixture of geometric is infinitely divisible.) By a limit argument prove that the density

$$f(x) = \int_0^\infty \lambda e^{-\lambda x} G\{d\lambda\},$$

where G is a distribution concentrated on $(0, \infty)$, is infinitely divisible.

21. If X and Y are independent random variables such that $X > 0$ and Y has an exponential density, then prove that XY is infinitely divisible.
22. If X and Y are independent random variables such that X is stable with exponent α, while Y is positive and stable with exponent $\beta < 1$, show that $XY^{1/\alpha}$ is stable with exponent $\alpha\beta$.

CHAPTER 2

Definition and Characteristics of a Stochastic Process

2.1 Introduction

A stochastic process is a family of random variables $X(t)$ depending on an index set $t \in T$. Here the index set T may, in general, be an abstract set, but we shall take it to be the real line $(-\infty, \infty)$ and interpret t as *time*. For each t the random variable $X(t)$ is defined on a sample space Ω, where we are given a Borel field B if subsets of Ω and a probability measure P on B. The set of values taken by $X(t)$ is called the *sample space* of the process. To emphasize its dependence on t and ω, we shall denote the stochastic process as $\{X(t, \omega), t \in T, \omega \in \Omega\}$. For each $\omega \in \Omega, X(t, \omega)$ is a function of t, called a *sample function*. For each $t \in T, X(t, \omega)$ is an ordinary random variable. Sometimes we shall simplify the notation to $\{X(t), t \in T\}$.

There are two different ways in which a stochastic process $\{X(t), t \in T\}$ can be formulated: (1) its sample function can be defined for each value of the index (analytical definition) or (2) its finite-dimensional distribution properties (namely, the distribution properties for each finite set of index values) can be defined. Once a stochastic process is defined, it is important in many situations to obtain its various characteristics such as moments. In this chapter we describe these aspects and give some examples. If the analytic definition of a stochastic process is given, we need to develop its continuity, differentiability and integrability properties. This leads to a stochastic calculus, which will be treated briefly in Chapter 4 for stationary processes.

2.2 Analytic Definition

Let $X(t) = \phi(t; A_1, A_2, \ldots, A_k)$, where ϕ is an ordinary function of t and of a set of random variables $A_1, A_2, \ldots, A_k$ $(k \geq 1)$ defined on a probability space independent of t, and ϕ is a random variable on this space. Then $X(t)$ is a stochastic process. The following are examples of such a definition.

Example 2.1 For $t \geq 0$ let $X(t) = At + B$, where A and B are random variables. Here the sample functions are of the from $at + b$, where (a, b) is a sample pair of values of (A, B). For each t we have

$$P\{X(t) \leq x\} = \iint_{at+b\leq x} P\{A \in (a, a + da], B \in (b, b + db]\}.$$

In particular suppose that A, B are independent random variables, each having a uniform density in $(0, 1)$. If $u_\theta(x)$ is the uniform density in $(0, \theta)$, its distribution function (d.f.) can be written as

$$U_\theta(x) = \int_{-\infty}^{x} u_\theta(y)dy = \frac{x_+ - (x - \theta)_+}{\theta},$$

where for any real number x we denote $x_+ = \max(0, x)$. Using the convolution formula we find that for $t > 0$,

$$\begin{aligned} P\{X(t) \leq x\} &= \int_0^1 U_t(x - y)u_1(y)dy \\ &= \frac{1}{2t}[x_+^2 - (x - 1)_+^2 - (x - t)_+^2 - (x - 1 - t)_+^2]. \end{aligned}$$

Example 2.2 Let $X(t) = A\cos\theta t + B\sin\theta t$ $(t \geq 0)$, where θ is a constant, and A, B are independent random variables, each having a normal density with mean 0 and variance σ^2. The sample functions here are the trigonometric functions $a\cos\theta t + b\sin\theta t$. For each $t, X(t)$ is the weighted sum of normally distributed random variables, so $X(t)$ has a normal distribution with mean and variance given by

$$EX(t) = 0, \mathrm{Var}X(t) = \sigma^2(\cos^2\theta t + \sin^2\theta t) = \sigma^2.$$

2.3 Definition in Terms of Finite-Dimensional Distributions

We define $\{X(t), t \in T\}$ by specifying the joint d.f.'s

$$\begin{aligned} &F(x_1, x_2, \ldots, x_n; t_1, t_2, \ldots, t_n) \\ &\quad = P\{X(t_1) \leq x_1, X(t_2) \leq x_2, \ldots, X(t_n) \leq x_n\} \end{aligned} \tag{3.1}$$

for $t_1, t_2, \ldots, t_n \in T$, and $n \geq 1$. These d.f.'s satisfy the following *consistency conditions*:

(a) Symmetry: For every permutation $(j_1, j_2, \ldots, j_n)$ of $(1, 2, \ldots, n)$ we must have

$$F(x_{j_1}, x_{j_2}, \ldots, x_{j_n}; t_{j_1}, t_{j_2}, \ldots, t_{j_n})$$
$$= F(x_1, x_2, \ldots, x_n; t_1, t_2, \ldots, t_n). \quad (3.2)$$

(b) Compatibility: For $m < n$ we must have

$$F(x_1, x_2, \ldots, x_m, \infty, \infty, \ldots, \infty; t_1, t_2, \ldots, t_n)$$
$$= F(x_1, x_2, \ldots, x_m; t_1, t_2, \ldots, t_m). \quad (3.3)$$

Instead of the d.f. F we can also consider the characteristic function (c.f.) ϕ where

$$\phi(\theta_1, \theta_2, \ldots, \theta_n; t_1, t_2, \ldots, t_n)$$
$$= E\{\exp[i\theta_1 X(t_1) + i\theta_2 X(t_2) + \cdots + i\theta_n X(t_n)]\} \quad (3.4)$$

where $i = \sqrt{-1}$ and $\theta_1, \theta_2, \ldots, \theta_n$ are real numbers. In view of the uniqueness theorem connecting d.f.'s and c.f.'s we can verify the consistency conditions (a) and (b) for ϕ. These are, respectively

$$\phi(\theta_{j_1}, \theta_{j_2}, \ldots, \theta_{j_n}; t_{j_1}, t_{j_2}, \ldots, t_{j_n})$$
$$= \phi(\theta_1, \theta_2, \ldots, \theta_n; t_1, t_2, \ldots, t_n) \quad (3.5)$$

and

$$\phi(\theta_1, \theta_2, \ldots, \theta_m, 0, 0, \ldots, 0; t_1, t_2, \ldots, t_n)$$
$$= \phi(\theta_1, \theta_2, \ldots, \theta_m; t_1, t_2, \ldots, t_m). \quad (3.6)$$

As an example of the above definition we consider the following.

Example 3.1 (The Gaussian Process) This is a process $X(t)$ such that for $t_1, t_2, \ldots, t_n \in T, n \geq 1$, the random variables $X(t_1), X(t_2), \ldots, X(t_n)$ have an n-variate Gaussian (normal) distribution with

$$EX(t_p) = m(t_p), \quad \mathrm{Var}X(t_p) = \Gamma(t_p, t_p)$$
$$\mathrm{Covar}\{X(t_p), X(t_q)\} = \Gamma(t_p, t_q), (p \neq q) \quad (p, q = 1, 2, \ldots, n). \quad (3.7)$$

The variance–covariance matrix is $\Gamma = (\Gamma_{pq})$, where $\Gamma_{pq} = \Gamma(t_p, t_q)$. We assume that Γ is non-singular, and denote by $\Gamma^{-1} = \Lambda = (\Lambda_{pq})$ the inverse matrix. Then the joint density function of $X(t_1), X(t_2), \ldots, X(t_n)$ is given by

$$f(x_1, x_2, \ldots, x_n; t_1, t_2, \ldots, t_n)$$
$$= \frac{|\Lambda|^{1/2}}{(2\pi)^{n/2}} \exp - \left\{ \frac{1}{2} \sum\sum_{p,q=1}^{n} \Lambda_{pq}[x_p - m(t_p)][x_q - m(t_q)] \right\}. \quad (3.8)$$

It is more convenient to work with the c.f.

$$\phi(\theta_1, \theta_2, \ldots, \theta_n; t_1, t_2, \ldots, t_n)$$
$$= \exp \left\{ i \sum_{1}^{n} \theta_p m(t_p) - \frac{1}{2} \sum\sum_{p,q=1}^{n} \Gamma_{pq}[\theta_p - m(t_p)][\theta_q - m(t_q)] \right\}. \quad (3.9)$$

It is easily verified that the right side of (3.9) satisfies the conditions (3.5) and (3.6).

The Daniell–Kolmogorov theorem states that given a family of distributions functions $F(x_1, x_2, \ldots, x_n; t_1, t_2, \ldots, t_n)$ satisfying the consistency conditions there exists a probability space (Ω, F, P) and a stochastic process $X(t, \omega)$ having F as the probability distribution of $X(t_1), X(t_2), \ldots, X(t_n)$.

If in the definition of a stochastic process $\{X(t), t \in T\}$ the index set T has only a finite number of points, then the theory of such processes belongs to the branch of mathematical statistics known as multivariate analysis. The really interesting case is therefore the one in which T is an infinte set. We shall deal with the two cases where (a) $T = \{0, 1, 2, \ldots\}$, and (b) $T = [0, \infty)$. We shall refer to (a) as discrete time and (b) as continuous time.

We also remark that in general $X(t)$ is a k-dimensional stochastic process, so that for each t the random variable $X(t)$ is a vector of the form $\{X_1(t), X_2(t), \ldots, X_k(t)\}$. We shall, however, be mostly concerned with the univariate stochastic processes. Further, we have assumed that $X(t)$ is a real stochastic process. If $X(t)$ and $Y(t)$ are two real stochastic processes, then

$$Z(t) = X(t) + iY(t), \quad (i = \sqrt{-1}) \quad (3.10)$$

defines a complex stochastic process; each process on the right side of (3.10) is here specified either analytically, or in terms of its finite-dimensional d.f.'s. The following is an example of a complex process.

Example 3.2 Let

$$Z(t) = \sum_{1}^{N} A_p e^{i\theta_p t},$$

where $\theta_1, \theta_2, \ldots, \theta_N$ are real constants and $A_1, A_2, \ldots, A_N$ are real random variables. This can be written as

$$Z(t) = \sum_{1}^{N} A_p \cos \theta_p t + i \sum_{1}^{N} A_p \sin \theta_p t.$$

2.4 Moments of Stochastic Processes

For $t_1, t_2, \ldots, t_n \in T, n \geq 1, r_1, r_2, \ldots, r_n$ real, we define the product moment

$$\mu_{r_1 r_2 \ldots r_n}(t_1, t_2, \ldots, t_n) = EX(t_1)^{r_1} X(t_2)^{r_2} \cdots X(t_n)^{r_n} \tag{4.1}$$

if it exists. In particular, for $t \in T$, let

$$m(t) = \mu_1(t) = EX(t) \tag{4.2}$$

$$\mu_2(t) = EX(t)^2, \quad \mu_{11}(s,t) = EX(s)X(t). \tag{4.3}$$

Then $m(t)$ is the mean of $X(t)$,

$$\sigma^2(t) = E[X(t) - m(t)]^2 = \mu_2(t) - [m(t)]^2 \tag{4.4}$$

is its variance, and

$$\begin{aligned} \Gamma(s,t) &= E[X(s) - m(s)][X(t) - m(t)] \\ &= \mu_{11}(s,t) - m(s)m(t), \end{aligned} \tag{4.5}$$

is the covariance function. Further,

$$\rho(s,t) = \frac{\Gamma(s,t)}{\sigma(s)\sigma(t)} \tag{4.6}$$

gives the correlation function of the process.

For a complex process $Z(t) = X(t) + iY(t)$, we define

$$EZ(t) = EX(t) + iEY(t) \tag{4.7}$$

$$\mu_2(t) = E|Z(t)|^2, \quad \mu_{11}(s,t) = EZ(s)\bar{Z}(t), \tag{4.8}$$

where $\bar{Z}(t) = X(t) - iY(t)$.

Example 4.1 Let $X(t) = At + B$, where A, B are uncorrelated random variables with

$$E(A) = \alpha, \quad E(B) = \beta$$
$$\mathrm{Var}(A) = \sigma_1^2 \quad \mathrm{Var}(B) = \sigma_2^2.$$

We find that $m(t) = EX(t) = t\alpha + \beta, \sigma^2(t) = \mathrm{Var}X(t) = t^2\sigma_1^2 + \sigma_2^2$. Moreover, since

$$\begin{aligned}\mu_{11}(s,t) &= EX(s)X(t) = E(sA+B)(tA+B)\\ &= stE(A^2) + (s+t)\alpha\beta + E(B^2),\end{aligned}$$

we obtain

$$\begin{aligned}\Gamma(s,t) &= \mathrm{Covar}\{X(s), X(t)\} = \mu_{11}(s,t) - (s\alpha+\beta)(t\alpha+\beta)\\ &= st\sigma_1^2 + \sigma_2^2.\end{aligned}$$

2.5 Some Problems in Stochastic Processes

Given a stochastic process $X(t)$ we may ask several questions concerning it, or more specifically, concerning random variables associated with $X(t)$. In the case of a discrete time process $\{X_n, n \geq 0\}$ some of these questions are the following.

(1) Given a Borel subset C of the state space, the random variables

$$N = \min\{n \geq 1 : X_n \in C\} \tag{5.1}$$

denote the first time the process $\{X_n\}$ enters the set C. This is called the *first entrance time* or *hitting time* of C. We may need to know whether the process is sure to enter the set C, that is, whether the event

$$\{X_n \in C \text{ for some } n \geq 1\} \tag{5.2}$$

has probability 1. If this is not the case, then the event

$$\{X_1 \notin C, X_2 \notin C, X_3 \notin C \ldots\} \tag{5.3}$$

has positive probability. According to the standard convention the event (5.2) is indicated by $\{N < \infty\}$ and the event (5.3) by $\{N = \infty\}$. We need to know the probabilities of these events. In the case where $P\{N < \infty\} = 1$, we may also be interested in the moments of N.

(2) With C as above, we may be interested in the time the process spends in C during a time span $[0, n]$. This is *called the occupation time* of the set C. To define this analytically, we introduce the indicator function

$$1_{\{X_n \in C\}} = 1 \quad \text{if } X_n \in C, \quad \text{and} = 0 \text{ otherwise.} \tag{5.4}$$

Then,

$$Y_n = \sum_{m=0}^{n} 1_{\{X_m \in C\}} \tag{5.5}$$

is the occupation time.

(3) Denote

$$M_n = \max\{X_0, X_1, \ldots, X_n\}, \quad m_n = \min\{X_0, X_1, \ldots, X_n\}. \tag{5.6}$$

These are the maximum and minimum functionals of the process $\{X_n\}$. Here again, our questions concern the properties of these functionals.

(4) We may be interested in the limit behavior of $\{X_n\}$ as $n \to \infty$, or in its weaker version, in the limit

$$\lim_{n \to \infty} P\{X_n \in C\}. \tag{5.7}$$

In the case of a continuous time process $\{X(t), t \in [0, \infty)\}$, we are interested in the random variables defined as above in an analogous manner. In addition we may also ask questions concerning the continuity, differentiability, and integrability properties (*sample function properties*) of $X(t)$ as a random function of t (see Sec. 2.7).

2.6 Probability Models

Probability models may be characterized as mathematical models that involve a random element. Probability modeling emerged during what J. Neyman calls the era of *dynamic indeterminism*, starting with Mendelism, statistical mechanics, epidemiology and other areas, and now extending to all branches of the natural, physical, and social sciences. Each topic within these diverse subject areas has its own technical background. The initial description of the phenomenon to be investigated is usually available in fairly nonmathematical terms. The probabilist then sets up reasonable hypotheses (assumptions) to translate this description into mathematical (probabilistic) terms. It is this latter description that constitutes the probability model, and this gives rise to a stochastic process

$X(t)$. Relevant questions concerning the phenomenon are then formulated in terms of $X(t)$, as indicated in Sec. 2.5. It is also necessary to test the appropriateness of the model (this includes statistical inference from stochastic processes).

2.7 Comments on the Definition of a Stochastic Process

We conclude this chapter by showing that the definition of a stochastic process based on its finite-dimensional distributions is not adequate in the continuous time case. For example, the set

$$A_t = \left\{ \omega : b < \sup_{0 \leq \tau \leq t} X(\tau, \omega) < a \right\}$$

is defined in terms of an uncountable set of points, namely, $[0, t]$ and is not an event within the framework of finite dimensional distributions. We are therefore unable to assign a probability measure to A_t. The following examples further illustrate difficulties of this type.

Example 7.1 Given the index set $T = [0, \infty]$ and a probability space (Ω, F, P) we define two stochastic processes $\{X(t)\}$ and $\{Y(t)\}$ as follows. For $t \in T$

$$X(t, \omega) = 0 \quad \text{for } \omega \in \Omega, \tag{7.1}$$

$$Y(t, \omega) = 0 \quad \text{for } t \neq Z(\omega), \quad \text{and} = 1 \quad \text{for } t = Z(\omega), \tag{7.2}$$

where $Z(\omega)$ is a non-negative random variable with a continuous distribution. We have $P\{X(t) = Y(t)\} = 1$, so the two processes have the same finite dimensional distributions: for $0 < t_1 < t_2 < \cdots < t_n$ $(n \geq 1)$, $\{X(t_1), X(t_2), \ldots, X(t_n)\}$ and $\{Y(t_1), Y(t_2), \ldots, Y(t_n)\}$ have the same distribution, being concentrated at the origin $(0, 0, \ldots, 0)$. However, the sample functions of $X(t)$ are continuous, while those of $Y(t)$ have a discontinuity at $t = Z(\omega)$.

Example 7.2 Let $T = [0, 1], \Omega = [0, 1], F$ the Borel field of subsets of Ω, and P the Lebesgue measure on F. Define two processes $\{X(t)\}$ and $\{Y(t)\}$ as follows.

$$X(t, \omega) = 0 \tag{7.3}$$

$$Y(t, \omega) = 0 \quad \text{for } t \neq \omega, \quad \text{and} = 1 \quad \text{for } t = \omega. \tag{7.4}$$

As in Example 7.1, the finite-dimensional distributions of $X(t)$ and $Y(t)$ are the same. However,

$$\sup_{0\le t\le 1} X(t,\omega) = 0, \qquad \sup_{0\le t\le 1} Y(t,\omega) = 1 \tag{7.5}$$

for all $\omega \in \Omega$.

One way to deal with situations of the type described in Examples 7.1 and 7.2 is to introduce the notion of *separability* of a process, which (roughly speaking) enables us to determine the sample function properties of the process $X(t)$ from a countable set of points of T, except perhaps for a set of measure zero. It can be proved that for a given system of finite-dimensional distributions there exists a separable process having these distributions. In practice we deal only with this process, and call it the separable version of $X(t)$. Sufficient conditions for $X(t)$ to be separable are the existence of left and right limits and the right-continuity of $X(t)$ (the cadlag property). In this book we shall assume that our processes satisfy these conditions.

Further Reading

Billingsley, P (1986). *Probability and Measure*, 2nd Edn. New York: John Wiley.

Books of Historical Interest

Doob, JL (1953). *Stochastic Processes.* New York: John Wiley.

Moyal JE, MS Bartlett and DG Kendall (1949). Symposium on stochastic processes. *Journal of the Royal Statistical Society B*, 11(2).

Wax, N (ed.) (1954). *Selected Papers on Noise and Stochastic Processes.* New York: Dover.

CHAPTER 3

Some Important Classes of Stochastic Processes

3.1 Stationary Processes

A stochastic process $\{X(t), t \in T\}$ is said to be *stationary in the strict sense* if for $n \geq 1$, t_1, $t_2, \ldots, t_n$, $\tau \in T$, the random variables

$$X(t_1), X(t_2), \ldots, X(t_n)$$

have the same joint distribution as

$$X(t_1 + \tau), X(t_2 + \tau), \ldots, X(t_n + \tau).$$

If this is the case, then $X(0)$ and $X(\tau)$ have the same distribution, that is, the distribution of $X(t)$ is independent of t. Therefore, if $X(t)$ has finite variance,

$$EX(t) = m, \quad \operatorname{Var} X(t) = \sigma^2, \tag{1.1}$$

where m and σ^2 are constants. Here, we can take $m = 0$, $\sigma^2 = 1$ without loss of generality. Then the covariance function is given by

$$\begin{aligned}\Gamma(s,t) &= EX(s)X(t) = EX(0)X(t-s) \\ &= \Gamma(0, t-s) = R(t-s) \text{ (say)};\end{aligned} \tag{1.2}$$

$R(t)$ is thus the correlation function of the process. Clearly, $R(0) = 1$. If only properties (1.1) and (1.2) hold, then $X(t)$ is said to be *stationary in the wide sense*. It is clear that we may extend the index set to $T = (-\infty, \infty)$.

Example 1.1 Let $X(t) = A\cos\theta t + B\sin\theta t$, where A and B are uncorrelated random variables, each having mean 0 and variance σ^2. Clearly

$EX(t) = 0$. We have

$$\begin{aligned}\Gamma(s,t) &= EX(s)X(t) \\ &= E(A\cos\theta s + B\sin\theta s)(A\cos\theta t + B\sin\theta t) \\ &= \sigma^2(\cos\theta s\cos\theta t + \sin\theta s\sin\theta t) \\ &= \sigma^2\cos\theta(t-s). \end{aligned} \tag{1.3}$$

In particular, $\mathrm{Var}X(t) = \Gamma(t,t) = \sigma^2$. Thus the process is stationary in the wide sense. The correlation function is given by

$$R(t) = \cos\theta t. \tag{1.4}$$

Example 1.2 (The Stationary Gaussian Process) If in the Gaussian process defined in Example 3.1 of Chap. 2, $m(t) = m$ (a constant), and the function $\Gamma(t_p, t_q)$ depends only on the difference $t_p - t_q$, then the process is seen to be stationary in the wide sense. However, since a Gaussian distribution is completely determined by the means $m(t_p)$ and the variance–covariance matrix Γ, it follows that the process is in fact stationary in the strict sense.

Example 1.3 Let

$$Z(t) = \sum_1^N A_p e^{i\theta_p t}, \tag{1.5}$$

where $\theta_1, \theta_2, \ldots, \theta_N$ are real constants, $A_1, A_2, \ldots, A_N$ are real random variables which are uncorrelated, and have mean 0 and $\mathrm{Var}(A_p) = \sigma_p^2$ ($p = 1, 2, \ldots, N$). We have

$$m(t) = EZ(t) = 0, \tag{1.6}$$

$$\Gamma(s,t) = EZ(s)\bar{Z}(t) = E\sum\sum_{p,q=1}^{N}(A_pA_q)e^{i\theta_p s - i\theta_q t} = \sum_{p=1}^{N}\sigma_p^2 e^{i\theta_p(s-t)} \tag{1.7}$$

and in particular

$$\mathrm{Var}Z(t) = E|Z(t)|^2 = \sum_1^N \sigma_p^2. \tag{1.8}$$

From (1.6)–(1.8) we see that the process $Z(t)$ is stationary in the wide sense.

3.2 Processes with Stationary Independent Increments

These are defined for continuous time and have the following properties:

(i) For $0 \leq t_1 < t_2 < \cdots < t_n (n \geq 2)$ the random variables

$$X(t_1), X(t_2) - X(t_1), X(t_3) - X(t_2), \ldots, X(t_n) - X(t_{n-1}) \tag{2.1}$$

are independent.

(ii) The distribution of the increment $X(t_p) - X(t_{p-1})$ depends only on the difference $t_p - t_{p-1}$.

For such a process $X(t)$ we can take $X(0) \equiv 0$ without loss of generality. For if $X(0) \neq 0$, then the process $Y(t) = X(t) - X(0)$ has stationary independent increments, and also $Y(0) \equiv 0$.

The c.f. of $X(t_1), X(t_2), \ldots, X(t_n)$ is given by

$$\begin{aligned}
&\phi(\theta_1, \theta_2, \ldots, \theta_n; t_1, t_2, \ldots, t_n) \\
&\quad = E\left\{\exp\left[i \sum_1^n \theta_p X(t_p)\right]\right\} \\
&\quad = E\left\{\exp\left[i \sum_1^n \theta_p \sum_1^p [X(t_q) - X(t_{q-1})]\right]\right\} \\
&\quad = E\left[\exp\left\{i \sum_1^n [X(t_q) - X(t_{q-1})][\theta_q + \theta_{q+1} + \cdots + \theta_n]\right\}\right] \\
&\quad = \prod_{q=1}^n \phi(\theta_q + \theta_{q+1} + \cdots + \theta_n; t_q - t_{q-1}).
\end{aligned} \tag{2.2}$$

From (2.2) it is easily verified that the consistency conditions are satisfied.

We have the representation

$$X(t+s) = X(t) + [X(t+s) - X(t)], \tag{2.3}$$

where the random variables $X(t)$ and $X(t+s) - X(t)$ are independent, and $X(t+s) - X(t)$ has the same distribution as $X(s)$. If $X(t)$ has finite mean $m(t)$, then (2.3) yields the relation

$$m(t+s) = m(t) + m(s). \tag{2.4}$$

The only bounded solution of the functional equation (2.4) is known to be $m(t) = \mu t$, where obviously $\mu = EX(1)$. Similarly if the variance is finite, then $\operatorname{Var} X(t) = \sigma^2 t$, where $\sigma^2 = \operatorname{Var} X(1)$. Thus

$$EX(t) = \mu t, \quad \operatorname{Var} X(t) = \sigma^2 t. \tag{2.5}$$

For $s < t$ the covariance function is given by

$$\begin{aligned}
\Gamma(s,t) &= EX(s)X(t) - m(s)m(t) \\
&= E\{X(s)[X(t) - X(s)] + X(s)^2\} - \mu^2 st \\
&= EX(s) \cdot E[X(t) - X(s)] + EX(s)^2 - \mu^2 st \\
&= \mu s \cdot \mu(t-s) + \sigma^2 s + \mu^2 s^2 - \mu^2 st \\
&= \sigma^2 s.
\end{aligned}$$

Similarly $\Gamma(s,t) = \sigma^2 t$ if $s > t$. Thus

$$\Gamma(s,t) = \sigma^2 \min(s,t). \tag{2.6}$$

[Incidentally, the results (2.5) and (2.6) show that a process with stationary independent increments is not stationary as defined above.]

If we write

$$X(t) = \sum_{k=1}^{n} \left[X\left(\frac{k}{n}t\right) - X\left(\frac{k-1}{n}t\right) \right], \tag{2.7}$$

then $X(t)$ is seen to be the sum of n independent random variables all of which are distributed as $X\left(\frac{t}{n}\right)$. Thus a process with stationary independent increments is the generalization to continuous time of sums of independent and identically distributed random variables (random walks). In particular the relations (2.5) are continuous time analogues of the well-known discrete time results.

The following are the two important special cases of processes with stationary independent increments.

Example 2.1 (Brownian Motion) This is a real process with stationary independent increments such that $X(t) - X(s)$ has a Gaussian distribution with mean 0 and

$$\operatorname{Var}[X(t) - X(s)] = \sigma^2 |t - s|. \tag{2.8}$$

The equations

$$Y(t_p) = \sum_{q=1}^{p} [X(t_q) - X(t_{q-1})] \ \ (p = 1, 2, \ldots, n) \tag{2.9}$$

define a linear transformation of Gaussian random variables with a non-singular matrix of transformation. Since $Y(t_p) = X(t_p)$ it follows that for $n \geq 1$, the random variables $X(t_1), X(t_2), \ldots, X(t_n)$ have an n-variate Gaussian distribution. Therefore $X(t)$ is a Gaussian process.

Relation (2.3) expresses the well-known additive property of Gaussian distributions.

Example 2.2 (The Poisson Process) This is a real process with stationary independent increments such that the distribution of the increment $X(t) - X(s)$ $(s < t)$ is given by

$$P\{X(t) - X(s) = j\} = e^{-\lambda(t-s)} \frac{\lambda^j}{j!} (t-s)^j \quad (j = 0, 1, 2, \ldots). \tag{2.10}$$

We have

$$E[X(t) - X(s)] = \text{Var}[X(t) - X(s)] = \lambda(t - s). \tag{2.11}$$

From relation (2.3) we find that

$$e^{-\lambda(t+s)} \frac{\lambda^j}{j!} (t+s)^j = \sum_{k=0}^{j} e^{-\lambda t} \frac{(\lambda t)^{j-k}}{(j-k)!} e^{-\lambda s} \frac{(\lambda s)^k}{k!}. \tag{2.12}$$

3.3 Markov Processes

These are defined for discrete or continuous time, and have the property that for $0 \leq t_1 < t_2 < \cdots < t_n (n \geq 2)$ the conditional distribution of $X(t_n)$ given $X(t_1), X(t_2), \ldots, X(t_{n-1})$ depends only on $X(t_{n-1})$. Thus

$$\begin{aligned} P\{X(t_n) \leq x | X(t_1) = x_1,\ X(t_2) = x_2, \ldots, X(t_{n-1}) = x_{n-1}\} \\ = P\{X(t_n) \leq x | X(t_{n-1}) = X_{n-1}\}. \end{aligned} \tag{3.1}$$

The conditional d.f.

$$F(x, \tau; y, t) = P\{X(t) \leq y | X(\tau) = x\} \quad (\tau < t) \tag{3.2}$$

is called the *transition d.f.* of $X(t)$. It satisfies the Chapman–Kolmogorov equation

$$F(x, \tau; y, t+s) = \int_{-\infty}^{\infty} F(x, \tau; du, t) F(u, t; y, t+s) \quad (\tau < t < t+s). \tag{3.3}$$

To prove (3.3) we note that

$$
\begin{aligned}
&P\{X(t+s) \le y | X(\tau) = x\} \\
&\quad = \int_{-\infty}^{\infty} P\{X(t) \in du | X(\tau) = x\} P\{X(t+s) \le y | X(\tau) = x, X(t) = u\} \\
&\quad = \int_{-\infty}^{\infty} P\{X(t) \in du | X(\tau) = x\} P\{X(t+s) \le y | X(t) = u\},
\end{aligned}
$$

where we have used the Markov property (3.1) to simplify the expression for the last conditional probability.

Suppose now that the transition d.f. (3.2) is given for all τ, t with $\tau < t$. Also let

$$F_0(x) = P\{X(0) \le x\} \tag{3.4}$$

be the *initial d.f.* Then for $0 < t_1 < t_2 < \cdots < t_n$ $(n \ge 1)$ the joint distribution of $X(t_1), X(t_2), \ldots, X(t_n)$ can be determined in terms of (3.2) and (3.4). Thus for $t > 0$ the unconditional (or absolute) d.f. of $X(t)$ is given by

$$
\begin{aligned}
F(x,t) &= \int_{-\infty}^{\infty} P\{X(0) \in dx_0\} P\{X(t) \le x | X(0) = x_0\} \\
&= \int_{-\infty}^{\infty} F_0(dx_0) F(x_0, 0; x, t).
\end{aligned}
\tag{3.5}
$$

Therefore

$$
\begin{aligned}
&F(x_1, x_2, \ldots, x_n; t_1, t_2, \ldots, t_n) \\
&\quad = P\{X(t_1) \le x_1, X(t_2) \le x_2, \ldots, X(t_n) \le x_n,\} \\
&\quad = \int P\{X(t_1) \in dy_1\} P\{X(t_2) \in dy_2 | X(t_1) = y_1\} \\
&\qquad \times P\{X(t_3) \in dy_3 | x(t_1) = y_1, X(t_2) = y_2\} \cdots \\
&\qquad \times P\{X(t_n) \in dy_n | X(t_1) = y_1, \ldots, X(t_{n-1}) = y_{n-1}\} \\
&\quad = \int F(dy_1, t_1) F(y_1, t_1; dy_2, t_2) F(y_2, t_2; dy_3, t_3) \cdots \\
&\qquad \times F(y_{n-1}, t_{n-1}; dy_n, t_n),
\end{aligned}
\tag{3.6}
$$

where the integral is taken over the region $(y_1 \le x_1, y_2 \le x_2, \ldots, y_n \le x_n)$. It is found that these finite dimensional d.f.s satisfy the symmetry and compatibility conditions.

If the above d.f.s have a density then (3.6) yields the relation

$$\begin{aligned} f(x_1, x_2, \ldots, x_n; t_1, t_2, \ldots, t_n) \\ = f(x_1, t_1) f(x_1, t_1; x_2, t_2) \cdots f(x_{n-1}, t_{n-1}; x_n, t_n) \end{aligned} \tag{3.7}$$

which characterizes the Markov property in this case.

If the transition d.f. $F(x, \tau; y, t)$ depends only on $t - \tau$, we say that the process $X(t)$ is *time-homogeneous*. In this case we shall write

$$F(x, \tau; y, t) = F(x; y, t - \tau) \tag{3.8}$$

for the transition d.f. and $f(x, \tau; y, t) = f(x; y, t - \tau)$ for the density when it exists.

As examples of Markov processes we consider the following.

Example 3.1 (Processes with Stationary Independent Increments) If $X(t)$ is such a process, then for $0 \le t_1 < t_2 < \cdots < t_n$ $(n \ge 2)$ we can write

$$X(t_n) = X(t_n) - X(t_{n-1}) + \sum_{p=1}^{n-1} [X(t_p) - X(t_{p-1})], \tag{3.9}$$

where $t_0 = 0$ and we have assumed that $X(0) = 0$. On the right side of (3.9), $X(t_n) - X(t_{n-1})$ is independent of the random variables making up the sum, which reduces to $X(t_{n-l})$ and so the distribution of $X(t_n)$ for given $X(t_1), X(t_2), \ldots, X(t_{n-1})$ depends only on $X(t_{n-1})$. Therefore $X(t)$ is a Markov process.

From the identity $X(t) = X(\tau) + [X(t) - X(\tau)]$ $(0 < \tau < t)$ we find the transition d.f. of $X(t)$ as

$$\begin{aligned} F(x, \tau; y, t) &= P\{X(t) \le y | X(\tau) = x\} \\ &= P\{X(t) - X(\tau) \le y - x | X(\tau) = x\}. \end{aligned}$$

Since the increment $X(t) - X(\tau)$ is independent of $X(\tau)$ and has the same distribution as $X(t - \tau)$ we obtain

$$\begin{aligned} F(x, \tau; y, t) &= P\{X(t - \tau) \le y - x | X(0) = 0\} \\ &= F(0, 0; y - x, t - \tau). \end{aligned} \tag{3.10}$$

Thus the process $X(t)$ is homogeneous in time as well as space.

In particular, the Brownian Motion and the Poisson process are both Markovian. The transition density of the Brownian Motion is

$$f(x,\tau;y,t)=\frac{1}{\sqrt{2\pi(t-\tau)}\sigma}e^{-\frac{1}{2\sigma^2(t-\tau)}(y-x)^2}\quad(0<t<\tau),$$

while for the Poisson process we have

$$\begin{aligned}P_{jk}(\tau,t)&=P\{X(t)=k|X(\tau)=j\}\\&=e^{-\lambda(t-\tau)}\lambda^{k-j}(t-\tau)^{k-j}/(k-j)!\quad\text{for } k\geq j\\&=0\quad\text{elsewhere.}\end{aligned}$$

Example 3.2 (The Gauss–Markov Process) This is a Gaussian process which is also Markovian. We use the notation of Example 9.3 of Chap. 1, but assume that $m(t_p)=0$ and $\Gamma(t_p,t_p)=1$. The density of $X(t_1)$ is given by

$$f(x_1,t)=\frac{1}{\sqrt{2\pi}}e^{-\frac{1}{2}x_1^2},\tag{3.11}$$

while the joint density of $X(t_1)$ and $X(t_2)$ is given by

$$f(x_1;x_2;t_1,t_2)=\frac{1}{2\pi\sqrt{1-\gamma_{12}^2}}e^{-\frac{1}{2(1-\gamma_{12}^2)}(x_1^2-2\gamma_{12}x_1x_2+x_2^2)},\tag{3.12}$$

where we have denoted $\gamma_{12}=\Gamma(t_1,t_2)$ for convenience. The conditional density of $X(t_2)$ for a given $X(t_1)$ is therefore given by

$$\begin{aligned}f(x_1,t_1;x_2,t_2)&=\frac{f(x_1,x_2;t_1,t_2)}{f(x_1,t_1)}\\&=\frac{1}{\sqrt{2\pi(1-\gamma_{12}^2)}}e^{-\frac{1}{2(1-\gamma_{12}^2)}(x_2-\gamma_{12}x_1)^2}.\end{aligned}\tag{3.13}$$

Thus the transition density is Gaussian with mean $\gamma_{12}x_1$ and variance $(1-\gamma_{12}^2)$. We can therefore write down the joint density of $X(t_1)$, $X(t_2),\ldots,X(t_n)$ $(n\geq 1)$ as in (3.7). In particular for $0\leq t_1<t_2<t_3$ we have

$$\begin{aligned}\Gamma(t_1,t_3)&=E[X(t_1)X(t_3)]=EE[X(t_1)X(t_3)|X(t_2)]\\&=E\{X(t_1)E[X(t_3)|X(t_2)]\}=E[X(t_1)\Gamma(t_2,t_3)X(t_2)]\\&=\Gamma(t_2,t_3)\Gamma(t_1,t_2).\end{aligned}$$

Therefore the covariance function of the Gauss–Markov process has the important property

$$\Gamma(t_1, t_3) = \Gamma(t_1, t_2)\Gamma(t_2, t_3) \quad (0 \le t_1 \le t_2 \le t_3). \tag{3.14}$$

Example 3.3 (The Ornstein–Uhlenbeck Process) This is a Gauss–Markov process that is stationary. Thus for $0 < t_1 < t_2 < \cdots < t_n$ $(n \ge 1)$ the joint distribution of $X(t_1), X(t_2), \ldots, X(t_n)$ is Gaussian with

$$E[X(t_p)] = m, \quad \Gamma(t_p, t_q) = \sigma^2 R(t_q - t_p). \tag{3.15}$$

From (3.14) we see that $R(s+t) = R(s)R(t)$. Since $|R(t)| \le 1$ we must have either $R(t) = 0$ (degenerate case) or there exists a constant β $(0 \le \beta < \infty)$ such that

$$R(t) = e^{-\beta|t|}. \tag{3.16}$$

3.4 Problems for Solution

1. Let $X(t) = \cos(At + U)$, where A and U are independent random variables, U uniformly distributed in $(0, 2\pi)$ and A normally distributed with mean 0 and variance σ^2. Show that the process $X(t)$ is stationary in the wide sense.
2. *Random Binary Transmission.* Let us toss an ideal coin at times $0, \Delta, 2\Delta, \ldots$, and define a process $X(t) (t \ge 0)$ as follows:

$$\begin{aligned} X(t) &= +1 \quad \text{if heads at the } n\text{th toss} \\ &= -1 \quad \text{if tails at the } n\text{th toss,} \end{aligned}$$

where $n = \left[\frac{t-U}{\Delta}\right]$, and U is a random variable uniformly distributed in $(0, \Delta)$, $[x]$ being the largest integer contained in x. Show that $X(t)$ is stationary in the wide sense, and its correlation function is given by

$$\begin{aligned} R(t) &= 1 - \frac{|t|}{\Delta} \quad \text{if } |t| < \Delta \\ &= 0 \qquad\qquad \text{if } |t| \ge \Delta. \end{aligned}$$

3. *Random Telegraph Signal.* Let $X(t)$ be a Poisson process with parameter λ, and define a process $Y(t)$ $(t \ge 0)$ as follows: $Y(0)$ is

independent of $X(t)$ $(t \geq 0)$, and

$$\Pr\{Y(0) = +1\} = \frac{1}{2}, \quad \Pr\{Y(0) = -1\} = \frac{1}{2}.$$

$$\text{For } t > 0, Y(t) = Y(0)(-1)^{X(t)}, \text{ so that}$$
$$\begin{aligned} Y(t) &= Y(0) \quad \text{if } X(t) \text{ is even} \\ &= -Y(0) \quad \text{if } X(t) \text{ is odd.} \end{aligned}$$

Show that $Y(t)$ is stationary in the wide sense, with the correlation function

$$R(t) = e^{-2\lambda|t|}.$$

4. Suppose that $\{X_n, n \geq 0\}$ is a sequence of independent and identically distributed random variables. A second sequence $\{Y_n, n \geq 1\}$ is defined by

$$Y_n = X_n + \alpha X_{n-1} \quad (n \geq 0),$$

where α is a real constant. Show that $\{Y_n\}$ is strictly stationary.

5. Let $\{X_n, n \geq 0\}$ be a sequence of independent random variables, each having normal density with mean 0 and variance σ^2. For each $n \geq 1$ define

$$Y_n = a + bX_n + cX_{n-1},$$

where a, b, c are real constants. Show that $\{Y_n, n \geq 1\}$ is a stationary Gaussian process and find its means, variances, and covariance function.

6. Let $X(t)$ be a Brownian Motion and $Y(t)$ be defined as in (a) or (b) below. Show that in each case $Y(t)$ is a Brownian Motion.

   (a) $Y(0) = 0, Y(t) = \frac{1}{\sqrt{c}} X(ct)$ $(t > 0)$, where $c > 0$.

   (b) $Y(0) = 0, Y(t) = tX\left(\frac{1}{t}\right)$ $(t > 0)$.

7. Let $X(t)$ be a Brownian Motion, and

$$Y(t) = e^{-\beta t} X(e^{2\beta t}) \quad (t \geq 0),$$

where $\beta > 0$. Show that $Y(t)$ is the Ornstein–Uhlenbeck process.

8. *The Compound Poisson Process.* Let $\{X_n, n \geq 1\}$ be a sequence of independent random variables with a common d.f. $F(x)$ $(-\infty < x < \infty)$

and $N(t)$ a Poisson process with parameter λ, which is independent of the X_n. Let $X(0) = 0$, and

$$X(t) = X_1 + X_2 + \cdots + X_{N(t)} \ (t > 0).$$

Prove the following:

(a) The process $X(t)$ has stationary independent increments.
(b) If X_n has a finite second moment, then

$$EX(t) = \lambda t\alpha, \quad \mathrm{Var}X(t) = \lambda t\beta,$$

where $\alpha = E(X_n)$, $\beta = E(X_n^2)$.

CHAPTER 4
Stationary Processes

4.1 Examples of Real Stationary Processes

Suppose that the real process $\{X(t), t \in T\}$ is stationary (in the strict or wide sense) and has finite mean m and variance σ^2. From the inequality

$$|EX(t)X(t')| \leq \sqrt{EX(t)^2 \cdot EX(t')^2}, \quad t, t' \in T \tag{1.1}$$

it follows that

$$|EX(t)X(t')| < \infty, \tag{1.2}$$

so that the product moment $EX(t)X(t')$ is also finite. Without loss of generality we may take $EX(t)X(t')$ to be the covariance function $\Gamma(t, t')$ of the process. This amounts to assuming that $m = 0$. Let η be a real random variable independent of the process $X(t)$ and with mean 0 and variance unity. Then for the process

$$Y(t) = \eta X(t), \tag{1.3}$$

we have

$$EY(t) = E(\eta)EX(t) = 0 \tag{1.4}$$

since $E(\eta) = 0$, and

$$\begin{aligned} \text{Covar}\{Y(t), Y(t')\} &= EY(t)Y(t') = E(\eta^2)EX(t)X(t') \\ &= EX(t)X(t'). \end{aligned} \tag{1.5}$$

Thus, $EX(t)X(t')$ can be taken to be the covariance function of the process $Y(t)$, which proves our assertion. (Incidentally, this also proves that two or more processes may have the same covariance.) We thus denote

$$\Gamma(t, t') = EX(t)X(t') \quad \text{for } t, t' \in T. \tag{1.6}$$

On account of stationarity we have

$$\Gamma(t,t') = \Gamma(0,t'-t) = \Gamma(t-t',0), \tag{1.7}$$

with $\Gamma(t,t) = \sigma^2 > 0$. The correlation function is given by $\{R(t), t \in T\}$, where

$$R(t) = \frac{\Gamma(s,s+t)}{\sigma^2}. \tag{1.8}$$

The stationarity of the process implies that we may define the process for negative values of t. Accordingly we shall take the index set to be $T = \{\ldots -1,0,1,2,\ldots\}$ or $T = (-\infty,\infty)$.

From definition (1.8) we see that $R(0) = 1$. Also $\sigma^2 R(-t) = \Gamma(s,s-t) = \Gamma(s-t,s) = \sigma^2 R(t)$. Thus

$$R(-t) = R(t), \quad R(0) = 1. \tag{1.9}$$

From (1.1) we also find that

$$|R(t)| \leq 1. \tag{1.10}$$

The above definition of $R(t)$ and its properties are for the case where $X(t)$ is a real process. For complex processes we shall give the analogous definition of $R(t)$ and derive further properties. It will turn out that $R(t)$ determines several important characteristics of the stationary process.

The following are some important examples of real stationary processes.

Example 1.1 Let $\{X_n\}$ be a sequence of independent and identically distributed random variables. Since $(X_{n_1+m}, X_{n_2+m}, \ldots, X_{n_r+m})$ have the same joint distribution as $(X_{n_1}, X_{n_2}, \ldots, X_{n_r})$ for $r \geq 1$, it follows that $\{X_n\}$ is a strictly stationary process.

Example 1.2 (White Noise) This is a sequence $\{X_n\}$ of uncorrelated random variables with mean zero and variance σ^2. We have

$$\begin{aligned} E(X_m X_n) &= \sigma^2 \quad \text{if } m = n \\ &= 0 \quad \text{if } m \neq n. \end{aligned}$$

Therefore, $\{X_n\}$ is stationary in the wide sense. Its correlation function is given by

$$R_n = 1 \quad \text{if } n = 0, \quad \text{and} \quad R_n = 0 \quad \text{if } n \neq 0. \tag{1.11}$$

Example 1.3 (Moving Average) Let

$$X_n = \frac{\xi_n + \xi_{n-1} + \cdots + \xi_{n-p}}{p+1}, \quad (n = \cdots -1, 0, 1, 2, \ldots), \tag{1.12}$$

where $\{\xi_n\}$ is a white noise with variance σ^2. We have

$$E(X_n) = 0.$$

The covariance function is given by

$$\begin{aligned}
\Gamma_{mn} = EX_m X_n &= \frac{1}{(p+1)^2} E\left(\sum_{j=0}^{p} \xi_{m-j} \cdot \sum_{k=0}^{p} \xi_{n-k}\right) \\
&= \frac{1}{(p+1)^2} \sum_{j=0}^{p}\sum_{k=0}^{p} E(\xi_{m-j}\xi_{n-k}) \\
&= \frac{1}{(p+1)^2} \sum_{\substack{0\le j\le p \\ 0\le n-m+j\le p}}^{p} \sigma^2 = \frac{\sigma^2}{p+1}\left(1 - \frac{|n-m|}{p+1}\right).
\end{aligned} \tag{1.13}$$

In particular, the variance of the process is given by

$$\sigma_X^2 = \Gamma_{mn} = \frac{\sigma^2}{p+1}. \tag{1.14}$$

Therefore, $\{X_n\}$ is stationary in the wide sense. Its correlation function is given by

$$\begin{aligned}
R_n = \frac{\Gamma_{m,m+n}}{\sigma_X^2} &= 1 - \frac{|n|}{p+1} \quad \text{for } |n| < p+1 \\
&= 0 \quad \text{otherwise.}
\end{aligned} \tag{1.15}$$

Example 1.4 (Linear Autoregression) Let $\{X_n\}$ be a wide sense stationary process, and

$$X_n = \beta X_{n-1} + \xi_{n-1}, \tag{1.16}$$

where $\{\xi_n\}$ is a white noise with variance σ^2. From (1.16), we find that

$$\begin{aligned}
X_n &= \xi_{n-1} + \beta(\xi_{n-2} + \beta X_{n-2}) \\
&= \xi_{n-1} + \beta\xi_{n-2} + \beta^2 X_{n-2}.
\end{aligned}$$

Proceeding as in this manner we obtain after k steps,

$$X_n = \sum_{j=0}^{k} \beta^j \xi_{n-j-1} + \beta^{k+1} X_{n-k-1}. \tag{1.17}$$

Here the term $\beta^{k+1} X_{n-k-1}$ represents the influence of the process $k+1$ time units before n. Since $E(X_{n-k-1})^2$ does not depend on k we find that as $k \to \infty$

$$E(\beta^{k+1} X_{n-k-1})^2 = \beta^{2k+2} E(X_{n-k-1})^2 \to 0 \tag{1.18}$$

if $|\beta| < 1$, in which case the influence of the infinitely distant past is zero and

$$X_n = \sum_{j=0}^{\infty} \beta^j \xi_{n-j-1}. \tag{1.19}$$

Thus the process $\{X_n\}$ is a weighted moving average of white noise terms with weights β^j. We have

$$\begin{aligned} E(X_n) &= 0 \\ \Gamma_{mn} &= E(X_m X_n) = \sum_{j=0}^{\infty} \sum_{k=0}^{\infty} \beta^{j+k} E(\xi_{m-j-1} \xi_{n-k-1}) \\ &= \sigma^2 \sum_{j \geq 0, j \geq m-n} \beta^{2j+n-m} = \frac{\sigma^2}{1-\beta^2} \beta^{|n-m|}. \end{aligned} \tag{1.20}$$

This shows that $\{X_n\}$ is stationary in the wide sense. For its variance we have

$$\sigma_X^2 = \Gamma_{mm} = \frac{\sigma^2}{1-\beta^2}. \tag{1.21}$$

The correlation function is given by

$$R_n = \frac{\Gamma_{m,m+n}}{\sigma_X^2} = \beta^{|n|} \quad (n = \cdots -1, 0, 1, 2, \ldots). \tag{1.22}$$

4.2 The General Case

We shall now consider stationary processes $\{X(t), t \in T\}$ that are not necessarily real. We assume that

$$E|X(t)|^2 < \infty \quad \text{for all } t \in T \tag{2.1}$$

(such a process is said to be of *second order*). If this condition is satisfied, then

$$|EX(t)\bar{X}(t')| < \infty \quad \text{for all } t, t \in T. \tag{2.2}$$

This follows from the *Schwarz inequality*

$$E|X(t)\bar{X}(t')| \leq \sqrt{E|X(t)(|^2 E|X(t')|^2} \tag{2.3}$$

for $t, t' \in T$. To prove (2.3), let λ be a real constant, and consider

$$\{\lambda|X(t)| + |\bar{X}(t')|\}^2 \geq 0. \tag{2.4}$$

Taking the mean value of both sides of (2.4) we obtain

$$\lambda^2 E|X(t)|^2 + 2\lambda E|X(t)\bar{X}(t')| + E|X(t')|^2 \geq 0; \tag{2.5}$$

the quadratic expression in λ on the left side on (2.5) thus remains non-negative for all λ. Therefore,

$$\{E|X(t)\bar{X}(t')|\}^2 \leq E|X(t)|^2 E|X(t')|^2,$$

which is the required inequality.

We denote

$$\Gamma(t, t') = EX(t)\bar{X}(t'). \tag{2.6}$$

As in the case of a real process we may take $\Gamma(t, t')$ to be the covariance function of the process. Conversely, if $\Gamma(t, t')$ is finite for all $t, t' \in T$, then

$$E|X(t)|^2 = \Gamma(t, t) < \infty \quad \text{for all } t \in T. \tag{2.7}$$

Since the process is stationary, (1.7) holds, with $\Gamma(t, t) = E|X(t)|^2 = \sigma^2 > 0$, unless $X(t)$ is identically zero. The correlation function $\{R(t), t \in T\}$ is defined by (1.8), but it should be noted that $R(t)$ is a complex function of t. We shall take index set to be $T = \{\cdots -1, 0, 1, 2, \ldots\}$ or $T = (-\infty, \infty)$.

The following theorem lists some of the important properties of the correlation function, extending (1.9) and (1.10). We take $\sigma^2 = 1$.

Theorem 2.1 *The correlation function $R(t)$ of a stationary process has the following properties.*

(a) $\bar{R}(t) = R(-t)$
(b) $|R(t)| \leq 1$
(c) $|R(t+h) - R(t)|^2 \leq 2 \text{ Re}[R(0) - R(h)]$

(d) *$R(t)$ is a non-negative definite function, that is, for every $t_1, t_2, \dots, t_n \in T$ and for every function $\theta(t)$ on T we have*

$$\sum_{p=1}^{n}\sum_{q=1}^{n} R(t_q - t_p)\theta(t_p)\bar{\theta}(t_q) \geq 0. \tag{2.8}$$

Proof. (a) We have

$$\bar{R}(t) = E\bar{X}(s)X(s+t) = EX(s+t)\bar{X}(s) = R(-t).$$

(b) From the Schwarz inequality (2.3) we obtain

$$|EX(s)\bar{X}(s+t)|^2 \leq E|X(s)|^2 E|X(s+t)|^2.$$

This gives $|R(t)|^2 \leq R(0)^2$ or $|R(t)| \leq 1$.

(c) Again using the Schwarz inequality (2.3) we find that

$$|EX(0)[\bar{X}(t+h) - \bar{X}(t)]|^2 \leq E|X(0)|^2 E|X(t+h) - X(t)|^2.$$

This gives

$$|R(t+h) - R(t)|^2 \leq E|X(t+h) - X(t)|^2.$$

Now

$$\begin{aligned} E|X(t+h) - X(t)|^2 &= E[X(t+h) - X(t)][\bar{X}(t+h) - \bar{X}(t)] \\ &= 2R(0) - R(-h) - R(h) = 2\ \mathrm{Re}[R(0) - R(h)], \end{aligned}$$

where we have used (a). We thus arrive at the desired result.

(d) We have

$$\begin{aligned} \sum_{p=1}^{n}\sum_{q=1}^{n} R(t_q - t_p)\theta(t_p)\bar{\theta}(t_q) &= \sum_{p=1}^{n}\sum_{q=1}^{n} \theta(t_p)\bar{\theta}(t_q)EX(t_p)\bar{X}(t_q) \\ = E\sum_{p=1}^{n}\theta(t_p)X(t_p)\sum_{q=1}^{n}\bar{\theta}(t_q)\bar{X}(t_q) &= E\left|\sum_{p=1}^{n}\theta(t_p)X(t_p)\right|^2 \geq 0. \end{aligned}$$

□

4.3 A Second Order Calculus for Stationary Processes

We now develop a stochastic calculus for stationary processes. Considering first the broader class of second order processes (stationary or not), we say that $X(t)$ converges to a random variable X in the mean square as $t \to t_0$ and write

$$X(t) \to X(\text{m.s.}) \quad \text{as } t \to t_0 \tag{3.1}$$

if

$$E|X(t) - X|^2 \to 0 \quad \text{as } t \to t_0. \tag{3.2}$$

From the general properties of m.s. convergence (see Sec. 4.5) it follows that $E|X|^2 < \infty$, and moreover $EX(t) \to E(X)$ as $t \to t_0$, that is,

$$\lim_{t \to t_0} EX(t) = E \lim_{t \to t_0} X(t) \quad (\text{m.s.}) \tag{3.3}$$

The following theorem states the necessary and sufficient condition for m.s. convergence in terms of the covariance function.

Theorem 3.1 *Let $X(t)$ be a second order process. Then $X(t) \to X$ (m.s.) as $t \to t_0$, where X is a random variable iff the covariance function $\Gamma(t, t')$ converges to a finite limit as $t, t' \to t_0$. Then*

$$\Gamma(t, t') \to E|X|^2. \tag{3.4}$$

Proof. Let $X(t) \to X$ (m.s.) as $t \to t_0$. We have

$$\begin{aligned} E\{X(t)\bar{X}(t') - X\bar{X}\} &= E\{X(t) - X\}\{\bar{X}(t') - \bar{X}\} \\ &\quad + E\{X(t) - X\}\bar{X} + E\{\bar{X}(t') - \bar{X}\}X. \end{aligned} \tag{3.5}$$

Using the Schwarz inequality we find that

$$\begin{aligned} |E\{X(t) - X\}\{\bar{X}(t') - \bar{X}\}|^2 &\leq E|X(t) - X|^2 E|X(t') - X|^2 \\ &\to 0 \quad \text{as } t, t' \to t_0, \end{aligned}$$

and similarly for the other two terms on the right side of (3.5). Therefore,

$$\Gamma(t, t') = EX(t)\bar{X}(t') \to E|X|^2 \quad \text{as } t, t' \to t_0.$$

Conversely, let $\Gamma(t, t') = \gamma$ as $t, t' \to t_0$, where γ is finite. We have

$$\begin{aligned} E|X(t) - X(t')|^2 &= E\{X(t) - X(t')\}\{\bar{X}(t) - \bar{X}(t')\} \\ &= \Gamma(t, t) - \Gamma(t, t') - \Gamma(t', t) + \Gamma(t', t') \\ &\to \gamma - \gamma - \gamma - \gamma = 0 \quad \text{as } t, t' \to t_0. \end{aligned}$$

By the m.s. criterion of convergence it follows that $X(t) \to X$ (m.s.) as $t \to t_0$, where X is a random variable. □

For a stationary second order process $\{X(t), -\infty < t < \infty\}$ we proceed to define continuity, differentiability and Riemann-integrability, and derive necessary and sufficient conditions for these properties to hold.

Definitions. Let $\{X(t)\}$ be a stationary second order process.

3.1. We say that $X(t)$ is continuous in the mean square at t if

$$X(t+h) \to X(t)(\text{m.s.}) \quad \text{as } h \to 0. \tag{3.6}$$

3.2. We say that $X(t)$ has a derivative $X'(t)$ in the mean square at t if

$$\frac{X(t+h) - X(t)}{h} \to X'(t)(\text{m.s.}) \quad \text{as } h \to 0. \tag{3.7}$$

3.3. Consider an interval $a \leq t \leq b$. Let $a = t_0 < t_1 < \cdots < t_n = b$ be a subdivision of $[a, b], t_{p-1} \leq \xi_p < t_p$ $(p = 1, 2, \ldots, n)$, and denote the sum

$$Y_n = \sum_{p=1}^{n} (t_p - t_{p-1}) X(\xi_p). \tag{3.8}$$

If, as $n \to \infty$ in such a way that $\max_{1 \leq p \leq n} |t_p - t_{p-1}| \to 0$, Y_n converges (m.s.) to a limit, we say that this limit is the Riemann integral of $X(t)$ over $[a, b]$, and denote it by

$$I = \int_a^b X(t)dt. \tag{3.9}$$

Clearly, I is a random variable. If (3.9) exists, we say that $X(t)$ is Riemann-integrable over $[a, b]$.

Theorem 3.2 *A stationary second order process $X(t)$ is continuous at t iff its correlation function $R(t)$ is continuous at $t = 0$. In this case $R(t)$ is continuous at all t.*

Proof. For fixed t, let $Y_h = X(t+h)$. We have

$$EY_h\bar{Y}_{h'} = EX(t+h)\bar{X}(t+h') = R(h'-h).$$

Applying Theorem 3.1 to the process $\{Y_h\}$ we find that $Y_h \to X(t)$ (m.s.) as $h \to 0$ iff $R(t)$ is continuous at the origin. If this condition is satisfied, then from Theorem 2.1(c) it follows that $R(t)$ is continuous at all t. □

Theorem 3.3 *A stationary second order process $X(t)$ has a derivative $X'(t)$ at t iff $R''(t)$ exists and is finite at $t = 0$. If this condition is satisfied,* *then $R''(t)$ exists and is finite at all t, and $-R''(t)$ is the correlation function of $X'(t)$.*

Proof. (i) For fixed t, let us denote

$$Y_h = \frac{X(t+h) - X(t)}{h}$$

The process $\{Y_h\}$ is of second order and by Theorem 3.1, $Y_h \to X'(t)$ (m.s.) iff its covariance function $EY_h\bar{Y}_{h}$, converges to a finite limit as $h, h' \to 0$. Since

$$\begin{aligned} EY_h\bar{Y}_{h'} &= \frac{1}{hh'}E\{X(t+h) - X(t)\}\{\bar{X}(t+h') - \bar{X}(t)\} \\ &= \frac{R(h'-h) - R(-h) - R(h') + R(0)}{hh'}, \end{aligned}$$

the desired result follows.

(ii) Suppose $R''(0)$ exists and is finite. Then by (i)$X'(t)$ exists and is finite at all t. Now

$$\begin{aligned} R'(t'-t) &= \lim_{h\to 0} \frac{R(t'-t+h) - R(t'-t)}{h} \\ &= \lim_{h\to 0} EX(t)\frac{\bar{X}(t'+h) - \bar{X}(t')}{h} = EX(t)\bar{X}(t') \end{aligned}$$

where we have used the commutative property (4.3). Again

$$\begin{aligned} -R''(t'-t) &= \lim_{h\to 0} \frac{R'(t'-t-h) - R'(t'-t)}{h} \\ &= \lim_{h\to 0} E\frac{X(t+h) - X(t)}{h} \cdot \bar{X}'(t') = EX'(t)\bar{X}'(t'), \end{aligned}$$

which shows that $-R''(t)$ exists and is indeed the correlation function of $X'(t)$. □

Theorem 3.4 *A stationary second order process $X(t)$ with the correlation function $R(t)$ is Riemann-integrable over $[a, b]$ iff the double-integral*

$$\int_a^b \int_a^b R(t'-t)dtdt' \tag{3.10}$$

exists. In this case

$$E\left\{\int_a^b X(t)dt \cdot \int_a^b \bar{X}(t')dt'\right\} = \int_a^b \int_a^b R(t'-t)dtdt'. \tag{3.11}$$

Proof. Let Y_n be defined by (3.8). Then $Y_n \to$ a limit I (m.s.) iff

$$EY_n\bar{Y}_m \to E|I|^2, \tag{3.12}$$

where Y_m corresponds to a second subdivision $a = t'_0 < t'_1 < \cdots < t'_m = b$ of $[a, b]$, and $t'_{q-1} < \xi'_q < t'_q (q = 1, 2, \ldots, m)$. Now we have

$$EY_n\bar{Y}_m = \sum_{p=1}^{n}\sum_{q=1}^{n}(t_p - t_{p-1})(t'_q - t'_{q-1})R(\xi'_q - \xi_p) \tag{3.13}$$

and if the limit of the double sum exists as

$$n \to \infty, \quad m \to \infty, \quad \max_{1\le p\le n} |t_p - t_{p-1}| \to 0,$$
$$\max_{1\le q\le n} |t'_q - t'_{q-1}| \to 0,$$

then it is the double integral (3.10). This proves the first part of the theorem. The second part follows from (3.12). □

We also note that if $m(t) = EX(t)$, then from (3.8)

$$EY_n = \sum_{p=1}^{n}(t_p - t_{p-1})m(\xi_p) \to \int_a^b m(t)dt,$$

and since

$$\lim E(Y_n) = E \lim Y_n = E\int_a^b X(t)dt,$$

we obtain the identity

$$E\int_a^b X(t)\,dt = \int_a^b EX(t)dt. \tag{3.14}$$

Finally, we prove the following inportant result.

Theorem 3.5 *A function $R(t)(t \in T)$, with $R(0) = 1$, is the correlation function of a stationary second order process which is continuous (m.s.) iff there exists a distribution function $F(\lambda)(-\infty < \lambda < \infty)$ such that*

$$R(t) = \int_{-\infty}^{\infty} e^{it\lambda}dF(\lambda) \quad (t \in T). \tag{3.15}$$

Proof. (i) Suppose that $R(t)$ is the correlation function of a stationary second order process which is continuous (m.s.). Then by Theorem 2.1(d), $R(t)$ is non-negative definite and by Theorem 3.2,

$R(t)$ is continuous at all t. The result (3.15) follows from a theorem of S. Bochner.

(ii) Conversely, suppose that $R(t)$ can be represented as in (3.15). Let Λ be a random variable with the distribution function F. Also let η be a random variable independent of Λ and with $E(\eta) = 0, E|\eta|^2 = 1$. For $t \in T$ define $X(t) = \eta e^{it\Lambda}$. Then the process $\{X(t), t \in T\}$ has the correlation function $R(t)$, since

$$\begin{aligned} EX(t)\bar{X}(t') &= E|\eta|^2 E e^{i(t-t')\Lambda} \\ &= \int_{-\infty}^{\infty} e^{i(t-t')\lambda} dF(\lambda) = R(t-t'). \end{aligned}$$

It is clear that $X(t)$ is continuous (m.s.) □

In the case of a discrete index set it can be verified that (3.15) reduces to

$$R_n = \int_{-\pi+}^{\pi} e^{in\lambda} dF_1(\lambda) \quad (n = \ldots -2, -1, 0, 1, \ldots), \tag{3.16}$$

where $F_1(\lambda)$ can be expressed in terms of $F(\lambda)$.

If the process is real, then $R(t)$ is real, and (3.15), reduces to

$$R(t) = \int_{-\infty}^{\infty} \cos t\lambda \, dF(\lambda). \tag{3.17}$$

Since $R(t) = R(-t)$, it turns out that F is symmetric about the origin.

Theorem 3.5 exhibits the correlation function $R(t)$ of a stationary second order process as the characteristic function of a distribution function $F(\lambda)$. We call $F(\lambda)$ the *spectral distribution function* of the process. For a given $R(t)$ we can obtain $F(\lambda)$ by using the inversion formula for characteristic functions. In particular, if

$$\int_{-\infty}^{\infty} |R(t)| dt < \infty, \tag{3.18}$$

then the *spectral density* $f(\lambda)$ exists and is given by

$$f(\lambda) = \frac{1}{2\pi} \int_{-\infty}^{\infty} e^{-it\lambda} R(t) dt. \tag{3.19}$$

In the general case

$$F(\lambda_1) - F(\lambda_2) = \lim_{T\to\infty} \frac{1}{2\pi} \int_{-T}^{T} \frac{e^{-i\lambda_1 t} - e^{-i\lambda_2 t}}{it} R(t) dt, \tag{3.20}$$

where λ_1 and λ_2 are continuity points of $F(\lambda)$.

Example 3.1 (White Noise) In Example 1.2 we showed that the correlation function of the white noise is given by $R_n = 1$ if $n = 0$ and $R_n = 0$ otherwise. We can write

$$R_n = \int_{-\pi}^{\pi} \cos n\lambda \, f(\lambda) d\lambda, \tag{3.21}$$

where

$$\begin{aligned} f(\lambda) &= \frac{1}{2\pi} \quad \text{for} - \pi < \lambda < \pi \\ &= 0 \text{ elsewhere.} \end{aligned} \tag{3.22}$$

Thus the spectral density is uniform in the interval $(-\pi, \pi)$.

Example 3.2 Let $X(t) = A\cos\theta t + B\sin\theta t$, as in Example 1.1 of Chap. 3. We saw that $X(t)$ is stationary in the wide sense and its correlation function is given by $R(t) = \cos\theta t$. We can write this as

$$R(t) = \int_{-\infty}^{\infty} \cos t\lambda \, dF(\lambda), \tag{3.23}$$

where

$$\begin{aligned} F(\lambda) &= 0 \quad \text{for } \lambda < -\theta \\ &= \frac{1}{2} \quad \text{for } - \theta \leq \lambda < \theta. \\ &= 1 \quad \text{for } \lambda \geq \theta. \end{aligned} \tag{3.24}$$

Thus the spectral distribution consist of two atoms at $-\theta, +\theta$, with weight $1/2$ at each atom.

Example 3.3 (The Ornstein–Uhlenbeck Process) This was defined in Example 3.3 of Chap. 3 as a stationary Gauss–Markov process $X(t)$. Its correlation function was found to be $R(t) = e^{-\beta|t|}$ with $0 \leq \beta < \infty$. Since $R(t)$ is continuous, the process is continuous (m.s.) by Theorem 3.2, and the spectral representation exists by Theorem 3.5. Since $e^{-\beta|t|}$ is known to be the characteristic function of the Cauchy density

$$f(\lambda) = \frac{1}{\pi} \frac{\beta}{\beta^2 + \lambda^2} \quad (-\infty < \lambda < \infty); \tag{3.25}$$

its follows that (3.25) gives the spectral density of the process.

Since

$$R'(t) = -\beta e^{-\beta t} \text{ for } t > 0, \quad \text{and} \quad = \beta e^{\beta t} \text{ for } t < 0,$$

the derivative $R'(0)$ does not exist. By Theorem 3.3 the derivative $X'(t)$ does not exist (m.s.). Since the condition (3.10) is clearly satisfied, the Riemann-integral

$$Y(t) = \int_0^t X(s)ds \tag{3.26}$$

exists (m.s.) by Theorem 3.4. Therefore, we can put

$$Y(t) = \lim_{n\to\infty} \sum_{j=1}^{n} \frac{t}{n} X\left(j\frac{t}{n}\right). \tag{3.27}$$

Since $X\left(\frac{t}{n}\right), X\left(2\frac{t}{n}\right), \ldots, X\left(n\frac{t}{n}\right)$ have an n-variate Gaussian distribution it follows that $Y(t)$ also has a Gaussian distribution if it can be shown that its variance is nonzero. We have assumed that $EX(t) = 0$ and $\mathrm{Var}X(t) = 1$. Using (3.14) we find that

$$EY(t) = \int_0^t EX(s)ds = 0 \tag{3.28}$$

and

$$\begin{aligned} EY(t)Y(t') &= \int_0^t EX(s)ds \int_0^{t'} X(s')\,ds' \\ &= \int_0^t \int_0^{t'} EX(s)X(s')dsds' = \int_0^{t'} \int_0^t e^{-\beta|s-s'|}dsds'. \end{aligned}$$

For $t < t'$ this gives

$$\begin{aligned} EY(t)Y(t') &= \int_0^t \left\{ \int_0^s e^{-\beta(s-s')}ds' + \int_s^t e^{\beta(s-s')}ds' \right\} ds \\ &= \frac{1}{\beta} \int_0^t [2 - e^{-\beta(t'-s)} - e^{-\beta s}]ds \\ &= \frac{1}{\beta^2}[2\beta t - 1 + e^{-\beta t} + e^{-\beta t'} - e^{-\beta(t'-t)}], \end{aligned}$$

and similarly for $t \geq t'$. Thus

$$EY(t)Y(t') = \frac{1}{\beta^2}[2\beta \min(t,t') - 1 + e^{-\beta t} + e^{-\beta t'} - e^{-\beta|t-t'|}]. \tag{3.29}$$

The variance is therefore given by

$$\mathrm{Var}Y(t) = EY(t)^2 = \frac{2}{\beta^2}[\beta t - 1 + e^{-\beta t}] \tag{3.29a}$$

which is positive for $\beta > 0$. We note that

$$\begin{aligned} \mathrm{Var}Y(t) &\sim \frac{2t}{\beta} \quad \text{for large } t \\ &\sim t^2 \quad \text{for small } t. \end{aligned} \tag{3.30}$$

The Ornstein–Uhlenbeck process $X(t)$ is taken as a representation of the velocity of small particles suspended in fluids, whose erratic movement was observed by the botanist Roberet Brown in 1827. The integrated process $Y(t)$ is then the position of the particles at time t. The above results shown that for small t, $Y(t)$ does not behave like the Brownian motion.

4.4 Time Series Models

A time series is a series of observations made in time. Examples of such series occur in several branches of science and engineering. If the observations are made at fixed epochs of time, we get a discrete time series. The use of mechanical or electronic equipment makes it possible to record data continuously in time, yielding a continuous time series. We shall use the notation $\{X_n\}$ for a discrete time series and $\{X(t)\}$ for a continuous time series.

The raw data obtained usually consists of a long-term trend and seasonal fluctuations which are purely deterministic in nature and a random residual. In the classical approach these three components are assumed to be additive, so that

$$X(t) = m(t) + s(t) + Y(t), \tag{4.1}$$

where $m(t)$ is the trend, $s(t)$ represents the seasonal effect, and $Y(t)$ is the residual. As a first step in the analysis $m(t)$ and $s(t)$ are estimated and eliminated from the observed data, with the expectation that the residual $Y(t) = X(t) - M(t) - s(t)$ represents a stationary process. In modern approaches, other techniques are used to eliminate $m(t)$ and $s(t)$.

The analysis of time series based on models such as moving average, autoregression, and mixed models that combine these two, is called an analysis in the *time-domain*. The choice of the model often starts with the study of the properties of the correlation function. The analysis based on the spectral distribution is called an analysis in the *frequency domain.*

We now discuss the two important models (moving average and autoregression) that are frequently used in time series analysis. Special cases of these were discussed in Examples 1.3 and 1.4.

Example 4.1 (Moving Average) Let

$$X_n = \sum_{j=0}^{p} a_j \, \xi_{n-j} \quad (n = \cdots - 2, -1, 0, 1, \ldots), \tag{4.2}$$

where $\{\xi_n\}$ is a white noise with variance σ^2 and a_j are complex constants. We have $E(X_n) = 0$ and

$$\begin{aligned} E(X_m \bar{X}_n) &= \sum_{j=0}^{p} \sum_{k=0}^{p} a_j \bar{a}_k E(\xi_{m-j} \xi_{n-k}) \\ &= \sigma^2 \sum_{\substack{0 \le j \le p \\ 0 \le n-m+j \le p}} a_j \bar{a}_{n-m+j}. \end{aligned} \tag{4.3}$$

In particular, the variance is given by

$$\sigma_X^2 = E|X_m|^2 = \sigma^2 \sum_{0}^{p} |a_j|^2. \tag{4.4}$$

Therefore, the process is stationary in the wide sense and its correlation function is given by

$$R_n = \frac{\sigma^2}{\sigma_X^2} \sum_{\substack{0 \le j \le p \\ 0 \le n+j \le p}} a_j \bar{a}_{n+j}. \tag{4.5}$$

It is clear that $R_n = 0$ for $|n| > p$. Since

$$\frac{\sigma^2}{\sigma_X^2} \left| \sum_{-\infty}^{\infty} \sum_{j} a_j \bar{a}_{n+j} \right| \le \left| \sum_{0}^{p} a_j \right|^2 < \infty,$$

the spectral density exists and is given by

$$\begin{aligned} f(\lambda) = \frac{1}{2\pi} \sum_{-\infty}^{\infty} e^{-in\lambda} R_n &= \frac{\sigma^2}{2\pi\sigma_X^2} \sum_{j=0}^{p} a_j e^{ij\lambda} \sum_{n=-j}^{p-j} \bar{a}_{n+j} e^{-i(n+j)\lambda} \\ = \frac{\sigma^2}{2\pi\sigma_X^2} \sum_{0}^{p} a_j e^{ij\lambda} \sum_{0}^{p} \bar{a}_k e^{-ik\lambda} &= \frac{\sigma^2}{2\pi\sigma_X^2} \left| \sum_{0}^{p} a_j e^{ij\lambda} \right|^2. \end{aligned} \tag{4.6}$$

Example 4.2 (Multiple Autoregression) Let $\{X_n\}$ be a wide sense stationary process and

$$X_n = \beta_1 X_{n-1} + \beta_2 X_{n-2} + \cdots + \beta_p X_{n-p} + \xi_{n-p}, \tag{4.7}$$

where $\{\xi_n\}$ is a white noise with variance σ^2. In the linear case ($p = 1$) of Example 3.1 we solved the *difference equation* (4.7) recursively. In the general case ($p \geq 1$) we introduce the difference operator E by setting $EX_n = X_{n+1}$. The operators $E^2, E^3, \ldots$ are defined in the usual manner, namely, $E^2 = E \cdot E, E^3 = E \cdot E \cdot E$, and so on. We define the operator E^{-1} by similar properties. We can then write (4.7) as

$$(E^p - \beta_1 E^{p-1} - \beta_2 E^{p-2} - \cdots - \beta_p I) X_n = \xi_n. \tag{4.8}$$

The complementary solution of (4.8) is given by

$$X_n = c_1 \lambda_1^n + c_2 \lambda_2^n + \cdots + c_p \lambda_p^n, \tag{4.9}$$

where $\lambda_1, \lambda_2, \ldots \lambda_p$ are the roots of the *characteristic equation*

$$\lambda^p - \beta_1 \lambda^{p-1} - \beta_2 \lambda^{p-2} - \cdots - \beta_p = 0,$$

assumed to be distinct. For X_n to be stationary we must have $|\lambda_i| < 1$ ($i = 1, 2, \ldots, p$). If $\lambda_1 = \lambda_2$ is a double root, then the first term on the right side of (4.9) is replaced by $c_1 n \lambda_1^n$ and again we must have $|\lambda_1| < 1$. We assume this to be the case and so the complementary solution does not contribute to the stationary solution.

To obtain the particular solution of (4.8) we write for $|\lambda| > 1$,

$$\begin{aligned}
&(\lambda^p - \beta_1 \lambda^{p-1} - \beta_2 \lambda^{p-2} - \cdots - \beta_p)^{-1} \\
&\quad = \sum_{r=1}^{p} \frac{A_r}{\lambda - \lambda_r} = \sum_{r=1}^{p} A_r \sum_{j=0}^{\infty} \lambda_r^j \lambda^{-j-1} = \sum_{j=0}^{\infty} a_j \lambda^{-j-1}.
\end{aligned} \tag{4.10}$$

To obtain the coefficients a_j we note that for $|z| < 1$,

$$\begin{aligned}
\sum_0^{\infty} a_j z^{j+1} &= (z^{-p} - \beta_1 z^{-p+1} - \beta_2 z^{-p+2} - \cdots - \beta_p)^{-1} \\
&= z^p \left(1 - \sum_1^{\infty} \beta_j z^j\right)^{-1},
\end{aligned}$$

so that

$$\sum_{0}^{\infty} a_j z^j = \frac{z^{p-1}}{1 - \sum_1^p \beta_j z^j} \quad (|z| < 1). \tag{4.11}$$

Applying (4.10) to (4.8) we find that the particular solution is given by

$$X_n = (E^p - \beta_1 E^{p-1} - \beta_2 E^{p-2} - \cdots - \beta_p I)^{-1} \xi_n = \sum_{0}^{\infty} a_j \xi_{n-j-1}. \tag{4.12}$$

We have thus obtained a moving average with infinite number of terms. It can be easily verified that $\sum |a_j|^2 < \infty$. Proceeding as in Example 4.1 we obtain

$$\sigma_X^2 = E|X_m|^2 = \sigma^2 \sum_{0}^{\infty} |a_j|^2 \tag{4.13}$$

and

$$R_n = \frac{\sigma^2}{\sigma_X^2} \sum_{j \geq \max(0,-n)} a_j \bar{a}_{n+j}. \tag{4.14}$$

The spectral density is given by

$$f(\lambda) = \frac{\sigma^2}{2\pi\sigma_X^2} \left| \sum_{0}^{\infty} a_j e^{ij\lambda} \right|^2. \tag{4.15}$$

4.5 Mean Square Convergence

The sequence $\{X_n\}$ with $E|X_n|^2 < \infty$ converges to X in the mean square (m.s.) if

$$E|X_n - X|^2 \to 0 \quad \text{as } n \to \infty. \tag{5.1}$$

Convergence in the m.s. implies convergence in probability; for, by Chebyshev's inequality

$$P\{|X_n - X| \geq \varepsilon\} \leq \frac{1}{\varepsilon^2} E|X_n - X|^2 \to 0 \quad \text{as } n \to \infty \tag{5.2}$$

if $X_n \to X$ (m.s.). Conversely, convergence in probability does not always imply convergence (m.s.).

If $X_n \to X$ (m.s.), then since

$$|X_m - X_n|^2 \leq 2|X_m - X|^2 + 2|X_n - X|^2,$$

we find that

$$E|X_m - X_n|^2 \leq 2E|X_m - X|^2 + 2E|X_n - X|^2 \to 0 \tag{5.3}$$

as $m, n \to \infty$. Conversely, if $E|X_m - X_n|^2 \to 0$ as $m, n \to \infty$, then it can be proved that $X_n \to X$ (m.s.). Thus we have the m.s. criterion of convergence: $X_n \to X$ (m.s.) iff

$$E|X_m - X_n|^2 \to 0 \quad \text{as } m, n \to \infty. \tag{5.4}$$

Let $E|X_n|^2 < \infty$ for each n. If $E|X_n - X|^2 \to 0$ as $n \to \infty$, then $E|X_n - X|^2$ remains finite for all $n >$ some N. Since

$$E|X|^2 \leq 2E|X_n|^2 + 2E|X_n - X|^2,$$

if follows that $E|X|^2 < \infty$. Also, since

$$|E(X_n - X)| \leq E|X_n - X| \leq \sqrt{E|X_n - X|^2} \to 0$$

as $n \to \infty$, we find that

$$\lim_{n\to\infty} E(X_n) = E \lim_{n\to\infty} X_n \text{ (m.s.)}. \tag{5.5}$$

4.6 Problems for Solution

1. Suppose that $\{X(t), t \in T\}$ and $\{Y(t), t \in T\}$ are stationary processes which are uncorrelated; that is, $X(t)$ and $Y(t')$ are uncorrelated for every t, t'. Denote $Z(t) = X(t) + Y(t)$ for $t \in T$. Show that the process $\{Z(t)\}$ is also stationary and that its covariance function is the sum of the covariance functions of $\{X(t)\}$ and $\{Y(t)\}$.
2. Suppose that $\{X_n, n \geq 0\}$ is a sequence of independent and identically distributed random variables with mean 0 and variance σ^2. A second sequence $\{Y_n, n \geq 1\}$ it defined as in (i) or (ii) below. In each case verify whether $\{Y_n\}$ is stationary, and if it is, find the covariance function.

   (i) $Y_n = X_n X_{n-1}$
   (ii) $Y_n = X_n \cos n\theta + X_{n-1} \sin n\theta$, where θ is a constant.

3. *Chebyshev's Inequality.* If $X(t)$ is a real stationary process with the correlation function $R(t)$, shown that for any constant a,

$$P\{|X(s+t) - X(s) \geq a\} \leq 2\frac{R(0) - R(t)}{a^2}.$$

4. For the moving average of Example 1.3, show that the spectral density is given by

$$f(\lambda) = \frac{1}{2\pi(p+1)} \cdot \frac{1 - \cos(p+1)\lambda}{1 - \cos\lambda} \quad (-\infty < \lambda < \infty).$$

5. For the linear autoregression of Example 1.4, show that the spectral density is given by

$$f(\lambda) = \frac{1}{2\pi} \cdot \frac{1 - \beta^2}{1 - 2\beta\cos\lambda + \beta^2} \quad (-\infty < \lambda < \infty).$$

6. *Autoregression of Order 2.* Let $\{X_n\}$ be a wide sense stationary process and

$$X_n = \alpha X_{n-1} + \beta X_{n-2} + \xi_{n-2},$$

where $\{\xi_n\}$ is a white noise with variance σ^2. Assuming that $-1 < \beta < -\alpha^2/4$, show that the correlation function is given by

$$R_n = r^n \frac{\sin(n\theta + \psi)}{\sin\psi},$$

where $r = \sqrt{-\beta}, \theta = \text{arc tan}(-1 - 4\beta\alpha^{-2})$, and $\psi = \text{arc tan}[(1 - \beta)\tan\theta/(1+\beta)]$.

7. *Linear Filter.* Let $\{X_n\}$ be a wide sense stationary process with mean 0, variance σ_X^2, and spectral density $f_X(\lambda)$. A second sequence $\{Y_n\}$ is defined by

$$Y_n = \sum_{p=0}^{N} a_p X_{n-p},$$

where $a_0, a_1, \ldots, a_N$ are complex constants. Show that $\{Y_n\}$ is also wide sense stationary, and its spectral density is given by

$$f_Y(\lambda) = \frac{\sigma_X^2}{\sigma_Y^2} \left| \sum_0^N a_p e^{ip\lambda} \right|^2 f_X(\lambda),$$

where $\sigma_Y^2 = E|Y_n|^2$.

8. Let

$$X(t) = \sum_{p=1}^{N} c_p \cos(A_p t + U_p),$$

where $c_1, c_2, \ldots, c_N$ are real constants, and $A_1, A_2, \ldots, A_N, U_1, U_2, \ldots, U_N$ are independent random variables, with A_p having d.f. $F_p(x)$ and U_p uniformly distributed in $(0, 2\pi)(p = 1, 2, \ldots, N)$. Show that the process $\{X(t)\}$ is stationary in the wide sense, with spectral d.f.

$$F(\lambda) = \sum_{p=1}^{N} d_p^2 \tilde{F}_p(\lambda),$$

where $d_p^2 = c_p^2/(c_1^2 + c_2^2 + \cdots + c_N^2), (p = 1, 2, \ldots, N)$, and

$$\tilde{F}_p(\lambda) = \frac{1}{2}[F_p(\lambda) + 1 - F_{p(-\lambda-)}].$$

9. Let

$$X(t) = \sum_{p=1}^{N} (A_p \cos \theta_p t + B_p \sin \theta_p t),$$

where $\theta_p > 0, A_1, A_2, \ldots, A_N,\ B_1, B_2, \ldots, B_N$ are uncorrelated random variables with mean 0 and

$$\mathrm{Var}(A_p) = \mathrm{Var}(B_p) = \sigma_p^2 \quad (p = 1, 2, \ldots, N).$$

Show that $\{X(t)\}$ is stationary in the wide sense, and its spectral distribution consists of weights $\frac{1}{2}\sigma_p^2/\sigma^2$ at the atoms $\pm\theta_p$ $(p = 1, 2, \ldots, N)$, where $\sigma^2 = \sum_1^N \sigma_p^2$.

10. Let

$$X(t) = \sum_{1}^{N} A_p e^{i\theta_p t},$$

where $\theta_1, \theta_2, \ldots, \theta_N$ are real constants, $A_1, A_2, \ldots, A_N$ are real random variables which are uncorrelated, and have mean 0 and $\mathrm{Var}(A_p) = \sigma_p^2 (p = 1, 2, \ldots, N)$. This is Example 1.3 or Chap. 2, where we proved that $\{X(t)\}$ is a wide sense stationary process. Assume without loss of generality that $\theta_1, \theta_2, \ldots, \theta_N$ are all distinct, and $\theta_1 < \theta_2 < \cdots < \theta_N$. Prove that the spectral distribution consists of weights σ_p^2/σ^2 at the atoms θ_p $(p = 1, 2, \ldots, N)$, where $\sigma^2 = \sum_1^N \sigma_p^2$.

11. *Shot Noise.* Let $T_0 = 0$ and $T_1, T_2, \ldots$ be the epochs of successive events in a Poisson process $X(t)$ with parameter λ. Also, let $h(t)$ be an ordinary (non-random) function such that

$$\left| \int_{-\infty}^{\infty} h(s)h(s+t)ds \right| < \infty (-\infty < t < \infty).$$

Define

$$Y(t) = \sum_{n=0}^{\infty} h(t - T_n).$$

Show that $\{Y(t)\}$ is stationary in the wide sense and that its spectral distribution function $F(\lambda)$ is given by

$$F(\lambda_1) - F(\lambda_2) = \int_{-\infty}^{\infty} h(s)G(\lambda_1, \lambda_2, s)ds / \|h\|^2,$$

where

$$G(\lambda_1, \lambda_2, s) = \lim_{T \to \infty} \int_{-T}^{T} h(s+t) \frac{e^{-i\lambda_1 t} - e^{-i\lambda_2 t}}{it} dt$$

$$\|h\|^2 = \int_{-\infty}^{\infty} h(s)^2 ds,$$

and λ_1, λ_2 are points of continuity of $F(\lambda)$.

12. *Electron Transit.* Let $\{T_n, n = \cdots - 1, 0, 1, 2, \ldots\}$ be the epochs of successive events in a Poisson process with parameter λ. Define

$$Y(t) = \sum_{\{n: T_n \le t\}} (t - T_n),$$

where the sum if taken over n such that $t - t_0 \le T_n < t$ and $t_0 > 0$ is a given constant. Show that $\{Y(t), t \ge 0\}$ is stationary in the wide sense with

$$EY(t) = \frac{1}{2}\lambda t_0^2,$$

$$EY(t)Y(t') = \frac{1}{3}\lambda t_0^3 \left[\left(1 - \frac{|t'-t|}{t_0}\right)^3 + \frac{3}{2}\frac{|t'-t|}{t_0}\left(1 - \frac{|t'-t|}{t_0}\right)^2 \right].$$

13. Let T_n be as in Problem 12, and

$$Y(t) = \sum_{\{n:T_n \leq t\}} e^{-\beta(t-T_n)} (0 < \beta < \infty).$$

Show that $\{Y(t)\}$ is stationary in the wide sense and its correlation function is given by $R(t) = e^{-\beta|t|}$.

Further Reading

Abraham, B and J Ledolter, (1983). *Statistical Methods for Forecasting.* New York: John Wiley.

Brockwell, PJ and RA Davis, (1987). *Time Series: Theory and Methods.* New York: Springer-Verlag.

Chatfield, C, (1984). *The Analysis of Time Series: An Introduction.* 3rd Ed. London: Chapman and Hall.

CHAPTER 5

The Brownian Motion and the Poisson Process: Lévy Processes

5.1 The Brownian Motion

5.1.1 *Historical Remarks*

Robert Brown (1828) observed that pollen grains suspended in water perform a continual swarming motion.

Louis Bachelier (1900) defended his doctorate thesis, "Theory of Speculation" at the Sorbonne. In particular, he proved that the transition density function $f(x_0; x, t)$ of the Brownian Motion satisfies the heat equation

$$\frac{\partial f}{\partial t} = \frac{1}{2}\frac{\partial^2 f}{\partial x^2}.$$

Albert Einstein (1905) also derived this equation from considerations of statistical mechanics.

Norbert Wiener (1923) put the Brownian Motion on a firm mathematical foundation, after the ideas of Borel, Lebesgue, and Daniell appeared. Specifically, Wiener considered the space of continuous functions $\omega : t \in [0, \infty) \to \mathbb{R}$ to attach a precise meaning to Bachelier's statement that Brownian sample functions are continuous.

Paul Lévy (1939) found another construction of the Brownian Motion and in a 1948 monograph gave a profound description of the fine structure of individual Brownian sample functions.

A Note on Terminology. The Brownian Motion is also called the Wiener process, the Wiener–Einstein process, and the Wiener–Bachelier process. We shall continue to call it the Brownian Motion.

5.1.2 *Introduction*

(1) The Brownian Motion is a process $\{X(t), t \geq 0\}$ with stationary independent increments such that for $s < t$ the increment $X(t) - X(s)$ has a normal density with mean 0 and variance $\sigma^2|t - s|$.

If $\sigma = 1$ the Brownian Motion is standard.

(2) A Brownian Motion is Gaussian with

$$EX(t) = 0, \quad \Gamma(t, t') = EX(t)X(t') = \sigma^2 \min(t, t'). \tag{1.1}$$

(3) A Brownian Motion is Markov, with transition d.f.

$$\begin{aligned} F(x; y, t) &= P\{X(s+t) \le y \mid X(s) = x\} \\ &= P\{X(s+t) - X(s) \le y - x \mid X(s) = x\} \\ &= P\{X(t) \le y - x\} = N\left(\frac{y-x}{\sigma\sqrt{t}}\right). \end{aligned} \tag{1.2}$$

Thus the transition density is given by

$$f(x; y, t) = \frac{1}{\sigma\sqrt{t}} n\left(\frac{y-x}{\sigma\sqrt{t}}\right). \tag{1.3}$$

(4) We are interested in the following:

(i) The extremum functionals:

$$M(t) = \max_{0 \le s \le t} X(s), \quad m(t) = \min_{0 \le s \le t} X(s). \tag{1.4}$$

(ii) The hitting times

$$T_{ab} = \min\{t \colon X(t) = b\} \text{ on } \{X(0) = a\} \tag{1.5}$$

(that is, on the subclass of sample functions that have initial state a). We have

$$\begin{aligned} T_{ab} &= \min\{t \colon X(t) - X(0) = b - a\} \text{ on } \{X(0) = a\} \\ &\stackrel{d}{=} \min\{t \colon X(t) = b - a\} \text{ on } \{X(0) = 0\} \\ &= T_{0,b-a}. \end{aligned}$$

By symmetry we also find that $T_{0x} = T_0, -x$. Therefore it suffices to consider the random variable T_{0x} $(x > 0)$.

(5) *Stopping Times.* A random variable τ taking values on $[0, \infty]$ is called a stopping time with respect to a stochastic process $\{X(t), t \ge 0\}$ if the event $\{\tau \le t\}$ depends on the sample path of the process up to time t and nothing beyond.

For the hitting time $\tau_a = T_{0a}$ we have

$$\{\tau_a \le t\}^c = \{\tau_a > t\} = \{X(s) < a \ (0 \le s \le t)\}. \tag{1.6}$$

This shows that hitting times of a Brownian Mation are stopping times. (This holds for all Lévy processes.)

(6) *The Strong Markov Property.* The Markov property states that

$$P\{X(\tau+t) \leq y \mid X(s)(0 \leq s \leq \tau)\} = P\{X(\tau+t) \leq y \mid X(\tau)\} \quad (1.7)$$

for a fixed time τ. Frequently we have to calculate probabilities of the type (1.7) where τ is a random variable. The Markov property need not necessarily hold for τ, because, for example, τ may involve a knowledge of the future. However, G. A. Hunt proved that the Markov property holds for the Brownian Motion when τ is a stopping time. We characterize this as the strong Markov property of the Brownian Motion. (Hunt's proof can be extended to Lévy processes.)

(7) *The Set of Zeros.* Let $Z = \{t \geq 0 \colon X(t) = 0\} =$ set of zeros of the Brownian Motion. It is known that Z is a Cantor set (closed, uncountable, of Lebesgue measure zero).

5.1.3 *Properties of the Brownian Motion*

We state the following result without proof.

Theorem 1.1 *Brownian sample functions are* (i) *everywhere continuous, but* (ii) *nowhere differentiable. (In each case the exception set has measure zero.)*

Theorem 1.2 *Brownian sample functions are of unbounded variation with probability* 1.

Proof. It suffices to consider the interval $[0,1]$. Divide this interval into 2^n subintervals of equal length and consider the random variable

$$v_n = \sum_{k=1}^{2^n} \left| X\left(\frac{k}{2^n}\right) - X\left(\frac{k-1}{2^n}\right) \right|.$$

Clearly, $v_n \geq 0$. If we divide the subinterval $((k-1)2^{-n}, k2^{-n}]$ into two subintervals $\left((k-1)2^{-n}, (2k-1)2^{-n-1}\right]$ and $\left((2k-1)2^{-n-1}, k2^{-n}\right]$ we find that

$$\left| X\left(\frac{k}{2^n}\right) - X\left(\frac{k-1}{2^n}\right) \right| \leq \left| X\left(\frac{2k-1}{2^{n+1}}\right) - X\left(\frac{k-1}{2^n}\right) \right| + \left| X\left(\frac{k}{2^n}\right) - X\left(\frac{2k-1}{2^{n+1}}\right) \right|.$$

This shows that $v_n \leq v_{n+1}$, so that $\{v_n\}$ is a non-decreasing sequence. We shall show that $E(e^{-v_n}) \to 0$ as $n \to \infty$. It will then follow that $v_n \to \infty$ with probability 1. Now

$$\begin{aligned} E(e^{-v_n}) &= Ee^{-\sum_1^{2^n} X(\frac{k}{2^n})-X(\frac{k-1}{2^n})} \\ &= E\prod_1^{2^n} e^{-|X(\frac{k}{2^n})-X(\frac{k-1}{2^n})|} \\ &= \prod_1^{2^n} Ee^{-|X(\frac{k}{2^n})-X(\frac{k-1}{2^n})|} \\ &= \prod_1^{2^n} Ee^{-|X(\frac{1}{2^n})|} = \left[Ee^{-|X(\frac{1}{2^n})|}\right]^{2^n} \end{aligned}$$

where we have used the fact that $X(t)$ has stationary independent increments. For $x \geq 0$ we have $e^{-x} \leq 1 - x + \frac{x^2}{2}$, so that

$$\begin{aligned} Ee^{-|X(\frac{1}{2^n})|} &\leq E\left[1 - \left|X\left(\frac{1}{2^n}\right)\right| + \frac{1}{2}\left|X\left(\frac{1}{2^n}\right)\right|^2\right] \\ &= 1 - \sqrt{\frac{2}{\pi}} \cdot \frac{1}{2^{n/2}} + \frac{1}{2} \cdot \frac{1}{2^n}. \end{aligned}$$

Thus

$$E\left(e^{-v_n}\right) \leq \left(1 - \sqrt{\frac{2}{\pi}}, 2^{-\frac{n}{2}} + \frac{1}{2}2^{-n}\right)^{2^n} \to 0 \text{ as } n \to \infty.$$

□

Theorem 1.3 (Reflection Principle of D. André) *For $x > 0$ and $I \subset (0, \infty)$ we have*

$$P\{X(t) \in I, m(t) > 0 \mid X(0) = x\} = \int_I {}^\circ f(x; y, t)dy, \tag{1.8}$$

where

$${}^\circ f(x; y, t) = f(x; y, t) - f(-x; y, t). \tag{1.9}$$

Proof. We have

$$\begin{aligned} P\{X(t) \in I \mid X(0) = x\} &= P\{X(t) \in I, m(t) > 0 \mid X(0) = x\} \\ &\quad + P\{X(t) \in I, m(t) \leq 0 \mid X(0) = x\}. \end{aligned}$$

We proceed to evaluate this last probability. Given a sample function $X(s)$ $(0 \leq s \leq t)$ with $X(0) = x > 0$, $X(t) \in I, m(t) \leq 0$ we define a new sample function $Y(s)$ $(0 \leq s \leq t)$ as follows:

$$\begin{aligned} Y(s) &= -X(s) \quad \text{for } 0 \leq s \leq \tau \\ &= X(s) \quad \text{for } \tau < s \leq t \end{aligned}$$

where $\tau = \min\{t\colon X(t) = 0\}$ and $\tau \leq t$ since $m(t) \leq 0$. Clearly, there is a one-to-one correspondence between the sample functions of $X(s)$ of this type and those of $Y(s)$ $(0 \leq s \leq t)$. Note that τ is a stopping time and $X(t)$ is strongly Markov; the process $Y(s)$ $(0 \leq s \leq \tau)$ inherit these properties. Therefore,

$$P\{X(t) \in I, m(t) \leq 0 \mid X(0) = x\} = P\{Y(t) \in I \mid X(0) = -x\}.$$

and so

$$\begin{aligned} &P\{X(t) \in I, m(t) > 0 \mid X(0) = x\} \\ &\quad = P\{X(t) \in I \mid X(0) = x\} - P\{Y(t) \in I \mid Y(0) = -x\} \\ &\quad = \int_I f(x; y, t)dy - \int_I f(-x; y, t) \end{aligned}$$

as desired. □

Theorem 1.4 *The hitting time T_{0a} $(a > 0)$ has the stable d.f. $G_{1/2}\left(\frac{t}{a^2}\right)$. We have $T_{0a} < \infty$ with probability 1, but $E(T_{0a}) = \infty$.*

Proof. Since $T_{0a} \stackrel{d}{=} T_{a0}$, it suffices to consider T_{a0}. Choosing $I = (0, \infty)$ and $x = a > 0$ in Theorem 1.3 we find that

$$\begin{aligned} P\{T_{a0} > t\} &= P\{m(t) > 0 \mid X(0) = a\} \\ &= \int_0^\infty [f(a; y, t) - f(-a; y, t)]dy \\ &= [1 - F(a; 0, t)] - [1 - F(-a; 0, t)] \\ &= F(-a; 0, t) - F(a; 0, t) \\ &= N\left(\frac{a}{\sqrt{t}}\right) - N\left(\frac{-a}{\sqrt{t}}\right) = 2\left[N\left(\frac{a}{\sqrt{t}}\right) - \frac{1}{2}\right] \\ &= 2N\left(\frac{a}{\sqrt{t}}\right) - 1. \end{aligned}$$

Therefore,

$$P\{T_{a0} \le t\} = 2\left[1 - N\left(\frac{a}{\sqrt{t}}\right)\right] = G_{1/2}\left(\frac{t}{a^2}\right). \tag{1.10}$$

This gives

$$P\{T_{a0} < \infty\} = \lim_{t\to\infty} G_{1/2}\left(\frac{t}{a^2}\right) = 1.$$

□

Theorem 1.5 *Let $X(0) = 0$. Then we have the following.*

(i) $$P\{M(t) < x\} = N_+\left(\frac{x}{\sqrt{t}}\right) \tag{1.11}$$

$$EM(t) = \sqrt{\frac{2t}{\pi}}, \quad \mathrm{Var}M(t) = \left(1 - \frac{2}{\pi}\right)t \tag{1.12}$$

(ii) $$P\{m(t) < x\} = N_-\left(\frac{x}{\sqrt{t}}\right) \tag{1.13}$$

$$Em(t) = -\sqrt{\frac{2t}{\pi}}, \quad \mathrm{Var}m(t) = \left(1 - \frac{2}{\pi}\right)t \tag{1.14}$$

(iii) *As $t \to \infty, M(t) \to \infty$ and $m(t) \to -\infty$ with probability* 1.

Proof. (i) We have from Theorem 1.4

$$P\{M(t) < x\} = P\{T_{0x} > t\} = 1 - G_{1/2}\left(\frac{t}{x^2}\right)$$
$$= 1 - 2\left[1 - N\left(\frac{x}{\sqrt{t}}\right)\right] = 2N\left(\frac{x}{\sqrt{t}}\right) - 1 = N_+\left(\frac{x}{\sqrt{t}}\right).$$

(ii) We have

$$m(t) = \min_{0\le s\le t} X(s) \stackrel{d}{=} \min_{0\le s\le t} [-X(s)] = -\max_{0\le s\le t} [X(s)] = -M(t)$$

and the results for $m(t)$ follow from those of $M(t)$.

(iii) Clearly, when $M(t) \ge 0, M(t)$ is non-decreasing. Therefore, $M(t) \to M \le \infty$ with probability 1 and

$$P\{M(t) < x\} \to P\{M < x\} \quad \text{as } t \to \infty$$

But, from (i)

$$P\{M(t) < x\} \to 0 \quad \text{as } t \to \infty.$$

This gives $P\{M < x\} = 0$ which implies that $M = \infty$ with probability 1. Similarly, $m = -\infty$ with probability 1. □

Theorem 1.6 *Let Z be the set of zeros of the Brownian Motion, and denote*

$$\begin{aligned} z_-(t) &= \sup\{t' < t \colon t' \in Z\} \\ &= \textit{largest zero not exceeding } t \end{aligned} \tag{1.15}$$

$$\begin{aligned} z_+(t) &= \inf\{t' > t \colon t' \in Z\} \\ &= \textit{smallest zero exceeding } t. \end{aligned} \tag{1.16}$$

Then for $t_1 < t \leq t_2$ we have

$$P\{z_-(t) \leq t_1, z_+(t) > t_2\} = \frac{2}{\pi} \sin^- \sqrt{\frac{t_1}{t_2}}. \tag{1.17}$$

Proof. The required probability equals the probability that there is no zero in the interval $(t_1, t_2]$. Therefore, it is

$$\begin{aligned} &= \int_0^\infty f(0; x, t_1) P\{T_{x0} > t_2 - t_1\} dx \\ &\quad + \int_0^\infty f(0; -x, t_1) P\{T_{-x,0} > t_2 - t_1\} dx. \end{aligned}$$

Now since $f(0; x, t_1) = f(0; -x, t_1)$ and $T_{x0} \stackrel{d}{=} T_{-x,0}$ the two integrals add upto

$$\begin{aligned} &2 \int_0^\infty f(0; x, t_1) P\{T_{x0} > t_2 - t_1\} dx \\ &= 2 \int_0^\infty \frac{1}{\sqrt{2\pi t_1}} e^{-\frac{1}{2}\frac{x^2}{t_1}} dx \int_{t_2 - t_1}^\infty \frac{x}{\sqrt{2\pi s^3}} e^{-\frac{1}{2}\frac{x^2}{s}} ds \\ &= \frac{1}{\pi} \int_{t_2 - t_1}^\infty \frac{ds}{\sqrt{t_1 s^3}} \int_0^\infty x e^{-\frac{1}{2}\left(\frac{1}{t_1} + \frac{1}{s}\right) x^2} dx \\ &= \frac{1}{\pi} \int_{t_2 - t_1}^\infty \frac{ds}{\sqrt{t_1 s^3}} \cdot \frac{1}{\frac{1}{t_1} + \frac{1}{s}} = \frac{\sqrt{t_1}}{\pi} \int_{t_2 - t_1}^\infty \frac{ds}{\sqrt{s}(t_1 + s)} \quad (s = t_1 \tan^2 \theta) \\ &= \frac{\sqrt{t_1}}{\pi} \int_{\cos^{-1}\sqrt{\frac{t_1}{t_2}}}^{\frac{\pi}{2}} \frac{2d\theta}{\sqrt{t_1}} = \frac{2}{\pi}\left(\frac{\pi}{2} - \cos^{-1}\sqrt{\frac{t_1}{t_2}}\right) = \frac{2}{\pi} \sin^{-1} \sqrt{\frac{t_1}{t_2}}. \end{aligned}$$

□

Theorem 1.7 *Let $X(0) = 0$. The conditional process*

$$\{X(t)(t_1 \leq t \leq t_2) \mid X(t_1), X(t_2)\} \tag{1.18}$$

is Gaussian with

$$E[X(t) \mid X(t_1), X(t_2)] = X(t_1) + \frac{t - t_1}{t_2 - t_1}[X(t_2) - X(t_1)]$$

$$(P.\ \textit{Lévy's interpolation formula}) \tag{1.19}$$

$$\mathrm{Var}[X(t) \mid X(t_1), X(t_2)] = \frac{(t - t_1)(t_2 - t)}{(t_2 - t_1)} \tag{1.20}$$

$$\mathrm{Covar}\{X(t), X(t') \mid X(t_1), X(t_2)\} = \frac{(t - t_1)(t_2 - t')}{(t_2 - t_1)}$$

$$(t_1 \leq t < t' \leq t_2). \tag{1.21}$$

Proof. For $t \geq t_1$ we denote $X(t) = X(t_1) + Y(t - t_1)$, where Y is the Brownian Motion with $Y(0) = 0$ and $Y(t_2 - t_1) = X(t_2) - X(t_1)$. Then for $t_1 \leq t \leq t_2, X(t)$ is also Brownian with $X(t_1), X(t_2)$ given. Now let $X_1 = Y(t_2 - t_1), X_2 = Y(t - t_1), X_3 = Y(t' - t_1)$. From the Gaussian property of the Brownian it follows that (X_1, X_2, X_3) have a joint normal distribution with zero means and

$$\mathrm{Var}(X_1) = t_2 - t_1,\ \ \mathrm{Var}(X_2) = t - t_1,\ \ \mathrm{Var}(X_3) = t' - t_1,$$
$$\mathrm{Covar}(X_1, X_2) = t - t_1,\ \ \mathrm{Covar}(X_1, X_3) = t' - t_1,\ \ \mathrm{Covar}(X_2, X_3) = t - t_1.$$

The results (1.19)–(1.21) follow from the relations (in the standard notation)

$$E(X_2 \mid X_1) = \rho_{12}\frac{\sigma_2}{\sigma_1}X_1, \quad \mathrm{Var}(X_2|X_1) = \sigma_2^2(1 - \rho_{12}^2).$$

Using these we also find that

$$E(X_2X_3 \mid X_1) = EE[X_2X_3 \mid X_3, X_1] = \rho_{23}\frac{\sigma_2}{\sigma_3}E\left(X_3^2 \mid X_1\right)$$
$$= \rho_{23}\frac{\sigma_2}{\sigma_3}\sigma_3^2(1 - \rho_{13}^2) = \mathrm{Covar}(X_2, X_3)(1 - \rho_{13}^2).$$

□

Remark. The conditional stochastic process $\{X(t)(0 \leq t \leq 1) \mid X(1) = 0\}$ is called the *Brownian Bridge*. Theorem 1.7 shows that the Brownian Bridge is Gaussian with

$$E[X(t) \mid X(1) = 0] = 0, \quad \mathrm{Var}[X(t) \mid X(1) = 0] = t(1 - t) \tag{1.22}$$

$$\mathrm{Covar}\{X(s), X(t) \mid X(1) = 0\} = s(1 - t) \quad (0 \leq s < t < 1). \tag{1.23}$$

5.2 The Poisson Process

5.2.1 *Introduction*

(1) The Poisson process is a stochastic process $\{X(t), t \geq 0\}$ with stationary independent increments such that the increment $X(t)-X(s)$ $(0 \leq s < t)$ has the distribution

$$P\{X(t) - X(s) = j\} = e^{-\lambda(t-s)} \frac{\lambda^j (t-s)^j}{j!} \quad (j = 0, 1, 2, \ldots). \quad (2.1)$$

Taking $X(0) = 0$ we see that

$$EX(t) = \lambda t, \quad \mathrm{Var}X(t) = \lambda t.$$

The sample functions are step functions with steps (jumps) equal to 1. We identify the jumps as occurrences of Poisson events and refer to λ as the rate of the process.

(2) The Poisson process is a time-homogeneous Markov process. For its transition probabilities we have

$$\begin{aligned} P_{ij}(t) &= P\{X(\tau + t)\} = j \mid X(\tau) = i\} \\ &= P\{X(\tau + t) - X(\tau) = j - i \mid X(\tau) = i\} \\ &= P\{X(\tau + t) - X(\tau) = j - i\} \quad \text{(ind. increments)} \\ &= P\{X(t) = j - i \mid X(0) = 0\} \quad \text{(stationary increments)} \\ &= e^{-\lambda t} (\lambda t)^{j-i} / (j - i)! \quad (j \geq i). \end{aligned} \quad (2.2)$$

Thus $\{X(t)\}$ is homogeneous in time as well as space. We have

$$P_{ii+1}(t) = e^{-\lambda t} \lambda t = \lambda t - (1 - e^{-\lambda t}) \lambda t$$

$$\sum_{j>i+1} P_{ij}(t) = 1 - P_{ii}(t) - P_{ii+1}(t) = 1 - e^{-\lambda t} - e^{-\lambda t} \lambda t.$$

From this we find that

$$P_{ii+1}(t) = \lambda t + o(t), \quad \sum_{j>i+1} P_{ij}(t) = 0(t). \quad (2.3)$$

Conversely we ask whether a time-homogeneous Markov chain on the state space $\{0, 1, 2, \ldots\}$ satisfying (2.3) is indeed Poisson. This will be discussed under Markov chains (Chap. 10).

(3) *Hitting times.* For non-negative integers $m, n (n \geq m)$ we denote

$$T_{mn} = \inf\{t\colon X(t) = n\} \text{ on } \{X(0) = m\}. \tag{2.4}$$

We have $T_{mn} \stackrel{d}{=} T_{0,n-m}$ $(n > m)$ and $T_{mm} = T_{00} = 0$. We shall write T_n for T_{0n} and call T_n the *waiting time* for the nth Poisson event. The random variables $T_n - T_{n-1}$ $(n \geq 1)$ are called *sojourn times.*

(4) Stopping times and the strong Markov property are as defined in Sec.5.1 for the Brownian Motion.

5.2.2 *Properties of the Poisson Process*

Theorem 2.1 *Let $X(0) = 0$. Then the sojourn times $T_1, T_2 - T_1, T_3 - T_2, \ldots$ are IID random variables with density $\lambda e^{-\lambda t}$.*

Proof. (i) We have

$$P\{T_1 > t\} = P\{X(s) = 0 (0 \leq s \leq t)\} = P\{X(t) = 0\} = e^{-\lambda t}$$

which shows that T_1 has density $\lambda e^{-\lambda t}$.

(ii) For $k \geq 2$ we have

$$\begin{aligned} T_k - T_{k-1} &= \inf\{t - T_{k-1}\colon X(t) = k\} \\ &= \inf\{t - T_{k-1}\colon X(t) - X(T_{k-1}) = 1\} \\ &\stackrel{d}{=} \inf\{t - T_{k-1}\colon X(t - T_{k-1}) = 1\} \quad \text{on } \{X(T_{k-1}) = 0\} \\ &= \inf\{t'\colon X(t') = 1\} \quad \text{on } \{X(0) = 0\}. \end{aligned}$$

This shows that $T_1, T_k - T_{k-1}$ $(k \geq 2)$ are identically distributed. Independence follows from the strong Markov property. □

Remarks

2.1. The waiting time $T_n (n \geq 1)$ has the gamma density

$$e^{-\lambda x} \lambda^n x^{n-1}/(n-1)! \quad (x \geq 0). \tag{2.5}$$

2.2. From $\{T_n \leq t\} = \{X(t) \geq n\}$ we obtain

$$\int_0^t e^{-\lambda x} \lambda^n \frac{x^{n-1}}{(n-1)!} dx = 1 - \sum_{m=0}^{n-1} e^{-\lambda t} \frac{(\lambda t)^m}{m!}, \tag{2.6}$$

a formula connecting the gamma d.f. with Poisson probabilities.

2.3. We ask: starting from the sojourn times as in Theorem 2.1, can we construct the Poisson process in some manner? This is the subject of renewal theory.

2.4. We may view the Poisson process as a distribution $\pi = \{\pi(\omega) : \omega \in \Omega\}$ of points on $[0, \infty)$, even extending it to the real line, such that for $I = (a, b]$,

$$N\{I\} = X(b) - X(a) = \#\{\pi(\omega) \cap I\}. \tag{2.7}$$

Thus $N\{I\}$ is the counting process associated with π.

Theorem 2.2 *Given that n (≥ 1) Poisson events have occurred in $(0, t]$, the epochs of their occurrence have a joint uniform density in the set B_n, where*

$$B_n = \{(x_1, x_2, \ldots, x_n) \colon 0 \leq x_1 \leq x_2 \leq \cdots \leq x_n \leq t\} \tag{2.8}$$

Thus the density is given by

$$\begin{aligned} f(x_1, x_2, \ldots, x_n) &= \frac{n!}{t^n} \quad \text{for } (x_1, x_2, \ldots, x_n) \in B_n \\ &= 0 \qquad \text{otherwise.} \end{aligned} \tag{2.9}$$

To verify that this is a proper density, we note that the hypercube $(0, t]!$ can be partitioned into $n!$ subsets of equal volume corresponding to the $n!$ permutations of $(x_1, x_2, \ldots, x_n)$. Therefore, $n!|B_n| = t^n$ or $|B_n| = t^n/n!$

Proof. It is clear that the conditional distribution of $(T_1, T_2, \ldots, T_n)$, given $X(t) = n$ is concentrated on B_n. It suffices to consider the subset $B_n(s_1, s_2, \ldots, s_n)$ of B_n given by

$$\{0 \leq x_1 \leq s_1,\ x_1 \leq x_2 \leq s_2, \ldots, x_{n-1} \leq x_n \leq s_n \leq t\}.$$

Thus,

$$P\{T_1 \le s_1, T_2 \le s_2, \ldots, T_n \le s_n \mid X(t) = n\}$$
$$= \int_{B_n(s_1,s_2,\ldots,s_n)} P\{T_1 \in dx_1, T_2 \in dx_2, \ldots, T_n \in dx_n, T_{n+1} > t\}/P\{X(t) = n\}$$
$$= e^{\lambda t}\frac{n!}{(\lambda t)^n}\int_{B_n(s_1,s_2,\ldots,s_n)} P\{\tau_1 \in dx_1, \tau_1+\tau_2 \in dx_2, \ldots, \tau_1+\tau_2+\cdots+\tau_n \in dx_n, \tau_1+\tau_2+\cdots+\tau_{n+1} > t\},$$

where $\tau_1 = T_1, \tau_2 = T_2 - T_1, \ldots, \tau_{n+1} = T_{n+1} - T_n$ and by Theorem 2.1 $\tau_1, \tau_2, \ldots, \tau_{n+1}$ are IID with density $\lambda e^{-\lambda x}$. Therefore, the last probability becomes

$$e^{\lambda t}\frac{n!}{(\lambda t)}n\int_{B_n(s_1,s_2,\ldots,s_n)} P\{\tau_1 \in dx_1\}P\{\tau_2 \in dx_2 - x_1\} \cdots P\{\tau_n \in dx_n - x_{n-1}\}\cdot P\{\tau_{n+1} > t - x_n\}$$
$$= e^{\lambda t}\frac{n!}{(\lambda t)^n}\int_{B_n(s_1,s_2,\ldots,s_n)} \lambda e^{-\lambda x_1}dx_1\lambda e^{-\lambda(x_2-x_1)}dx_2, \ldots, \lambda e^{-\lambda(x_n - x_{n-1})} \times dx_n e^{-\lambda(t-x_1)}$$
$$= \frac{n!}{t^n}\int_{\beta_n(s_1,s_2,\ldots,s_n)} dx_1 dx_2, \ldots, d_{x_n} = \frac{n!}{t^n}\Big|B_n(s_1,s_2,\ldots,s_n)|.$$

□

Remarks

2.5 For $n=1$, Theorem 2.2 states that given $X(t)=1$, the time of occurrence is equally likely to be anywhere in $(0,t]$. On account of this we say that Poisson events occur completely at random.

2.6. For $n \ge 2$, the density (2.9) is identical with the order statistics $X_{[1]}, X_{[2]}, \ldots, X_{[n]}$ obtained from a random sample of size n $(X_1, X_2, \ldots, X_n)$ from a uniform population.

Theorem 2.3 *Let* $0 = a_0 < a_1 < a_2 < \cdots < a_{r-1} < a_r = t$ $(r \ge 2)$. *Given that* n *Poisson events occurred in* $(0,t]$, *the conditional distribution of the numbers of events occurring in* $(0,a_1], (a_1,a_2], \ldots, (a_{r-1},a_r]$ *is the* *multinomial given by*

$$\frac{n!}{k_1!k_2!\cdots k_r!}\left(\frac{t_1}{t}\right)^{k_1}\left(\frac{t_2}{t}\right)^{k_2}\cdots\left(\frac{t_r}{t}\right)^{k_r} \quad (k_i \geq 0, k_1+k_2+\cdots+k_r=n) \tag{2.10}$$

where $t_j = a_j - a_{j-1}$ $(j = 1, 2, \ldots, r)$.

Proof. Let $X(t)$ be a Poisson process with rate λ. For $k_i \geq 0, k_1 + k_2 + \cdots + k_r = n$ we have

$$\begin{aligned} &P\{X(a_1)-X(0)=k_1, X(a_2)-X(a_1)=k_2, \ldots, X(a_r)-X(a_{r-1}) \\ &\quad = k_r \mid X(t)=n\} \\ &\quad = e^{\lambda t}\frac{n!}{(\lambda t)^n}e^{-\lambda t_1}\frac{(\lambda t_1)^{k_1}}{k_1!}e^{-\lambda t_2}\frac{(\lambda t_2)}{k_2!}^{k_2}, \ldots, e^{-\lambda t_r}\frac{(\lambda t_r)}{k_r!}^{k_r} \end{aligned}$$

which reduces to the desired form. □

Remark 2.7 In the proof we have not used the fact that the intervals $(a_{i-1}, a_i]$ $(1 \leq i \leq r)$ are contiguous. Therefore Theorem 2.3 can be restated as follows:

If $I_1, I_2, \ldots, I_r$ *are non-overlapping sub-intervals of* $[0, \infty)$ *with a finite union (that is,* $I_i \cap I_j = \phi$ *(null), and* $I = \bigcup_1^r I_j$*, then*

$$\begin{aligned} &P\{X\{I_1\}=k_1, X\{I_2\}=k_2, \ldots, X\{I_n=k_r \mid X\{I\}=n\} \\ &= \frac{n!}{k_1!k_2!\ldots k_r!}P_1^{k_1}P_2^{k_2}\ldots P_r^{k_r} \ (k_i \geq 0,\ k_1+k_2+\cdots+k_r=n) \end{aligned} \tag{2.11}$$

where $P_j = |I_j|/|I|$ $(j = 1, 2, \ldots, r)$. □

Theorem 2.4 (Superposition of Poisson Processes) *Let* $\{X_i(t), t \geq 0, i=1,2,\ldots,r\}$ *be* r *independent Poisson processes with rates* $\lambda_1, \lambda_2, \ldots, \lambda_r$ *and* $X(t) = X_1(t) + X_2(t) + \cdots + X_r(t)$ $(t \geq 0)$. *Then* $\{X(t), t \geq 0\}$ *is also Poisson with rate* $\lambda = \lambda_1 + \lambda_2 + \cdots + \lambda_r$.

Proof. (i) For $0 = t_0 < t_1 < t_2 < \cdots < t_n$ $(n \geq 2)$ the random variables in the double array

$$\{X_i(t_p) - X_i(t_{p-1}),\ (1 \leq p \leq n\}\ (i = 1, 2, \ldots, r) \tag{2.12}$$

are mutually independent by our assumptions. This argument holds for any r processes with stationary independent increments. Therefore, the increments

$$X(t_p) - X(t_{p-1}) = \sum_{i=1}^{r} [X_i(t_p) - X_i(t_{p-1})]$$

$$X(t_q) - X(t_q - 1) = \sum_{i=1}^{r} [X_i(t_q) - X_i(t_{q-1})]$$

are independent and stationary.

(ii) For any fixed $t > 0$ we have

$$\begin{aligned} P\{X(t) = n\} &= \sum_{\Sigma k_i = n} P\{X_1(t) = k_1, X_2(t) = k_2, \ldots, X_r(t) = k_r\} \\ &= \sum_{\Sigma k_i = n} e^{-\lambda_1 t} \frac{(\lambda_1 t)^{k_1}}{k_1!} \cdot e^{-\lambda_2 t} \frac{(\lambda_2 t)^{k_2}}{k_2!} \cdots e^{-\lambda_r t} \frac{(\lambda_r t)^{k_r}}{k_r!} \\ &= e^{-\lambda t} \frac{(\lambda t)^n}{n!} \sum_{\Sigma k_i = n} \frac{n!}{k_1! k_2! \ldots k_r!} \left(\frac{\lambda_1}{\lambda}\right)^{k_1} \left(\frac{\lambda_2}{\lambda}\right)^{k_2} \cdots \left(\frac{\lambda_r}{\lambda}\right)^{k_r} \\ &= e^{-\lambda t} \frac{(\lambda t)^n}{n!} \quad (n \geq 0). \end{aligned}$$

□

Theorem 2.5 (The Coloring Theorem) *Let $\{X(t), t \geq 0\}$ be a Poisson process with rate λ. Suppose that the points T_n of this process are colored, randomly and independently of their location with $k(\geq 2)$ colors, the colors of different points being mutually independent. Let p_i be the probability that a point receives the ith color ($p_i \geq 0$, $p_1 + p_2 + \cdots + p_k = 1$). Let $X_i(t)$ be the number of points in $X(t)$ that receive the ith color. Then $\{X_i(t), (t \geq 0), i = 1, 2, \ldots, k\}$ are independent Poisson processes with rates $\lambda_1, \lambda_2, \ldots, \lambda_k$, where $\lambda_i = \lambda_{p_i}$ $(i = 1, 2, \ldots, k)$.*

Proof. Let C_n be the color received by T_n $(n \geq 1)$. We have

$$P\{C_n = i\} = p_i \quad (1 \leq i \leq k). \tag{2.13}$$

(i) The distribution of $\{X_i(\tau)\ (0 \leq \tau \leq s), i = 1, 2, \ldots, k\}$ involves the random variables

$$A = \{X(\tau)(0 \leq \tau \leq s), C_1, C_2, \ldots, C_{X(s)}\}$$

while the distribution of $\{X_i(s+t) - X_i(\tau)(s < \tau \leq s+t), i = 1, 2, \ldots, k\}$ involves the random variables

$$B = \{X(s+t) - X(\tau)\ (s < \tau \leq s+t), C_{X(s)+1}, \ldots, C_{X(s+t)}\}.$$

By our assumptions A and B are independent. Therefore the k-dimensional process $\{X_i(t), (t \geq 0), i = 1, 2, \ldots, k\}$ has independent increments.

(ii) For $n_i \geq 0, n_1 + n_2 + \cdots + n_k = n \geq 0$ we have

$$\begin{aligned} &P\{X_i(s+t) - X_i(s) = n_i \ (1 \leq i \leq k)\} \\ &= e^{-\lambda t}\frac{(\lambda t)^n}{n!} \cdot \frac{n!}{n_1!n_2!\cdots n_k!}p_1^{n_1}p_2^{n_2}\cdots p_k^{n_k} \\ &= \prod_{i=1}^{k} e^{-\lambda_i t}\frac{(\lambda_i t)^{n_i}}{n_i!}. \end{aligned}$$

This shows that the increments $X_i(s+t) - X_i(s)$ $(1 \leq i \leq r)$ are independently distributed as Poisson, with rates λ_i as was to be proved. □

Theorem 2.6 (The Poisson Process as a Renewal Process) *The renewal counting process $\{N(t), t \geq 0\}$ with exponential lifetimes is a Poisson process.*

Proof. Suppose that the successive lifetimes $\{X_k, k \geq 1\}$ have density $\lambda e^{-\lambda t}$. Denote $S_0 = 0, S_n = X_1 + X_2 + \cdots + X_n (n \geq 1)$. Then for $n \geq 1, S_n$ has density $e^{-\lambda x}\lambda^n x^{n-1}/(n-1)!$ and

$$P\{S_n \leq t\} = F_n(t) = \int_0^t e^{-\lambda x}\lambda^n \frac{x^{n-1}}{(n-1)!}dx \cdot \tag{2.14}$$

We recall that $N(t) = \max\{n\colon s_n \leq t\}$. Therefore

$$\begin{aligned} P\{N(t) = n\} &= F_n(t) - F_{n+1}(t) \\ &= \int_0^t e^{-\lambda x}\lambda^n \frac{x^{n-1}}{(n-1)!}dx - \int_0^t e^{-\lambda x}\lambda^{n+1}\frac{x^n}{n!}dx \\ &= \left[1 - \sum_0^{n-1} e^{-\lambda t}\frac{(\lambda t)^r}{r!}\right] - \left[1 - \sum_0^{n} e^{-\lambda t}\frac{(\lambda t)^r}{r!}\right] \\ &= e^{-\lambda t}\frac{(\lambda t)^n}{n!} \quad (n = 0, 1, 2, \ldots). \end{aligned}$$

This shows that for each $t > 0$, the random variable $N(t)$ has a Poisson distribution. To prove that the process $\{N(t), t \geq 0\}$ is Poisson, we need to prove that it has stationary independent increments. Consider the intervals $(0,t]$ and $(t,t+s]$. Then

$$N(t+s)-N(t)=\#\{n\colon X_1^{(t)}+X_2^{(t)}+\cdots+X_n^{(t)}\le s\},$$

where

$$X_1^{(t)}=S_{\nu(t)}-t,\ X_2^{(t)}=X_{\nu(t)+1}, X_3^{(t)}=X_{\nu(t)+2},\ldots$$

and $\nu(t)=N(t)+1=\min\{n\colon S_n>t\}=$ hitting time of the set (t,∞). Since

$$\{\nu(t)=n\}=\{S_1\le t,S_2\le t,\ldots,S_{n-1}\le t,S_n>t\}$$

it follows that the event $\{\nu(t)=n\}$ depends only on $X_1,X_2,\ldots,X_n$, and not on $X_{n+1},X_{n+2},\ldots$. Thus, $\nu(t)$ is a stopping time for $\{S_n\}$, the random variables $X_k^{(t)} (k\ge 2)$ are mutually independent, independent of $X_1^{(t)}$ and have density $\lambda e^{-\lambda t}$. We shall show that $X_1^{(t)}$ is independent of $\{X_1,X_2,\ldots,X_{N(t)},N(t)\}$ and has density $\lambda e^{-\lambda t}$. It will then follow that the increment $N(t+s)-N(t)$ is independent of $N(t)$ and has the same distribution as $N(s)$.

For $I\subset(0,\infty)$ we have

$$\begin{aligned}
&P\{N(t)=n,\ S_1<s_1,S_2\le s_2,\ldots,S_n\le s_n,X_1^{(t)}\in I\}\\
&\quad=\int_{B_n(s_1,s_2,\ldots,s_n)XI} P\{S_1\in dx_1,S_2\in dx_2,\ldots,S_n\in dx_n,S_{n+1}-t\in dx_{n+1}\}\\
&\quad=\int_{B_n(s_1,s_2,\ldots,s_n)XI} \lambda e^{-\lambda x_1}dx_1\lambda e^{-\lambda(x_2-x_1)}dx_2\cdots\lambda e^{-\lambda(x_n-x_{n-1})}\\
&\qquad\times dx_n\lambda e^{-\lambda(t+x_{n+1}-x)}dx_{n+1}\\
&\quad=\int_{B_n(s_1,s_2,\ldots,s_n)} e^{-\lambda t}\lambda^n dx_1dx_2,\ldots,dx_n\cdot\int_I\lambda e^{-\lambda x_{n+1}}dx_{n+1}.
\end{aligned}$$

This shows that $\{N(t),S_1,S_2,\ldots,S_{N(t)}\}$ is independent of $X_1^{(t)}$, and $X_1^{(t)}$ has density $\lambda e^{-\lambda x}$. □

5.2.3 *The Compound Poisson Process*

Let $\{N(t),t\ge 0\}$ be a Poisson process with parameter λ, and $\{Y_k,k\ge 1\}$ a sequence of IID random variables with d.f. $F(x)$, independent of $N(t)$. Denote $S_0=0,S_n=X_1+X_2+\cdots+X_n$ $(n\ge 1)$. We define a new process $\{X(t),t\ge 0\}$ by setting $X(t)=S_{N(t)}$. (We are observing $\{S_n\}$ at values of

n given by $N(t)$, that is, by the Poisson clock.) For $0 = t_0 < t_1 < t_2 < \cdots < t_n$ $(n \geq 2)$ we have

$$X(t_p) - X(t_{p-1}) = \sum_{k=N(t_{p-1})+1}^{N(t_p)} Y_k \quad (p = 1, 2, \ldots, r). \tag{2.15}$$

By our assumptions the random variables $X(t_p) - X(t_{p-1})(p = 1, 2, \ldots, r)$ are mutually independent, and moreover

$$\begin{aligned} X(t_p) - X(t_{p-1}) &\stackrel{d}{=} Y_1 + Y_2 + \cdots + Y_{N(t_p)-N(t_{p-1})} \\ &\stackrel{d}{=} Y_1 + Y_2 + \cdots + Y_{N(t_p - t_{p-1})}. \end{aligned} \tag{2.16}$$

That is, the increments $X(t_p) - X(t_{p-1})$ are stationary. Thus $\{X(t)\}$ is a process with stationary independent increments. It is called a *compound Poisson* process. In the special case where $Y_k = 1$ with *probability* 1, $X(t) = N(t)$. We call $\{N(t)\}$ a *simple Poisson* process, to distinguish it from the general case.

For the d.f. of $X(t)$ we have

$$P\{X(t) \leq x\} = \sum_{n=0}^{\infty} e^{-\lambda t} \frac{(\lambda t)^n}{n!} F_n(x), \tag{2.17}$$

where $F_n(x) = P\{S_n \leq x\}$ $(n \geq 0)$. The mean and variance of $X(t)$ are given by

$$EX(t) = \lambda bt, \quad \mathrm{Var} X(t) = Xct, \tag{2.18}$$

where $E(Y_k) = b$ and $E(Y_k^2) = c$, assuming these moments exist. We have the additional property as stated in the following.

Theorem 2.7 *The compound Poisson process $\{X(t), t \geq 0\}$ on the state space $(-\infty, \infty)$ can be expressed as*

$$X(t) = X_1(t) - X_2(t) \ (t \geq 0), \tag{2.19}$$

where $\{X_i(t), t \geq 0\}$ $(i = 1, 2)$ are independent compound Poisson processes on the state space $[0, \infty)$.

Proof. Let $N_1(t)$ be the number of terms in $S_{N(t)} = Y_1 + Y_2 + \cdots + Y_{N(t)}$ with $Y_k > 0$, and $N_2(t)$ the number of terms with $Y_k \leq 0$. Clearly, $N(t) = N_1(t) + N_2(t)(t \geq 0)$. By the Coloring Theorem 2.5 we find that $\{N_1(t)\}$

and $\{N_2(t)\}$ are independent Poisson processes with parameters λ_p and λ_q, respectively, where

$$p = P\{Y_k > 0\} = 1 - F(0), \quad q = P\{Y_k \leq 0\} = F(0). \tag{2.20}$$

Thus (2.19) holds with

$$X_1(t) = \sum_{k=1}^{N(t)} Y_k 1_{\{Y_k > 0\}}, \quad X_2(t) = -\sum_{k=1}^{N(t)} Y_k 1_{\{Y_k \leq 0\}}. \tag{2.21}$$

The sums in (2.21) are similar to those in $X(t)$, with the first sum containing $N_1(t)$ terms and the second containing $N_2(t)$ terms. It follows that $\{X_1(t)\}$ and $\{X_2(t)\}$ are compound Poisson processes. Since the sums do not contain any common terms and since $Y_1, Y_2, \ldots$ are independent it follows that $X_1(t)$ and $X_2(t)$ are independent. □

5.3 Lévy Processes

In the study of the Brownian and Poisson our starting point was their basic property of stationary independent increments. The sample functions of the Brownian are continuous with probability 1, while those of the Poisson are step functions. These properties were used to derive important properties of these processes.

We now consider the broad class of Lévy Processes, of which the Brownian and the Poisson are important special cases. For convenience we repeat the definition of stationary independent increments.

A process $\{X(t), t \geq 0\}$ is called a process with stationary independent increments if it satisfies the following conditions:

(i) For $0 \leq t_1 < t_2 < \cdots < t_n$ $(n \geq 2)$ the random variables

$$X(t_1), X(t_2) - X(t_1), X(t_3) - X(t_2), \ldots, X(t_n) - X(t_{n-1})$$

are independent.

(ii) The distribution of the increment $X(t_p) - X(t_{p-1})$ depends only on the difference $t_p - t_{p-1}$.

For such a process we can take $X(0) = 0$ without loss of generality. For if $X(0) \neq 0$, then the process $Y(t) = X(t) - X(0)$ has stationary independent increments, and $Y(0) = 0$.

If we write

$$X(t) = \sum_{k=1}^{n} \left[X\left(\frac{k}{n}t\right) - X\left(\frac{k-1}{n}t\right) \right] \tag{3.1}$$

then $X(t)$ is seen to be the sum of n independent random variables all of which are distributed as $X\left(\frac{t}{n}\right)$. Thus a process with stationary independent increments is the generalization to continuous time of sums of independent and identically distributed random variables (random walks). If follows, in particular, that $\{X(t)\}$ is a Markov process which is homogeneous in time and space.

We now impose two further assumptions on these processes to generate the class of Lévy processes. These are by way of regularity conditions which are satisfied in the special cases described here and also in Secs. 5.1 and 5.2. Lévy processes have the *strong Markov* property, a fact which will be used extensively.

A Lévy process $\{X(t), t \geq 0\}$ is a process with stationary independent increments, which in addition to (i) and (ii) above satisfies the following regularity conditions:

(iii) $X(t)$ is continuous in probability; that is, for each $\varepsilon > 0$,

$$P\{|X(t)| > \varepsilon\} \to 0 \quad \text{as } t \to 0^+.$$

(iv) There exist left and right limits $X(t^-)$ and $X(t^+)$ and we assume that $X(t) = X(t^+)$, that is, $X(t)$ is right continuous. The quantity $X(t) - X(t^-)$ is called the jump of the process at time t.

In Secs. 5.4–5.6 we describe additional examples of important Lévy processes. In Sec. 5.7 we provide a more complete description of a Lévy process in terms of its c.f.

5.4 The Gaussian Process

This is the process $\{X(t), t \geq 0\}$, where $X(t) = \sigma B(t) + dt (\sigma > 0,\ d \text{ real})$. For $\sigma = 1, d = 0$ this reduces to the standard Brownian studied in Sec. 5.1. In the general case $X(t)$ is a Gaussian process with

$$EX(t) = dt, \quad \text{Var}X(t) = \sigma^2 t \tag{4.1}$$

$$\text{Covar}\{X(t), X(t')\} = \sigma^2 \min(t, t'). \tag{4.2}$$

We denote the density of $X(t)$ as $f(x,t)$, so that

$$f(x,t) = \frac{1}{\sigma\sqrt{t}} n\left(\frac{x-dt}{\sigma\sqrt{t}}\right). \tag{4.3}$$

The corresponding d.f. is

$$F(x,t) = N\left(\frac{x-dt}{\sigma\sqrt{t}}\right). \tag{4.4}$$

The maximum and minimum functionals of the process will be denoted by $M(t)$ and $m(t)$, respectively. Also, let $T(x)$ denote the hitting time of the level x, so that

$$T(x) = \inf\{t\colon X(t) = x\} \quad (x \neq 0\}, T(0) = 0. \tag{4.5}$$

To simplify calculations we sometimes denote $X(t)$ as $X_d(t)$ and the corresponding quantities with a suffix d. Then since $B(t) \stackrel{d}{=} -B(t)$ we find that

$$X_{-d}(t) = \sigma B(t) - dt \stackrel{d}{=} -\sigma B(t) - dt = -X_d(t). \tag{4.6}$$

Consequently

$$f_{-d}(x,t) = f_d(-x,t) \quad \text{and} \quad F_{-d}(x,t) = 1 - F_d(-x,t) \tag{4.7}$$

and

$$M_{-d}(t) \stackrel{d}{=} -m_d(t), \quad T_{-d}(x) \stackrel{d}{=} T_d(-x). \tag{4.8}$$

We first prove the following result, which is an extension of the reflection principle of D. André, used in Sec. 5.1.

Lemma 4.1 *For $y < x$ denote*

$$P\{T(x) > t, X(t) \in dy\} = {}^{x}f(y,t)dy. \tag{4.9}$$

Then, for $y > 0$

$${}^{x}f(x-y,t) = f(x-y,t) - e^{-\frac{2dy}{\sigma^2}} f(x+y,t). \tag{4.10}$$

Proof. Using the continuity of the sample functions we find that

$$f(x-y,t) = {}^x f(x-y,t) + \int_0^t P\{T(x)\,\epsilon\, ds\} f(-y,t-s) \tag{4.11}$$

$$f(x+y,t) = \int_0^t P\{T(x)\ \epsilon\ ds\} f(y,t-s) \tag{4.12}$$

since on account of the strong Markov property $X(t) - X \circ T(x)$ is independent of $X \circ T(x)$ and has the same distribution as $X(t - T(x))$. Now it can be verified that

$$f(y,t-s) = e^{\frac{2dy}{\sigma^2}} f(-y,t-s). \tag{4.13}$$

This leads to the desired result. □

Theorem 4.1 *For the maximum and minimum functionals of the process* $X(t)$ *we have the following:*

(i) $$P\{M(t) < x\} = N\left(\frac{x-dt}{\sigma\sqrt{t}}\right) - e^{\frac{2dx}{\sigma^2}} N\left(\frac{-x-dt}{\sigma\sqrt{t}}\right) \quad (x \geq 0) \tag{4.14}$$

$$P\{M(t) \leq x\} = N\left(\frac{x-dt}{\sigma\sqrt{t}}\right) + e^{\frac{2dx}{\sigma^2}} N\left(\frac{x+dt}{\sigma\sqrt{t}}\right) \quad (x \leq 0). \tag{4.15}$$

(ii) *As* $t \to \infty, M(t) \to M \leq \infty, m(t) \to m \geq -\infty$ *with probability* 1. *If* $d > 0$, *then* $M = \infty$ *with probability* 1, *while*

$$P\{m \leq x\} = e^{\frac{2dx}{\sigma^2}} \quad (x \leq 0). \tag{4.16}$$

If $d < 0$, *then* $m = -\infty$ *with probability* 1, *while*

$$P\{M < x\} = 1 - e^{\frac{2dx}{\sigma^2}} \ (x \geq 0). \tag{4.17}$$

If $d = 0$, *then* $M = \infty, m = -\infty$, *with probability* 1.

Proof. On account of (4.8) we need to consider only $M(t)$. Integrating (4.10) over $(0 < y < \infty)$ we obtain

$$\begin{aligned} P\{T(x) > t\} &= \int_{-\infty}^{x} f(u,t)du - e^{\frac{2dx}{\sigma^2}} \int_x^{\infty} f(-u,t)du \\ &= N\left(\frac{x-dt}{\sigma\sqrt{t}}\right) - e^{\frac{2dx}{\sigma^2}} N\left(\frac{-x-dt}{\sigma\sqrt{t}}\right). \end{aligned} \tag{4.18}$$

From this the result (4.14) follows, since

$$P\{M(t) < x\} = P\{T(x) > t\}.$$

As $t \to \infty$, $M(t) \to M \leq \infty$ on account of monotonicity of $M(t)$. From (4.14) we find that

$$\begin{aligned} P\{m < x\} &= N(-\infty) - e^{\frac{2dx}{\sigma^2}} N(-\infty) = 0 && \text{if } d > 0 \\ &= N(0) - N(0) = 0 && \text{if } d = 0 \\ &= N(\infty) - e^{\frac{2dx}{\sigma^2}} N(\infty) = 1 - e^{\frac{2dx}{\sigma^2}} && \text{if } d < 0. \end{aligned}$$

We have thus proved the results for $M(t)$ as $t \to \infty$. □

Lemma 4.1 also leads to results for $T(x)$. If suffices to consider the case $x > 0$.

Theorem 4.2 (i) *For $x > 0$ the hitting time $T(x)$ has the density*

$$g(t,x) = \frac{x}{t} f(x,t) \quad (t > 0) \quad \textit{and} \quad = 0 \quad (t \leq 0). \tag{4.19}$$

(ii) $P\{T(x) < \infty\} = 1$ *if* $d \geq 0$ *and* $= e^{\frac{2dx}{\sigma^2}}$ if $d < 0$. (4.20)

(iii) *If $d > 0$, then*

$$ET(x) = xd^{-1}, \quad \operatorname{Var} T(x) = \sigma^2 x d^{-3} \tag{4.21}$$

while if $d = 0$, then $ET(x) = \infty$.

Proof. From (4.18) we find that

$$P\{T(x) \leq t\} = N\left(\frac{dt - x}{\sigma\sqrt{t}}\right) + e^{\frac{2dx}{\sigma^2}} N\left(\frac{-x - dt}{\sigma\sqrt{t}}\right). \tag{4.22}$$

For the density of $T(x)$ we have

$$\begin{aligned} g(t,x) &= n\left(\frac{dt - x}{\sigma\sqrt{t}}\right)\left(\frac{1}{2}\frac{d}{\sigma}, \frac{1}{\sqrt{t}} + \frac{1}{2}\frac{x}{\sigma}, \frac{1}{t^{3/2}}\right) \\ &\quad + e^{\frac{2dx}{\sigma^2}} n\left(\frac{-x - dt}{\sigma\sqrt{t}}\right)\left(-\frac{1}{2}\frac{d}{\sigma}, \frac{1}{\sqrt{t}} + \frac{1}{2}\frac{x}{\sigma}, \frac{1}{t^{3/2}}\right) \\ &= \frac{x}{\sigma t^{3/2}} n\left(\frac{dt - x}{\sigma\sqrt{t}}\right) = \frac{x}{t} f(x,t). \end{aligned}$$

This proves (i). The results of (ii) follow by letting $t \to \infty$ in (4.22). To derive the moments of $T(x)$ for $d \geq 0$ we use its Laplace transform. It can be verified that

$$e^{-\theta t}g_d(t,x) = e^{-x\eta(\theta)}g_{d_1}(t,x) \quad (\theta > 0),$$

where $d_1 = \sqrt{d^2 + 2\sigma^2\theta} > 0$ and

$$\eta(\theta) = \frac{\sqrt{d^2 + 2\sigma^2\theta} - d}{\sigma^2}. \tag{4.23}$$

Therefore,

$$Ee^{-\theta T(x)} = \int_0^\infty e^{-\theta t}g_d(t,x)dt = e^{-x\eta(\theta)}\int_0^\infty g_{d_1}(t,x)dt,$$

where the last integral equals unity since $d_1 > 0$. The results (iii) follow from this transform since for $d \geq 0$

$$\eta(0+) = 0, \quad \eta'(0+) = |d|^{-1} \quad \text{and} \quad \eta''(0+) = -\sigma^2|d|^{-3}. \qquad \square$$

To understand the results of Theorem 4.2 more meaningfully we consider the family of random variables $\{T(x), x \geq 0\}$. For $y > 0$ we have

$$\begin{aligned} T(x+y) - T(x) &= \inf\{t - T(x)\colon X(t) = x + y\} \\ &= \inf\{t - T(x)\colon X(t) - X \circ T(x) = y\} \end{aligned}$$

since $X \circ T(x) = x$. As observed in the proof of Lemma 4.1, $X(t) - X \circ T(x)$ is independent of $X \circ T(x)$ and has the same distribution as $X(t - T(x))$. It follows that $T(x+y) - T(x)$ is independent of $T(x)$ and has the same distribution as

$$\inf\{t'\colon X(t') = y\} = T(y).$$

Extending this properly we see that the process $\{T(x)\}$ has stationary independent increments with the roles of time and space reversed. In the next section we study this inverse Gaussian process.

(We remark here that although the preferred term for the proces $\{X(t)\}$ is the Gaussian in applications, it is actually one of the family of Gaussian processes. A similar remark applies to the inverse Gaussian.)

The following result due to A.V. Skorokhod leads to the Wiener–Hopf factorization of the process $\{X(t)\}$ and is also important in applications.

Lemma 4.2 *Let $x(t)$ $(t \geq 0)$ be a real valued continuous function with $x(0) = 0$. Then there exists a unique pair (z, c) of continuous functions such that*

$$z(t) = x(t) + c(t) \tag{4.24}$$

$$z(t) \geq 0, c \textit{ is nondecreasing, } c(0) = 0 \tag{4.25}$$

and

$$\int_0^t 1_{\{z(s) \neq 0\}} dc(s) = 0 \quad (t \geq 0). \tag{4.26}$$

Moreover, the function c is given by

$$c(t) = -m(t), \quad m(t) = \inf_{0 \leq s \leq t} x(s). \tag{4.27}$$

Proof. We shall prove that (i) with c defined in (4.27) and $z(t) = x(t) + c(t)$, the pair (z, c) is a solution of (4.24) and (ii) this solution is unique.

(i) Clearly $z(t) \geq 0$, c is nondecreasing, $c(0) = 0$. Also, c increases only when x is at a new minimum, in which case $z = 0$. Therefore the condition (4.26) is satisfied.

(ii) Suppose there is a second solution (z_1, c_1) such that

$$z_1(t) = x(t) + c_1(t)$$

satisfying the conditions (4.25) and (4.26). Suppose that $z_1(t') > z(t')$ for some t', and set

$$t_0 = \sup\{s \leq t' \colon z_1(s) = z(s)\}.$$

Then for $t_0 \leq t \leq t'$ we have $z_1(t) > z(t) \geq 0$ by continuity, so that

$$0 = \int_{t_0}^{t'} 1_{\{z_1(s) \neq 0\}} dc_1(s) = c_1(t') - c_1(t_0).$$

Thus

$$\begin{aligned} 0 < z_1(t') - z(t') &= c_1(t') - c(t') \leq c_1(t_0) - c(t_0) \\ &= z_1(t_0) - z(t_0) \end{aligned}$$

which is a contradiction, Hence $z_1(t') \leq z(t')$ for all t', and similarly $z_1(t') \geq z(t')$ for all t'. We conclude that $z(t) = z_1(t)$. $\square$

We apply this lemma to the Gaussian process $\{X(t)\}$. Denote

$$Ee^{-\theta X(t)} = e^{-t\phi\{\theta\}} \qquad (\theta \text{ real}), \tag{4.28}$$

where $\phi(\theta) = d\theta - \frac{1}{2}\sigma^2\theta^2$. For fixed $s > 0$ it is found that the equation $s + \phi(\theta) = 0$ has two roots $\eta_1(s) > 0$ and $-\eta_2(s) < 0$ given by

$$\eta_1(s) = \frac{d + \sqrt{d^2 + 2\sigma^2 s}}{\sigma^2}, \quad \eta_2(s) = \frac{-d + \sqrt{d^2 + 2\sigma^2 s}}{\sigma^2} \tag{4.29}$$

and $s + \phi(\theta) > 0$ for $-\eta_2 < \theta < \eta_1$, where $\eta_1 = \eta_1(s)$ and $\eta_2 = \eta_2(s)$.

Theorem 4.3 (Wiener–Hopf factorization of the Gaussian Process) *For $s > 0$ let T_s denote the random variable independent of $X(t)$ and having density se^{-st} $(t > 0)$. Then*

$$E\left[e^{-\theta X(T_s)}\right] = E\left[e^{-\theta M(T_s)}\right] E\left[e^{-\theta m(T_s)}\right] (-\eta_2 < \theta < \eta_1) \tag{4.30}$$

where for fixed $s > 0$ the two factors on the right side of (4.30) represent the transforms of infinitely divisible distributions on $(0, \infty)$ and $(-\infty, 0)$, respectively. This factorization is unique among such factors upto a factor $a^{-\theta a}$, a being a constant.

Proof. We first note that the left side of (4.30) is

$$= \int_0^\infty se^{-st} e^{-t\phi(\theta)} dt = \frac{s}{s + \phi(\theta)} \quad (-\eta_2 < \theta < \eta_1). \tag{4.31}$$

We define process $\{Z(t), t \geq 0\}$ where $Z(t)$ is the solution of the equation

$$Z(t) = X(t) + I(t) \tag{4.32}$$

such that $Z(t) \geq 0$, $I(t)$ is nondecreasing, $I(0) = 0$ and

$$\int_0^\infty 1_{\{Z(s) \neq 0\}} dI(s) = 0 \quad (t \geq 0). \tag{4.33}$$

Since the process $\{X(t)\}$ has continuous sample functions it follows from Lemma 4.2 that the unique solution of (4.32) is given by the pair $\{Z(t), I(t)\}$, where

$$Z(t) = X(t) - m(t), \quad I(t) = -m(t). \tag{4.34}$$

We proceed to derive the joint distribution of $\{Z(t), I(t)\}$. It can be easily verified that

$$e^{-\theta I(t)} = 1 - \theta \int_0^t e^{-\theta I(\tau)} dI(\tau) \quad (\theta > 0).$$

For $\theta_1 > 0,\ \theta_2 > 0$ we therefore obtain

$$\begin{aligned} e^{-\theta_1 Z(t) - \theta_2 I(t)} &= e^{-\theta_1 X(t) - (\theta_1+\theta_2) I(t)} \\ &= e^{-\theta_1 X(t)} - (\theta_1 + \theta_2) \int_0^t e^{-\theta_1 [X(t)-X(\tau)] - \theta_1 Z(\tau) - \theta_2 I(\tau) dI(\tau)}. \end{aligned}$$

Here in the integrand $Z(\tau) = 0$ in view of the condition (4.33). Also, $X(t) - X(\tau)$ is independent of $I(\tau)$ and has same distribution as $X(t-\tau)$. Therefore,

$$Ee^{-\theta_1 Z(t) - \theta_2 I(t)} = e^{-t\phi(\theta_1)} - (\theta_1 + \theta_2) \int_0^t e^{-(t-\tau)\phi(\theta_1)} E\big[e^{-\theta_2 I(\tau)} dI(\tau)\big]$$

and

$$\begin{aligned} &\int_0^\infty se^{-s\tau} E\big[e^{-\theta_1 Z(t) - \theta_2 I(t)}\big] dt \\ &= \frac{s}{s+\phi(\theta_1)} \left[1 - (\theta_1 + \theta_2) J^*(\theta_2, s)\right] (-\eta_1 < \theta_1 < \eta_1), \end{aligned} \tag{4.35}$$

where

$$J^*(\theta_2, s) = \int_0^\infty e^{-st} E\big[e^{-\theta_2 I(t)} dI(t)\big]. \tag{4.36}$$

Since for fixed $s > 0,\ \theta_2 > 0$ the expression on the left side of (4.35) is a bounded analytic function of $\theta_1 > 0$ it follows that we must have

$$J^*(\theta_2, s) = \lim_{\theta \to \eta_1 -} \frac{1}{\theta_1 + \theta_2} = \frac{1}{\eta_1 + \theta_2}. \tag{4.37}$$

Substituting for $J^*(\theta_2, s)$ in (4.35) we obtain

$$E\big[e^{-\theta_1 Z(T_s) - \theta_2 I(T_s)}\big] = \frac{s}{s+\phi(\theta_1)} \cdot \frac{\eta_1 - \theta_1}{\eta_1 + \theta_2}.$$

Since $Z(t) \stackrel{d}{=} M(t)$ and $I(t) = -m(t)$ we can rewrite this last equation as

$$E\big[e^{-\theta_1 M(T_s) + \theta_2 m(T_s)}\big] = \frac{s}{s+\phi(\theta_1)} \cdot \frac{\eta_1 - \theta_1}{\eta_1 + \theta_2}. \tag{4.38}$$

For $\theta_1 = \theta$ and $\theta_2 \to 0+$ this gives

$$E\left[e^{-\theta m(T_s)}\right] = \frac{s}{s+\phi(\theta)} \cdot \frac{\eta_1 - \theta}{\eta_1} \quad (-\eta_2 < \theta < \eta_1) \tag{4.39}$$

and for $\theta_1 \to 0+$ and $\theta_2 = -\theta > 0$ it gives

$$E\left[e^{-\theta M(T_s)}\right] = \frac{\eta_1}{\eta_1 - \theta} \quad (\theta < \eta_1). \tag{4.40}$$

From (4.39) and (4.40) we arrive at the factorization (4.30). Now

$$\frac{s}{s+\phi(\theta)} = \frac{s}{-\frac{1}{2}\sigma^2(\theta-\eta_1)(\theta+\eta_2)} = \frac{\eta_2}{\eta_2+\theta} \cdot \frac{\eta_1}{\eta_1-\theta}, \tag{4.41}$$

where the factors in the last expression are Laplace transforms of exponential densities

$$\eta_2 e^{-x\eta_2} \quad (x > 0) \quad \text{and} \quad \eta_1 e^{x\eta_1} \quad (x < 0) \tag{4.42}$$

both of which are infinitely divisible. Uniqueness of (4.30) follows from the property of infinitely divisible distributions. □

5.4.1 *Application to Brownian Storage Models*

In inventory models that allow backlogs and return of merchandise it is convenient to view the net input as a quantity that varies continuously with time, taking positive as well as negative values. Accordingly we formulate a storage model in which the net input upto time t as a Gaussian process $\{X(t), t \geq 0\}$ described in this section, and the storage level $Z(t)$ of time t satisfies the equation

$$Z(t) = Z(0) + X(t) + I(t) \tag{4.43}$$

subject to the constraints $Z(t) \geq 0$, $I(t)$ is non-decreasing, $I(0) = 0$ and

$$\int_0^t 1_{\{Z(s) \neq 0\}} dI(s) = 0 \quad (t \geq 0). \tag{4.44}$$

Here $I(t)$ is the amount of unsatisfied demand during (0,t] and (4.44) states that $I(t)$ increases only at epochs at which the storage level is zero.

A slight extension of Lemma 4.2 yields the unique solution $\{Z(t), I(t)\}$ as

$$Z(t) = \max\{Z(0) + X(t), X(t) - m(t)\}, \ I(t) = [Z(0) + m(t)]. \tag{4.45}$$

Of additional interest in applications is the non-empty period T, namely,

$$T = \inf\{t\colon Z(t) = 0\} \quad \text{on } \{Z(0) = x\}. \tag{4.46}$$

From (4.45) we see that

$$T \stackrel{d}{=} \inf\{t\colon x + X(t) = 0\} = T(-x). \tag{4.47}$$

Theorems 4.1 and 4.2 contain results for $I(t)$ and T. If $Z(0) = 0$, then $Z(t) = X(t) - M(t) \stackrel{d}{=} M(t)$ and the results for this case are also given by Theorem 4.1. In the general case $Z(0) \geq 0$ we have the following.

Theorem 4.4 *For $z > 0$, $x \geq 0$ we have*

$$P\{Z(t) < z \mid Z(0) = x\} = N\left(\frac{z - x - dt}{\sigma\sqrt{t}}\right) - e^{\frac{2dz}{\sigma^2}} N\left(\frac{-z - x - dt}{\sigma\sqrt{t}}\right) \tag{4.48}$$

and

$$\lim_{t\to\infty} P\{Z(t) < z \mid Z(0) = x\}$$
$$= 1 - e^{\frac{2dz}{\sigma^2}} \text{ if } d < 0 \quad \textit{and} \quad = 0 \quad \textit{if} \quad d \geq 0. \tag{4.49}$$

Proof. Since $\{X(t), X(t) - m(t)\} \stackrel{d}{=} \{X(t), M(t)\}$ for any Lévy process it follows from (4.45) that $Z(t) \stackrel{d}{=} \max\{Z(0) + X(t), M(t)\}$ $(t \geq 0)$. Therefore

$$\begin{aligned} P\{Z(t) < z \mid Z(0) = x\} &= P\{X(t) < z - x, M(t) < z\} \\ &= P\{X(t) < z - x, T(z) > t\} \\ &= \int_x^\infty {}^z f(z - u, t) du, \end{aligned} \tag{4.50}$$

where ${}^z f(y, t)$ is defined as in (4.9). Using Lemma 4.1 we find that

$$\begin{aligned} {}^z f(z - u, t) &= f(z - u, t) - e^{-\frac{2du}{\sigma^2}} f(z + u, t) \\ &= f(z - u, t) - e^{\frac{2dz}{\sigma^2}} f(-z - u, t). \end{aligned}$$

Substituting this in (4.50) we arrive at the result (4.48). It leads directly to the limit result (4.49). □

5.5 The Inverse Gaussian Process

This is a process $\{X(t), t \geq 0\}$ with stationary independent increments, with density

$$f(x,t) = \frac{t}{\sigma\sqrt{2\pi x^3}} e^{-\frac{d^2}{2\sigma^2 x}\left(x - \frac{t}{d}\right)^2} \quad (0 < x < \infty), \tag{5.1}$$

where $\sigma > 0$ and d is real. We can write this as

$$f(x,t) = \frac{t}{\sigma x^{3/2}} n\left(\frac{dx - t}{\sigma\sqrt{x}}\right). \tag{5.2}$$

The corresponding d.f. is

$$F(x,t) = N\left(\frac{dx - t}{\sigma\sqrt{x}}\right) + e^{\frac{2dt}{\sigma^2}} N\left(\frac{-dx - \tau}{\sigma\sqrt{x}}\right). \tag{5.3}$$

From this we see that

$$P\{X(t) < \infty\} = 1 \quad \text{if } d \geq 0 \quad \text{and} \quad = e^{\frac{2dt}{\sigma^2}} \quad \text{if } d < 0. \tag{5.4}$$

Let $L = \inf\{t\colon X(t) = \infty\}$. Then L is defined to be the lifetime of the process in the sense that $X(t) = \infty$ for $t \geq L$. Here $L = \infty$ with probability 1, if $d \geq 0$ and

$$P\{L \leq t\} = 1 - e^{\frac{2dt}{\sigma^2}} \quad \text{if } d < 0. \tag{5.5}$$

For completeness we include the following results which are essentially a restatement of Theorem 4.2 (noting the time–space reversal). Thus

$$Ee^{-\theta X(t)} = e^{-t\eta(\theta)}, \quad \eta(\theta) = \frac{\sqrt{d^2 + 2\sigma^2\theta} - d}{\sigma^2} \quad (\theta > 0) \tag{5.6}$$

$$EX(t) = td^{-1}, \quad \mathrm{Var}X(t) = \sigma^2 t d^{-3} \quad (d > 0) \tag{5.7}$$

and

$$EX(t) = \infty \quad (d = 0). \tag{5.8}$$

5.6 The Randomized Bernoulli Random Walk

As a special case of the compound Poisson process of Sec. 5.2 we consider $\{X(t),\ t \geq 0\}$ where

$$X(t) = Y_1 + Y_2 + \cdots + Y_{N(t)} \quad (t \geq 0), \tag{6.1}$$

where $\{N(t),\ t \geq 0\}$ is a Poisson process with parameter ν and $\{Y_k, k \geq 1\}$ is a sequence of IID rondom variables with

$$P\{Y_k = +1\} = p, \quad P\{Y_k = -1\} = q, \tag{6.2}$$

where $0 < \nu < \infty,\ 0 < p < 1$ and $q = 1 - p$. The state space of $\{X(t)\}$ is $\{\ldots, -1, 0, 1, 2, \ldots\}$. The representation (6.1) shows that we may view $\{X(t)\}$ as a Bernoulli random walk where the coin is tossed at the epochs when a Poisson event occurs. Thus, $\{X(t)\}$ is a Bernoulli random walk.

As shown in Sec. 5.2 we may represent $X(t)$ as

$$X(t) = A(t) - D(t), \tag{6.3}$$

where $\{A(t)\}$ and $\{D(t)\}$ are independent Poisson processes with parameters $\lambda = \nu p$ and $\mu = \nu q$, respectively. Denote

$$\begin{aligned} k_j(t) &= P\{X(t) = j\} \\ &= \sum_{m-j-}^{\infty} e^{-(\lambda+\mu)t} \frac{(\mu t)^m (\lambda t)^{m+j}}{m!(m+j)!}. \end{aligned} \tag{6.4}$$

We can express $k_j(t)$ as

$$k_j(t) = e^{-(\lambda+\mu)t} \rho^{j/2} I_j(2\sqrt{\lambda\mu} t), \tag{6.5}$$

where $\rho = \lambda/\mu$ and I_j is the modified Bessel function defined as

$$I_j(x) = \sum_m \frac{(x/2)^{2m+j}}{m!(m+j)!}. \tag{6.6}$$

We note that $I_j(x) = I_{-j}(x)$, so $k_j(t) = \rho^{-j} k_j(t)$. Also, denote

$$K_j(t) = P\{X(t) \leq j\} = \sum_{i=-\infty}^{j} k_i(t). \tag{6.7}$$

Since this last expression is an infinite series it is useful to obtain asymptotic estimates for it. These are based on the known fact that

$$e^{-x} I_j(x) = \frac{1}{\sqrt{2\pi x}} + o\left(\frac{1}{\sqrt{x}}\right) \quad (x \to \infty). \tag{6.8}$$

Writing (6.5) as

$$k_j(t) = e^{-(\lambda+\mu-2\sqrt{\lambda\mu})t} \rho^{j/2} e^{-\sqrt{2\lambda\mu} t} I_j(2\sqrt{\lambda\mu} t)$$

and applying (6.8) we find that

$$k_j(t) = c_1 e^{-c_2 t_2} t^{-1/2} \rho^{j/2}[1 + o(1)] \quad (t \to \infty), \tag{6.9}$$

where $c_1^{-1} = 2\pi^{1/2}(\lambda\mu)^{1/4} > 0$ and $c_2 = \left(\sqrt{\lambda} - \sqrt{\mu}\right)^2 \geq 0$. The following results concern $K_j(t)$.

Theorem 6.1 *Let $K_j(t)$ be the cumulative probability function of the randomized Bernoulli random walk. Then for all j,*

$$\begin{aligned} 1 - K_j(t) &= c_1 e^{-c_2 t} t^{-1/2} \rho^{\frac{j+1}{2}} (1 - \sqrt{\rho})^{-1} + o(t^{-1/2}) \quad \textit{for } \rho < 1 \\ K_j(t) &= c_1 e^{-c_2 t} t^{-1/2} \rho^{\frac{j+1}{2}} (\sqrt{\rho} - 1)^{-1} + o(t^{-1/2}) \quad \textit{for } \rho > 1 \\ &= \frac{1}{2} + \left(j + \frac{1}{2}\right) \frac{t^{-1/2}}{2\sqrt{\lambda\pi}} + o(t^{-1/2}) \quad \text{for } \rho = 1. \end{aligned}$$

Proof. For $\rho < 1$ we obtain, using (6.9),

$$\begin{aligned} 1 - K_j(t) &= \sum_{j+1}^{\infty} k_\nu(t) = c_1 e^{-c_2 t} t^{-1/2} \sum_{j+1}^{\infty} \rho^{\nu/2}[1 + o(1)] \\ &= c_1 e^{-c_2 t} t^{-/2} \rho^{\frac{j+1}{2}} (1 - \sqrt{\rho})^{-}1 + o(t^{-/2}). \end{aligned}$$

For $\rho > 1$ we obtain the desired result by considering the reversed process $\{-X(t), t \geq 0\}$. For $\rho = 1$, the process is symmetric about the origin and so

$$K_{-1}(t) = \frac{1}{2} - \frac{1}{2} k_0(t).$$

For $j \geq 0$ this gives

$$K_j(t) = K_{-1}(t) + \sum_{0}^{j} k_\nu(t) = \frac{1}{2} - \frac{1}{2} k_0(t) + \sum_{0}^{j} k_\nu(t).$$

The relation (6.9) in this case becomes

$$k_j(t) = \frac{t^{-1/2}}{2\sqrt{\lambda\pi}}[1 + o(1)]$$

and so

$$K_j(t) = \frac{1}{2} - \frac{1}{2} \frac{t^{-1/2}}{\sqrt{\lambda\pi}} + (j + 1) \frac{t^{-1/2}}{2\sqrt{\lambda\pi}} + o(t^{-1/2})$$

which proves the desired result for $j \geq 0$. The proof for $j < 0$ is similar. □

Let $M(t)$ and $m(t)$ be the maximum and minimum functionals of the process $\{X(t)\}$. Also, let T_i be the hitting time of the state i, so that

$$T_i = \inf\{t: X(t) = i\} \quad (i \neq 0), \quad T_0 = 0. \tag{6.10}$$

The following results are given for the above random variables.

Theorem 6.2 (i) *For $x \geq 1$ we have*

$$P\{M(t) < i\} = K_i(t) - \rho^i k_{-i}(t) \tag{6.11}$$

(ii) *The asymptotic estimates for $M(t)$ as $t \to \infty$ are given by the following:*

$$\begin{aligned} P\{M(t) < i\} &= 1 - \rho^i + o(t^{-1/2}) && \text{if } \rho < 1 \\ &= o(t^{-1/2}) && \text{if } \rho > 1 \\ &= \frac{i}{\sqrt{\lambda\pi}} t^{-1/2} + o(t^{-1/2}) && \text{if } \rho = 1. \end{aligned} \tag{6.12}$$

Proof. For $i \geq 1$ denote

$$\begin{aligned} {}^i k_{j(t)} &= P\{T_i > t, X(t) = j\} && (j < i) \\ &= 0 && (j \geq i). \end{aligned}$$

An inspection of sample functions of $X(t)$ shows that for $j \geq 0$

$$\begin{aligned} k_{i-j}(t) &= -k_{i-j}(t) + \int_0^t P\{T_i \,\epsilon\, ds\} k_{-j}(t-s). \\ k_{i+j}(t) &= \int_0^t P\{T_i \,\epsilon\, ds\} k_j(t-s). \end{aligned}$$

Since $k_j(t) = \rho^j k_{-j}(t)$ it follows from these that

$${}^i k_{i-j}(t) = k_{i-j}(t) - \rho^i k_{-i-j}(t) \ (j \geq 0). \tag{6.13}$$

Adding this last relation over $j \geq 0$ we find that

$$P\{T_i > t\} = \sum_{-\infty}^{i} k_\nu(t) - \rho^i \sum_{-\infty}^{-i} k_\nu(t).$$

This leads to the result (i) since $\{M(t) < i\} = \{T_i > t\}$. The results in (ii) are based on the asymptotic estimates for $k_j(t)$ provided by Theorem 6.1. □

Theorem 6.3 (i) *For $i \geq 1$ the hitting time T_i has density $g_i(t)$ given by*

$$g_i(t) = \frac{i}{t} k_i(t), \tag{6.14}$$

(ii) $P\{T_i < \infty\} = 1$ *if* $\rho \geq 1$, *and* $= g^i$ *if* $\rho < 1$. *If* $\rho = 1$, *then* $E(T_i) = \infty$, *and if* $\rho > 1$, *then*

$$E(T_c) = i(\lambda - \mu)^{-1} \quad \textit{and} \quad \mathrm{Var}(T_i) = i(\lambda + \mu)(\lambda - \mu)^{-3}. \tag{6.15}$$

Proof. Since

$$P\{T_i \in dt\} = ik_{i-1}(t)\lambda dt$$

using (6.13) we find that the density of T_i is given by

$$\begin{aligned}
g_i(t) &= \lambda\left[k_{i-1}(t) - \rho^i k_{-i-1}(t)\right] \\
&= \lambda e^{-(\lambda+\mu)t}\left[\rho^{\frac{i-1}{2}} I_{i-1}(2\sqrt{\lambda\mu t}) - \rho^i \rho^{\frac{-i-1}{2}} I_{i+1}(2\sqrt{\lambda\mu t})\right] \\
&= \lambda e^{-(\lambda+\mu)t}\rho^{\frac{i-1}{2}}\left(\frac{2i}{2\sqrt{\lambda\mu t}}\right) I_i(2\sqrt{\lambda\mu t}) \\
&= \frac{i}{t} k_i(t)
\end{aligned}$$

which proves (i). To derive the moments of T_i we derive its Laplace transform. Proceeding as in Sec. 5.4 we find that $\{T_i - T_{i-1}(i \geq 1)\}$ is a sequence of IID random variables, so that

$$E(e^{-\theta T_i}) = \xi^i \quad (\theta > 0), \tag{6.16}$$

where $\xi \equiv \xi(\theta) = E(e^{-\theta T_i})$ is to be determined. The epoch of the first jump (increase or decrease) in $X(t)$ is given by $T = \min(T_1, T_{-1})$. For this we have

$$E[e^{-\theta T}; T = T_1] = \int_0^\infty e^{-\theta t} e^{-(\lambda+\mu)t}\lambda dt = \frac{\lambda}{\theta + \lambda + \mu}$$

and similarly

$$E[e^{-\theta T}; T = T_{-1}] = \frac{\mu}{\theta + \lambda + \mu}.$$

Clearly

$$T_i \overset{d}{=} T_1 + T_{i-1} \quad \text{if } T = T_1 \quad \text{and} \quad T_i = T_{-1} + T_{i+1} \quad \text{if } T = T_{-1},$$

where T_1 and T_{i-1} are independent and so are T_{-1} and T_{i+1}. Therefore,

$$\xi^i = \frac{\lambda\xi^{i-1} + \mu\xi^{i+1}}{\theta + \lambda + \mu}$$

which shows that ξ satisfies the quadratic equation $\mu\xi^2 - (\theta+\lambda+\mu)\xi + \lambda = 0$. The desired root of this equation is

$$\xi(\theta) = \frac{\theta + \lambda + \mu - \sqrt{(\theta + \lambda + \mu)^2 - 4\lambda\mu}}{2\mu} < 1. \tag{6.17}$$

The results (ii) follow from this transform, since

$$\xi(0+) = \min(1, \rho), \quad \xi'(0+) = -(\lambda - \mu)^{-1} \text{ if } \lambda > \mu, \text{ and } = -\infty \text{ if } \lambda = \mu$$

and

$$\xi''(0+) = 2\lambda(\lambda - \mu)^{-3} \quad \text{if } \lambda > \mu.$$

□

5.6.1 *Application to the Simple Queue*

Suppose customers arrive at a single server facility in a Poisson process $\{A(t)\}$ at a rate λ and their service times are IID random variables with density $\mu e^{-\mu t}(0 < \lambda < \infty, 0 < \mu < \infty)$. The quantity of interest is the number $Q(t)$ of customers waiting or being served at time t (the queue length). The classical approach to the study of this process is based on the fact that it is a birth and death process with parameters $\lambda_n = \lambda(n \geq 0), \mu_0 = 0, \mu_n = \mu(n \geq 1)$. A more fruitful approach is based on the following characterization. If a sufficient number of customers are present in the system, they will be served in a Poisson process $\{D(t)\}$ at a rate μ. Let $X(t) = A(t) - D(t)$ $(t \geq 0)$; then $\{X(t), t \geq 0\}$ is called the net input process. The server will be idle when no customers are present, so we see that

$$Q(t) = Q(0) + A(t) - \int_0^t 1_{\{Q(s)>0\}} D\{ds\}$$

or equivalently

$$Q(t) = Q(0) + X(t) + \int_0^t 1_{\{Q(s)>0\}} D\{ds\}. \tag{6.18}$$

We thus obtain an integral equation for $Q(t)$. Its unique solution is given by

$$Q(t) = \max\{Q(0) + X(t), X(t) - m(t)\} \quad (t \geq 0). \tag{6.19}$$

In particular, when $Q(0) = 0, Q(t) = X(t) - m(t) \stackrel{d}{=} M(t)$. Theorem 6.2 yields the distribution of $Q(t)$. In particular, if $\rho \geq 1, Q(t) \xrightarrow{d} \infty$, while if $\rho \geq 1, Q(t) \xrightarrow{d} Q$, where Q has the geometric distribution $(1-\rho)\rho^j (j \geq 0)$. Limit results for $\rho \geq 1$ can also be obtained. Of additional interest is the random variable

$$B_i = \inf\{t\colon Q(t) = 0\} \text{ on } \{Q(0) = i\} \tag{6.20}$$

which is the busy period initiated by i (≥ 1) customers. The representation (6.19) yields the fact

$$B_i = \inf\{t\colon i + X(t) = 0\} = T_{-i}. \tag{6.21}$$

The results for B_i are obtained from Theorem 6.3 by interchanging λ and μ.

5.7 Lévy Processes: Further Properties

Under the conditions (i)–(iv) of Sec. 5.3 the c.f. of $X(t)$ is given by the following.

Theorem 7.1 *We have*

$$E[e^{i\omega X'(t)}] = e^{-t\phi(\omega)} \quad (i = \sqrt{-1}, \omega \text{ real}), \tag{7.1}$$

where

$$\phi(\omega) = i\omega a - \int_{-\infty}^{\infty} \frac{e^{i\omega x} - 1 - i\omega\tau(x)}{x^2} M\{dx\} \tag{7.2}$$

a being a real constant, $\tau(x)$ a centering function given by

$$\tau(x) = -1 \text{ for } x < -1, \quad = x \text{ for } |x| \leq 1 \quad \text{and} \quad = +1 \text{ for } x > 1 \tag{7.3}$$

and M a measure such that $M\{I\} < \infty$ for each bounded interval I and

$$M^+(x) = \int_{[x,\infty)} y^{-2} M\{dy\} < \infty, \; M^-(-x) = \int_{(-\infty,-x]} y^{-2} M\{dy\} < \infty \tag{7.4}$$

for each $x > 0$.

Proof. Let $\phi_t(\omega) = E[e^{i\omega X(t)}]$. From (3.1) we find that for each $t > 0$, $\phi_t(\omega) = [\phi_{t/n}(\omega)]^n$ for all $n \geq 1$. This property is known as *infinite divisibility.* Now for infinitely divisible characteristic functions it is known that $\phi_t(\omega) = e^{\psi_t(\omega)}$, where $\psi_t(\omega)$ has the same form as $-t\phi(\omega)$, with a and

M depending on t. Now writing $X(t+s) = X(t) + [X(t+s) - X(t)]$ we find from (i) and (ii) that the increment $X(t+s) - X(t)$ has the same distribution as $X(s)$ and is independent of $X(t)$. Hence $\phi_{t+s}(\omega) = \phi_t(\omega)\phi_s(\omega)$ or equivalently, $\psi_{t+s}(\omega) = \psi_t(\omega) + \psi_s(\omega)$. On account of (iii), $\psi_t(\omega) \to 0$ as $t \to 0+$, so we must have $\psi_t(\omega) = t\psi_1(\omega)$. Thus $\phi_t(\omega) = e^{t\psi(\omega)}$, with $\psi(\omega) = \psi_1(\omega)$, is the desired form. □

It should be noted that at the origin the integrand in (7.2) is defined by continuity. Thus the contribution of the origin to the integral is

$$\lim_{x \to 0} \frac{e^{i\omega x} - 1 - i\omega\tau(x)}{x^2} M\{(-x, x)\} = -\frac{1}{2}\omega^2\sigma^2,$$

where $M\{0\} = \sigma^2 \geq 0$. Away from the origin $e^{2\omega(x)} - 1 - i\omega\tau(x)$ is bounded and so the integrand behaves like x^{-2}, which is integrable with respect to the measure M on account of the conditions (7.4). We can therefore write

$$\phi(\omega) = i\omega a + \frac{1}{2}\omega^2\sigma^2 - \int_{|x|>0} \frac{e^{i\omega x} - 1 - i\omega\tau(x)}{x^2} M\{dx\}. \tag{7.2a}$$

Example 7.1 Suppose that M is concentrated at the origin with weight $M\{0\} = \sigma^2 > 0$. Then (7.2a) gives $\phi(\omega) = i\omega a + \frac{1}{2}\omega^2\sigma^2$ and so

$$Ee^{i\omega X(t)} = e^{-i\omega at - \frac{1}{2}\sigma^2\omega^2 t}. \tag{7.5}$$

This shows that $X(t) = -at + \sigma B(t)$, where $B(t)$ is the Brownian motion. This is a Gaussian process [see Sec. 5.4].

Example 7.2 Suppose that M has no atom at the origin and the integral

$$\int_{-\infty}^{\infty} y^{-2} M\{dy\} = \lambda \quad (0 < \lambda < \infty). \tag{7.6}$$

We define a distribution function F by writing

$$F\{dy\} = \lambda^{-1} y^{-2} M\{dy\} \quad (y \neq 0). \tag{7.7}$$

We can then write $\phi(\omega)$ as

$$\phi(\omega) = -\lambda \int_{-\infty}^{\infty} (e^{i\omega x} - 1) F\{dx\} = \lambda[1 - \psi(\omega)],$$

where $\psi(\omega)$ is the c.f. of F, and we have chosen

$$a = -\lambda \int_{-\infty}^{\infty} \tau(x) F\{dx\}.$$

Therefore,

$$Ee^{i\omega X(t)} = e^{-\lambda t[1-\psi(\omega)]} \tag{7.8}$$

which shows that $\{X(t)\}$ is the compound Poisson process of Sec. 5.2.3.

Example 7.3 (The Stable Process with Exponent 1/2) Suppose the measure M is concentrated on $(0,\infty)$ with density $(2\pi)^{-1/2}x^{1/2}$. It is more convenient to write (7.2) as a Laplace transform in this case. Thus

$$Ee^{-\theta X(t)} = e^{-t\phi/\theta} \quad (\theta > 0), \tag{7.9}$$

where

$$\phi(\theta) = a\theta + \int_0^\infty \frac{1-e^{-\theta x} - \theta\tau(x)}{x^2} M\{dx\}. \tag{7.10}$$

Now

$$\begin{aligned}
&\int_0^\infty \frac{1-e^{-\theta x}}{x^2} M\{dx\} \\
&= \frac{1}{\sqrt{2\pi}} \int_0^\infty (1-e^{-\theta x})x^{-3/2}dx = \frac{1}{\sqrt{2\pi}} \int_0^\infty x^{-3/2}dx \int_0^x \theta e^{-\theta y}dy \\
&= \frac{1}{\sqrt{2\pi}} \int_0^\infty \theta e^{-\theta y}dy \int_y^\infty \frac{dx}{x^{3/2}} = \frac{\sqrt{2}\theta}{\sqrt{\pi}} \int_0^\infty e^{-\theta y}y^{-1/2}dy \\
&= \frac{\sqrt{2}\theta}{\sqrt{\pi}} \cdot \frac{r(1/2)}{\theta^{1/2}} = \sqrt{2\theta}
\end{aligned}$$

and

$$I = \sqrt{2\pi} \int_0^\infty \frac{\tau(x)}{x^2} M\{dx\} = \int_0^1 \frac{dx}{x^{1/2}} + \int_1^\infty \frac{dx}{x^{3/2}} < \infty.$$

Taking $\sqrt{2\pi}a = I$ we find that $\phi(\theta) = \sqrt{2\theta}$ and

$$Ee^{-\theta X(t)} = e^{-t\sqrt{2\theta}}. \tag{7.11}$$

The result (7.11) as a special case (with $d = 0, \sigma = 1$) of the Laplace transform of the inverse Gaussian process described in Sec. 5.5, as can be seen from (5.6). It follows from (5.2) and (5.3) that the density and the d.f. of $X(t)$ are given respectively by

$$g_{1/2}\left(\frac{x}{t^2}\right) = \frac{t}{x^{3/2}} n\left(\frac{t}{\sqrt{x}}\right) \tag{7.12}$$

and

$$G_{1/2}\left(\frac{x}{t^2}\right) = 2\left[1 - N\left(\frac{t}{\sqrt{x}}\right)\right] \quad (0 < x < \infty). \tag{7.13}$$

The result (7.11) also shows that $X(t)t^{-2} \stackrel{d}{=} X(1)$. This shows that the process $\{X(t), t \geq 0\}$ is stable with exponent $1/2$. More generally, the class of stable processes with exponent α $(0 < \alpha \leq 2)$ as characterized by the property

$$\frac{X(t)}{t^{1/\alpha}} \stackrel{d}{=} X(1) \tag{7.14}$$

(actually, this is true only for strictly stable processes, but this distinction matters only for $\alpha = 1$). The case $\alpha = 2$ corresponds to the Brownian Motion.

Example 7.4 (The Gamma Process) Consider the case where M is concentrated on $(0, \infty)$ with density $e^{-\lambda x}x$. Then

$$\begin{aligned}\int_0^\infty \frac{e^{i\omega x} - 1}{x^2} M\{dx\} &= \int_0^\infty \frac{e^{-(\lambda - i\omega)x} - e^{-\lambda x}}{x} dx \\ &= \log \frac{\lambda}{\lambda - iw}\end{aligned}$$

and

$$I = \int_0^\infty \frac{\tau(x)}{x^2} M\{dx\} = \int_0^\infty \frac{\tau(x)}{x} e^{-\lambda x} dx < \infty.$$

Choosing $a = -I$ we obtain $\phi(\omega) = -\log(\lambda/\lambda - i\omega)$ and

$$Ee^{i\omega X(t)} = \left(\frac{\lambda}{\lambda - i\omega}\right)^t, \tag{7.15}$$

which is characteristic function of the gamma density

$$\begin{aligned}f(x;t) &= e^{-\lambda x} x^t \frac{x^{t-1}}{\Gamma(t)} \quad &&\text{for } x > 0 \\ &= 0 &&\text{for } x \leq 0.\end{aligned} \tag{7.16}$$

We call $X(t)$ the gamma process. Its mean and variance are given by

$$EX(t) = \frac{t}{\lambda}, \quad \mathrm{Var} X(t) = \frac{t}{\lambda^2}. \tag{7.17}$$

The probabilistic interpretation of the measure M is the following. If $M\{0\} = \sigma^2 > 0$ then (7.2a) shows that $X(t)$ has a Gaussian component. Denote by $N_+(t,x)$ the number of jumps of size $\geq x > 0$ and by $N_-(t,x)$ the number of jumps of size $\leq x < 0$ that occur in a time-interval $(0,t]$. Then $N_+(t,x]$ and $N_-(t,x)$ are Poisson processes with parameters $M^+(x)$ and $M^-(x)$, respectively. The total number of positive jumps in $(0,t]$ is a Poisson process with parameter $M^+(0)$ and thus in each finite time-interval there are a finite number of positive jumps iff $M^+(0) < \infty$. A similar statement holds for the total number of negative jumps.

For the Brownian motion (see Sec. 5.1) since $M\{I\} = 0$ for every interval I not containing the origin we conclude from the above that the process has no jumps. This confirms the property that Brownian sample functions are continuous with probability 1.

For the compound Poisson process (see Sec 5.2), we have

$$M^+(x) = \lambda[1 - F(x-)], \quad M^-(-x) = \lambda F(-x) \ (x > 0). \tag{7.18}$$

Thus jumps of size $\geq x > 0$ form a Poisson process with parameter $\lambda[1 - F(x-)]$ and jumps of size $\leq -x < 0$ form a Poisson process with parameter $\lambda F(-x)$. This statement is an extension of the result proved in Sec. 5.2.3, where we also showed that $X(t)$ can be represented as $X(t) = X_+(t) - X_-(t)$, where $X_+(t) \geq 0, X_-(t) \geq 0$. Thus the compound Poisson process is of *bounded variation.*

In the general case a Lévy process $X(t)$ is of bounded variation iff $M\{0\} = 0$ and moreover,

$$\int_{|x|<1} \frac{1}{|x|} M\{dx\} < \infty. \tag{7.19}$$

In this case

$$a + \int_{-\infty}^{\infty} \frac{\tau(x)}{x^2} M\{dx\} = a + \int_{|x|<1} \frac{1}{|x|} M\{dx\} + \int_{|x|\geq 1} \frac{1}{x^2} M\{dx\} = -d$$

(the integrals being convergent), so we can write (7.2a) as

$$\begin{aligned}\phi(\omega) &= -i\omega d - \lambda \int_{0+}^{\infty} \frac{e^{i\omega x} - 1}{x^2} M\{dx\} - \int_{\infty}^{0} \frac{e^{i\omega x} - 1}{x^2} M\{dx\} \\ &= i\omega d + \phi_+(\omega) + \phi_-(\omega) \text{ (say).}\end{aligned}$$

Therefore,

$$Ee^{i\omega X(t)} = e^{i\omega dt - t\phi_+(\omega) - t\phi_-(\omega)},$$

which shows that

$$X(t) = dt + X_+(t) - X_-(t), \tag{7.20}$$

where $X_+(t)$ and $X_-(t)$ are Lévy processes with non-negative increments (and consequently with non-decreasing sample functions) and d is the *drift* of the process. We call $X_+(t), X_-(t)$ *subordinators.* The simple Poisson process of Sec. 5.2, inverse Gaussian of Example 5.5, and the gamma process of Example 7.4 are all subordinators. We note that for the gamma process

$$M^+(0) = M(0) = \int_\infty^0 e^{-\lambda x} x^{-1} dx = \infty$$

so that in every finite time-interval the process has an infinite number of jumps.

Remarks

7.1. Other possible centering functions that can be used in (7.2) are

$$\text{(i) } \tau(x) = \sin x, \quad \text{and (ii) } \tau(x) = \frac{x}{1+x^2}.$$

7.2. The measure ν defined by

$$\nu\{0\} = 0, \quad \nu(dx) = x^{-2} M(dx) \quad (x \neq 0)$$

is called a Lévy measure, We have

$$\int_{-\infty}^{\infty} \min(1, x^2)\nu(dx) < \infty,$$

as can be easily verifed from (7.3).

7.3. Let

$$M(t) = \sup_{0<\tau<t} X(\tau), \quad m(t) = \inf_{0\le\tau\le t} X(\tau). \tag{7.21}$$

On account of our assumptions on $X(t)$, the functionals $M(t)$ and $m(t)$ are both random variables.

The following characterization of a Lévy process confirms the important roles played by the Brownian Motion and the Poisson process.

Ito Representation. For a Lévy process $\{X(t), t \geq 0\}$ we have

$$X(t) = dt + \sigma B(t)$$
$$+ \lim_{\varepsilon \to 0+} \left[\Sigma\{J(s) \colon (0 \leq s \leq t)\}, |J(s)| > \varepsilon\} - t \int_{|y|>\varepsilon} \tau(y) \frac{M\{dy\}}{y^2} \right],$$

where $\sigma B(t)$ is independent of $X(t) - \sigma B(t)$.

Here $B(t)$ is the Brownian Motion and $J(s)$ are the jumps.

Now let I be a Borel subset of $\mathbb{R} - \{0\}$ and

$$N_I(t) = \#\{\text{jumps } J(s) \colon 0 \leq s \leq t, J(s) \in I\}.$$

Then $N_I(t)$ is a Poisson random variable with mean $t\nu\{I\}$.

Moreover, let $I_1, I_2, \ldots,$ be subsets of $\mathbb{R} - \{0\}$ such that $I_p \cap I_q = 0$ $(p \neq q)$. Then $N_{I_1}(t), N_{I_2}(t), \ldots$ are mutually independent.

5.8 Problems for Solution

In problems 1–3, $\{B(t), t \geq 0\}$ is the standard Brownian Motion.

1. Show that

$$P\{M(t) > a, M(t) - B(t) > b\}$$
$$= P\{B(t) - m(t) > a, -m(t) > b\} = 1 - N_+ \left(\frac{a+b}{\sqrt{t}} \right).$$

2. Let $X(t) = B(t) - tB(1)$. Show that $\{X(t), 0 \leq t \leq 1\}$ is Gaussian with the correlation function

$$\Gamma(t_1, t_2) = \sqrt{\frac{t_1(1-t_2)}{t_2(1-t_1)}} \quad (0 \leq t_1 \leq t_2 \leq 1).$$

3. (a) Find the probability that $B(t)$ has no zero in (t_0, t_2), given that it has no zero in (t_0, t_1) when $0 < t_0 \leq t_1 \leq t_2$,
   (b) Letting $t_0 \to 0+$ in (a) show that for $0 < t_1 \leq t_2, P\{B(t)$ has no zero in $(0, t_2) \mid B(t)$ has no zero in $(0, t_1)\} = \sqrt{\frac{t_1}{t_2}}$.
4. Let $\{X(t), t \geq 0\}$ and $\{Y(t), t \geq 0\}$ be two independent Poisson processes with rates λ and μ, respectively. For $n \geq 0$, let T_n be the waiting time for the nth event in $Y(t)$, with $T_0 \equiv 0$. Show that the random variables $X(T_n) - X(T_{n-1}) (n \geq 1)$ are IID with a geometric distribution.

5. Let $\{X(t), t \geq 0\}$ be a process with a stationary independent increments, having a finite variance. Denote $Y(t) = X(t+1) - X(t)$. Show that the process $\{Y(t), t \geq 0\}$ is stationary in the wide sense, with the correlation function

$$R(t) = (1 - |t|)^t.$$

Problems 6-10 are on the randomized Bernoulli random walk of Sec. 5.6.

6. Denote by

$$k_j^*(s) = \int_0^\infty e^{-st} k_j(t) dt \quad (s > 0)$$

the Laplace transform of the probability function $k_j(t)$. Show that for $j \geq 0$

$$k_j^*(s) = \frac{\eta^{-j}}{\lambda(\eta - \xi)}, \quad k_j^*(s) = \frac{\xi^{-j}}{\lambda(\eta - \xi)},$$

where $\xi \equiv \xi(s)$ and $\eta \equiv \eta(s)$ are the roots of the equation $\lambda\theta^2 - (s + \lambda + \mu)\theta + \mu = 0 \quad (\xi < \eta)$.

7. Prove the following.

   (i) If $\lambda > \mu$, the process drifts to $+\infty$, that is,

$$M(t) \to \infty, \quad m(t) \to m > -\infty.$$

   (ii) If $\lambda < \mu$, the process drifts to $-\infty$, that is,

$$M(t) \to M < \infty, \quad m(t) \to -\infty.$$

   (iii) If $\lambda = \mu$, the process oscillates in the sense that

$$M(t) \to \infty, \quad m(t) \to -\infty.$$

8. (i) If $\lambda > \mu$, show that as $i \to \infty$, the hitting time T_i has an asymptotically normal distribution with mean and variance given by Theorem 6.3.
   (ii) If $\lambda = \mu$, show that as $t \to \infty$, $2\lambda T_i / i^2$ has the limit distribution with stable density with exponent $1/2$.

9. In the notation of Sec 5.6, let

$$Z = T + T_{-1} \quad \text{if } T = T_1 \quad \text{and} \quad Z = T + T_1 \quad \text{if } T = T_{-1}.$$

Show that the random variable Z has a proper distribution iff $\lambda = \mu$, in which case $E(Z) = \infty$.

10. For the maximum functional $M(t)$ the following.

 (i) As $t \to \infty$

$$\begin{aligned} \frac{M(t)}{t} &\to \lambda - \mu \quad \text{if } \lambda > \mu \\ &\to 0 \qquad\quad \text{if } \lambda \leq \mu. \end{aligned}$$

 (ii) If $\lambda = \mu$, then

$$\lim_{t\to\infty} P\left\{\frac{M(t)}{\sqrt{2\lambda t}} < x\right\} = 2N(x) - 1.$$

11. Let $Y(t) = X(t) - t$ $(t \geq 0)$, where $\{X(t)\}$ is subordinator with

$$Ee^{-\theta X(t)} = e^{-t\phi(\theta)} \quad (\theta > 0).$$

 Denote

$$T_a = \inf\{t\colon Y(t) = -a\} \quad (a > 0),\ T_0 = 0.$$

 Show that $\{T_a, a \geq 0\}$ is a subordinator with $E(e^{-sT_a}) = e^{-a\eta(s)}$ where $\eta \equiv \eta(s)$ is the unique continuous root of the functional equation $\eta = s + \phi(\eta)$ with $\eta(\infty) = \infty$.
12. In problem 11, if $\{X(t)\}$ is the inverse Gaussian process of Sec. 5.5, show that $\{T_a\}$ is again an inverse Gaussian process with unit drift.
13. In problem 11, if $\{X(t)\}$ is a stable process with exponent 1/2, show that $\{T_a\}$ is an inverse Gaussian with unit drift and

$$P\{T_a < \infty\} = e^{-2a} < 1.$$

Further Reading

Durrett, R (1984). *Brownian Motion and Martingales in Analysis.* Wadsworth.

Ito, K and McKean, P Jr (1965). *Diffusion Processes and Their Sample Paths.* New York: Springer-Verlag.

Kingman, JFC. (1993). *Poisson Processes.* Oxford Science Publishers.

CHAPTER 6

Renewal Processes and Random Walks

6.1 Renewal Processes: Introduction

Let $\{X_k, k \geq 1\}$ be a sequence of mutually independent nonnegative random variables with the distributions

$$K(x) = P\{X_1 \leq x\}, \quad F(x) = P\{X_k \leq x\} \ (k \geq 2). \tag{1.1}$$

We ignore the trivial case where $P\{X_k = 0\} = 1$ $(k \geq 2)$. Denote

$$E(X_k) = \mu \leq \infty, \quad \mathrm{Var}(X_k) = \sigma^2 \leq \infty \tag{1.2}$$

for $k \geq 2$. Also, let

$$S_0 = 0, \quad S_n = X_1 + X_2 + \cdots + X_n \ (n \geq 1) \tag{1.3}$$

be the partial sums of the sequence $\{X_k\}$ and

$$F_n(x) = P\{S_n \leq x\} \quad (x \geq 0, n \geq 0) \tag{1.4}$$

be the d.f. of $S_n (n \geq 0)$. We have $F_0(x) = 1$ $(x \geq 0), F_1(x) = K(x)$, and

$$F_{n+1}(x) = \int_0^x F_n(x - y) F\{dy\} \quad (n \geq 1). \tag{1.5}$$

We call $\{S_n, n \geq 0\}$ the renewal process induced by the distributions (K, F). If $K = F$ we call it the process induced by F. For $t \geq 0$ we define the random variable

$$N(t) = \max\{n \colon S_n \leq t\} \tag{1.6}$$

and denote

$$U(t) = EN(t) \leq \infty, \quad V(t) = EN(t)^2 \leq \infty. \tag{1.7}$$

We call $\{N(t), t \geq 0\}$ the renewal-counting process and $U(t)$ the renewal function.

6.1.1 *Physical Interpretation*

Consider a population of individuals (or items such as electric bulbs, components of industrial machinery, etc.) such that when the individual dies (fails) it is replaced (renewed) by a new individual, and the individuals live and die independently of each other. Denote by X the lifetime (lifespan) of an individual, and assume that X is a random variable with the d.f. $F(x)$. Suppose that at time $t = 0$ the population consists of an individual of age x_0; then the remaining lifetime X_1 of this individual has the d.f. given by

$$K(x) = P\{X - x_0 \leq x \mid X > x_0\} = \frac{P\{x_0 < X \leq x + x_0\}}{P\{X > x_0\}}$$

$$= \frac{F(x + x_0) - F(x_0)}{1 - F(x_0)} \quad (x \geq 0). \tag{1.8}$$

If this individual is a new one, then $x_0 = 0$ and $K(x) = F(x)$; otherwise $K(x) \neq F(x)$. Suppose the initial individual is drawn from a population with age distribution $\Phi(x_0)$, then the d.f. of X_1 is given by

$$K(x) = \int_0^\infty \frac{F(x + x_0) - F(x_0)}{1 - F(x_0)} \Phi\{da\}. \tag{1.9}$$

Clearly, the replacements (renewals) take place at the epochs $X_1, X_1 + X_2$, $X_1 + X_2 + X_3, \ldots$, where $X_1, X_2, X_3, \ldots$ are the lifetimes of successive replacements. By our assumptions the lifetimes $X_2, X_3, \ldots$ are IID with the d.f. $F(x)$, and are independent of X_1. The random variable $N(t)$ defined by (1.6) is the number of replacements made in the interval $(0, t]$, including any made at t, but excluding the initial individual. We shall ignore the case $P\{X_k = 0\} = 1$, where the individuals die instantaneously. (We also remark that in reliability theory $X_1, X_2, \ldots$ are called failure times, and $1 - F(x)$ the survival function.) In general K can be an arbitrary distribution. We ask the following questions:

(1) How many replacements will be needed on the average?
(2) What is the rate of replacement?
(3) How old is the individual currently in use? When will the next replacement be made?

The age $X(t)$ of the individual alive at time t and $Y(t)$ its remaining lifetime are given by

$$X(t) = X(0) + t \text{ if } N(t) = 0, \quad \text{and} \quad t - S_{N(t)} \text{ if } N(t) > 0 \tag{1.10}$$

and

$$Y(t) = S_{N(t)+1} - t \ (t \geq 0). \tag{1.11}$$

Remark 1.1 Since $S_{n+1} = S_n + X_{n+1}, \{S_n, n \geq 0\}$ is a Markov process on the state space $[0, \infty)$. The random variable $\nu(t) = N(t) + 1$ can be expressed as

$$\nu(t) = \min\{n\colon S_n > t\}; \tag{1.12}$$

so $\nu(t)$ is the hitting time of the set (t, ∞) and is a stopping time relative to $\{S_n\}$. This important fact will be useful in many calculations.

We recall that a distribution F on $[0, \infty)$ is arithmetic if it is concentrated on a set of the form $\{0, d, 2d, \ldots\}$ and that the largest d with this property is called the span of F. We shall say that the renewal process $\{S_n\}$ induced by the distributions (K, F) is *discrete with span* d if F is arithmetic with span d and K is concentrated on $\{0, d, 2d, \ldots\}$. If F is not concentrated on any set of the form $\{0, d, 2d, \ldots\}$, we shall say that the renewal process induced by (K, F) is continuous.

Corresponding to a result for a continuous renewal process there is usually a version for the discrete process. However, we shall either not state this discrete version or else state it without proof.

Example 1.1 (The Terminating Process) Suppose that the distributions K and F are both defective so that

$$K(\infty) \equiv p_0, \quad F(\infty) = p \quad (0 < p_0 < 1, 0 < p < 1). \tag{1.13}$$

Then there is a random variable N such that

$$\{X_1 < \infty, X_2 < \infty, \ldots, X_N < \infty, X_{N+1} = \infty\} = \{S_N < \infty, S_{N+1} = \infty\}. \tag{1.14}$$

Clearly

$$P\{N = 0\} = 1 - p_0, \quad P\{N = n\} = p_0 p^{n-1}(1 - p) \ (n \geq 1) \tag{1.15}$$

and $P\{N < \infty\} = 1$. Thus the renewal process $\{S_n\}$ has a finite lifetime in the sense that

$$S_n \leq M \quad \text{for } n \leq N, \ \text{ and } = \infty \quad \text{for } n > N, \tag{1.16}$$

where $M = S_N$. This shows that

$$N(t) \leq N \quad \text{for } t \leq M, \quad \text{and } = N \quad \text{for } t > M. \tag{1.17}$$

We therefore call $\{S_n\}$ a terminating renewal process.

In the remainder of this chapter we shall only consider the case where the distributions (K, F) are both proper so that $N(t) \to \infty$ with probability 1 as $t \to \infty$.

Example 1.2 (The Binomial Process) For $k \geq 1$ let

$$P\{X_k = 0\} = q, \ \ P\{X_k = 1\} = p \quad (0 < p < 1, q = 1 - p). \tag{1.18}$$

Here the renewals occur at the epochs $0, 1, 2, \ldots$, and consequently

$$N(t) = N_1 + N_2 + \cdots + N_T, T = [t] \tag{1.19}$$

where N_j is the number of renewals made at j $(j \geq 1)$. We have

$$P\{N_j = n\} = pq^{n-1} \ (n \geq 1). \tag{1.20}$$

Thus N_j has a geometric distribution with mean and variance given by

$$E(N_j) = \frac{1}{p}, \quad \text{Var}(N_j) = \frac{q}{p^2}. \tag{1.21}$$

Since $N_1, N_2, \ldots, N_T$ are IID, $N(t)$ has a negative binomial distribution. Therefore $N(t)$ is a proper random variable with

$$U(t) = EN(t) = \frac{T}{p}, \quad \text{Var}N(t) = T\frac{q}{p^2}. \tag{1.22}$$

6.2 The Renewal-Counting Processes $\{N(t)\}$

Theorem 2.1 *The random variable $N(t)$ is proper, with finite moments of all orders; that is,*

(i) $$N(t) < \infty \text{ with probability } 1, \tag{2.1}$$

(ii) $$EN(t)^r < \infty \ (r > 0). \tag{2.2}$$

Proof. Since we have assumed that $P\{X_k = 0\} < 1$ $(k \geq 2)$ there exists an $\alpha > 0$ such that $P\{X_k \geq \alpha\} > 0$. Without loss of generality we may take $\alpha = 1$; then $P\{X_k \geq 1\} = p > 0$ $(k \geq 2)$. We now define a modified renewal process as follows. Let

$$\bar{X}_1 = 0 \text{ a.s.}$$
$$\bar{X}_k = 0 \text{ if } X_k < 1, \text{ and } = 1 \text{ if } X_k \geq 1 \ (k \geq 2).$$

Let $\overline{S_0} \equiv 0, \overline{S_n} = \bar{X}_1 + \bar{X}_2 + \cdots + \bar{X}_n (n \geq 1)$ and $\bar{N}(t) = \max\{n; \bar{S}_n \leq t\}$. It is clear that $\{\bar{S}_n\}$ is the Bernoulli process of Example 1.1 with $p_0 = 0$. Since $\bar{S}_n \leq S_n$ we have $N(t) \leq \bar{N}(t)$ with probability 1. Since $\bar{N}(t)$ is a proper random variable with finite moments of all orders, the desired results follow. □

Theorem 2.2 *The distribution of $N(t)$ is given by*

$$P_n(t) = P\{N(t) = n\} = F_n(t) - F_{n+1}(t) \ (n = 0, 1, 2, \ldots) \tag{2.3}$$

and its first two moments by

$$EN(t) = U(t) \equiv \sum_1^\infty F_n(t) \tag{2.4}$$

$$EN(t)^2 = V(t) = U(t) + 2\int_0^t U(t-\tau)U\{d\tau\}. \tag{2.5}$$

Proof. We have

$$P\{N(t) \geq n\} = P\{S_n \leq t\} = F_n(t)$$

so that

$$P_n(t) = P\{N(t) \geq n\} - P\{N(t) \geq n+1\} = F_n(t) - F_{n+1}(t)$$

which proves (2.3). This gives

$$EN(t) = \sum_{n=0}^\infty n[F_n(t) - F_{n+1}(t)] = \sum_1^\infty F_n(t)$$

which is the result (2.4), and

$$EN(t)^2 = \sum_0^\infty n^2[F_n(t) - F_{n+1}(t)] = \sum_1^\infty (2n-1)F_n(t).$$

Now

$$\sum_{1}^{\infty} nF_n(t) = \sum_{m=1}^{\infty}\sum_{n=0}^{\infty} F_{m+n}(t) = \sum_{m=1}^{\infty}\sum_{n=0}^{\infty}\int_0^t F_m(t-\tau)F_n\{d\tau\}$$
$$= U(t) + \int_0^t U(t-\tau)U\{d\tau\}.$$

This leads to the result (2.5). □

Example 2.1 (The Poisson Process) Suppose that the lifetime X has density $\lambda e^{-\lambda t}$ $(0 < \lambda < \infty)$. Then $F(x) = 1 - e^{-\lambda x}$ and

$$K(x) = \frac{[1 - e^{-\lambda(x+x_0)}] - [1 - e^{-\lambda x_0}]}{1 - (1 - e^{-\lambda x_0})} = 1 - e^{-\lambda x} = F(x) \tag{2.6}$$

(indicating age-independent failure times). Here $F_n(x)$ has the gamma density $e^{-\lambda x}\lambda^n x^{n-1}/(n-1)!$ and so

$$F_n(t) = 1 - \sum_{0}^{n-1} e^{-\lambda t}\frac{(\lambda t)^n}{n!} \quad (n \geq 1). \tag{2.7}$$

This gives

$$P\{N(t) = n\} = F_n(t) - F_{n+1}(t) = e^{-\lambda t}\frac{(\lambda t)^n}{n!} \quad (n \geq 0). \tag{2.8}$$

We have thus shown that the random variable $N(t)$ has a Poisson distribution. In order to show that the renewal-counting process $\{N(t), t \geq 0\}$ is a Poisson process we need to show that it has stationary independent increments. This is done in Chap. 5. The essential argument there is that since $\nu(t) = N(t) + 1$ is a stopping time relative to $\{S_n\}$, the random variables $S_\nu(t) - t, X_{\nu(t)+k}$ $(k \geq 1)$ are IID with density $\lambda e^{-\lambda t}$, and these are independent of $\{X_1, X_2, \ldots, X_{N(t)}, N(t)\}$. We note that $EN(t) = \lambda t$ and $\mathrm{Var}N(t) = \lambda t$.

Example 2.2 (The Stationary Process) Suppose that $0 < \mu < \infty$ and the initial age $X(0)$ has the density

$$\frac{1 - F(x_0)}{\mu} \quad (0 < x_0 < \infty). \tag{2.9}$$

Then the d.f. of X_1 is given by

$$K(t) = \int_0^\infty \frac{F(x_0+t)-F(x_0)}{1-F(x_0)} \cdot \frac{1-F(x_0)}{\mu}\, dx_0$$
$$= \int_0^t \frac{1-F(s)}{\mu}\, ds. \tag{2.10}$$

Thus K has density (2.9), and consequently the distribution of S_n also has density f_n, where for $n \geq 2$

$$f_n(t) = \int_0^t \frac{1-F(t-s)}{\mu} P\{X_2+X_3+\cdots+X_n \,\epsilon\, ds\}$$
$$= \frac{1}{\mu}[P\{X_2+X_3+\cdots+X_n \;\leq t\} - P\{X_2+X_3+\cdots+X_{n+1} \leq t\}] \tag{2.11}$$

and $f_1(t)$ is given by (2.9). The renewal function u is given by

$$U(t) = \int_0^t \sum_1^\infty f_n(s)ds = \int_0^t \frac{1}{\mu} ds = \frac{t}{\mu}. \tag{2.12}$$

From Theorem 2.2 we find that $EN(t) = t\mu^{-1}$ and

$$EN(t)^2 = U(t) + 2\int_0^t U(t-s)U\{ds\}$$
$$= \frac{t}{\mu} + \frac{2}{\mu^2}\int_0^t (t-s)ds = \frac{t}{\mu} + \frac{t^2}{\mu^2}.$$

This gives $\text{Var}N(t) = t\mu^{-1}$. This process is called stationary because it will be proved later (in Sec. 6.5) that for $s > 0$ the increment $N(s+t)-N(s)$ has a distribution that does not depend on s. These properties are shared by the Poisson process, which has the additional property that the increments are independent.

Theorem 2.3 *As $t \to \infty$*

$$\frac{N(t)}{t} \to \frac{1}{\mu} \tag{2.13}$$

with probability 1, *the limit being interpreted as zero if $\mu = \infty$.*

Proof. (i) Let $\mu < \infty$. Then by the strong law of large numbers we have

$$\frac{S_n}{n} = \frac{X_1}{n} + \frac{X_2 + \cdots + X_n}{n-1} . \frac{n-1}{n} \to \mu$$

with probability 1. Since $N(t) \to \infty$ as $t \to \infty$ we have $S_{N(t)}/N(t) \to \mu$ with probability 1. The inequalities

$$\frac{S_N(t)}{N(t)} \leq \frac{t}{N(t)} < \frac{S_{N(t)+1}}{N(t)+1} \cdot \frac{N(t)+1}{N(t)}$$

show that as $t \to \infty, N(t)t^{-1} \to \mu^{-1}$ with probability 1, as desired.

(ii) Let $\mu = \infty$. We define a second renewal process $\{\bar{S}_n, n \geq 0\}$ with lifetimes $\{\bar{X}_k,\ k \geq 1\}$ given by

$$\bar{X}_k = X_k \quad \text{if } X_k \leq A, \quad \text{and } = 0 \quad \text{if } X_k > A, \tag{2.14}$$

where $A > 0$. The mean lifetime of this process is given by

$$\mu_A = E(X_k;\ X_k \leq A)\ \leq A. \tag{2.15}$$

Let $\bar{N}(t) = \max\{n;\ \bar{S}_n \leq t\}$. Since $\bar{S}_n \leq S_n$ we have $N(t) \leq \bar{N}(t)$. Therefore

$$\limsup_{t\to\infty} \frac{\bar{N}(t)}{t} \leq \lim_{t\to\infty} \frac{\bar{N}(t)}{t} = \frac{1}{\mu_A}.$$

Since $\mu_A \to \infty$ as $A \to \infty$ the desired result follows for $\mu = \infty$. □

In the Poisson case (Example 2.1) we saw that $U(t) = \lambda t$, so that $U(t)/t = \mu^{-1}$, where $\mu = E(X_k)$, the mean lifetime. In the general case we shall prove that this relation holds asymptotically as $t \to \infty$. For convenience we assume that $K(x) = F(x)$.

Theorem 2.4 (The Elementary Renewal Theorem) *As* $t \to \infty$

$$\frac{U(t)}{t} \to \frac{1}{\mu}, \tag{2.16}$$

the limits being interpreted as: 0 *if* $\mu = \infty$.

Proof. (i) Let $\mu < \infty$. As already observed in Remark 1.1, the random variable $\nu(t) = N(t) + 1$ is a stopping time relative to $\{S_n\}$. Since $E\nu(t) = U(t) + 1 < \infty$, the Wald equation (see Theorem 7.7.1) states that

$ES_{\nu(t)} = \mu E\nu(t) = \mu[U(t)+1]$. For the remaining lifetime $Y(t) = S_{\nu(t)} - t$ we therefore obtain $E[Y(t)+t] = \mu[U(t)+1]$. Thus

$$\frac{U(t)}{t} + \frac{1}{t} = \frac{EY(t)}{\mu t} + \frac{1}{\mu}. \tag{2.17}$$

Since $EY(t) \geq 0$ we obtain

$$\frac{U(t)}{t} + \frac{1}{t} \geq \frac{1}{\mu}$$

which gives

$$\liminf_{t\to\infty} \frac{U(t)}{t} \geq \frac{1}{\mu}. \tag{2.18}$$

We now consider the renewal process $\{\bar{S}_n\}$ constructed in the proof of Theorem 2.3. Denote by $\bar{U}(t)$ the renewal function of this process and $\bar{Y}(t)$ the remaining lifetime at time t. Since $\bar{N}(t) \leq \bar{N}(t)$ we have $U(t) \leq \bar{U}(t)$. Also, $\bar{Y}(t) < A$. Thus applying (2.17) to this process we obtain

$$\frac{\bar{U}(t)}{t} + \frac{1}{t} \leq \frac{A}{\mu_A t} + \frac{1}{\mu_A},$$

where $\mu_A \leq \mu < \infty$. Hence

$$\limsup_{t\to\infty} \frac{\bar{U}(t)}{t} \leq \frac{1}{\mu_A}.$$

However, as already observed, $U(t) \leq \bar{U}(t)$. Therefore

$$\limsup_{t\to\infty} \frac{U(t)}{t} \leq \limsup_{t\to\infty} \frac{\bar{U}(t)}{t} \leq \frac{1}{\mu_A}. \tag{2.19}$$

Now as $A \to \infty$, $\mu_A \to \mu$ so that (2.19) gives

$$\limsup_{t\to\infty} \frac{U(t)}{t} \leq \frac{1}{\mu}. \tag{2.20}$$

From (2.18) and (2.20) the result (2.16) follows for $\mu < \infty$.

(ii) Let $\mu = \infty$. We again consider the process $\{\bar{S}_n\}$. Since $\mu_A \to \infty$ as $A \to \infty$ in this case, (2.19) gives $U(t)/t \to 0$ as $t \to \infty$. □

The proof of Theorem 2.3 suggests that any limit theorem concerning $\{S_n\}$ possibly yields a corresponding limit theorem for $N(t)$ in view of the relationship $\{N(t) < n\} = \{S_n > t\}$. It turns out that this is indeed the case. In particular the following result states that if $\{S_n\}$ obeys the central limit theorem, so does $\{N(t)\}$. Again, we shall assume $K = F$ for convenience.

Theorem 2.5 *Suppose* $0 < \mu < \infty$ *and* $\sigma^2 = \text{Var}(X_k) < \infty$. *Then as* $t \to \infty$

$$P\left\{\frac{N(t) - t\mu^{-1}}{\sqrt{\sigma^2\mu^{-3}t}} \le x\right\} \to N(x). \tag{2.21}$$

Proof. By the central limit theorem the sequence of random variables $\{(S_n - n\mu)/\sigma\sqrt{n}\}$ converges in distribution to the standard normal. Since $N(t) \to \infty$ as $t \to \infty$ with probability 1, it follows that $\{[S_{N(t)} - \mu N(t)]/\sigma\sqrt{N(t)}\}$ also converges in distribution to the standard normal. Now from

$$S_{N(t)} \le t < S_{N(t)} + X_{N(t)+1}$$

we obtain

$$\frac{S_{N(t)} - \mu N(t)}{\sigma\sqrt{N(t)}} \le \frac{t - \mu N(t)}{\sigma\sqrt{N(t)}} < \frac{S_{N(t)} - \mu N(t)}{\sigma\sqrt{N(t)}} + \frac{X_{N(t)+1}}{\sigma\sqrt{N(t)}},$$

where it can be proved that $X_{N(t)+1}/\sqrt{N(t)} \to 0$ with probability 1 as $t \to \infty$. So we conclude that as $t \to \infty$

$$P\left\{\frac{t - \mu N(t)}{\sigma\sqrt{N(t)}} \le x\right\} \to N(x).$$

Again, writing

$$\frac{t - \mu N(t)}{\sigma\sqrt{N(t)}} = \frac{t\mu^{-1} - N(t)}{\sqrt{\sigma^2\mu^{-3}t}} \cdot \sqrt{\frac{t}{N(t)\mu}}$$

and recalling that $N(t)t^{-1} \to \mu^{-1}$ with probability 1 we find that $[t\mu^{-1} - N(t)]/\sqrt{\sigma^2\mu^{-3}t}$ converges in distribution to the standard normal. The desired result follows from the symmetry of the standard normal density. □

We note that since the convergence in distribution does not always imply convergence of moments, Theorem 2.5 does not imply

$$EN(t) \sim \frac{t}{\mu}, \quad \text{Var}N(t) \sim \frac{t\sigma^2}{\mu^3}. \tag{2.22}$$

The first of these results is given by Theorem 2.4.

Remark 2.1 (The Renewal Measure) As in the case of the Poisson process (see Sec. 5.2.2) we may view the renewal-counting process $\{N(t)\}$ as a distribution of points on $[0, \infty)$ such that for $I = (a, b]$

$$N\{I\} \equiv \{n : a < S_n \leq b\}. \tag{2.23}$$

Then

$$U\{I\} = EN\{I\} = U(b) - U(a). \tag{2.24}$$

We call $U\{I\}$ the renewal measure associated with $\{S_n\}$; it is bounded over finite intervals I.

If K and F have densities (k and f), then F_n also has density f_n (say) so that

$$\infty > U\{I\} = \sum_{n=1}^{\infty} \int_I f_n(s)ds = \int_I \sum_1^{\infty} f_n(s)ds.$$

This shows that U has density u in the sense that

$$U\{I\} = \int_I u(s)ds, \quad u(s) = \sum_1^{\infty} f_n(s). \tag{2.25}$$

If we "confuse" expectation with probability, $U(t)\,dt$ can be interpreted as the probability that a replacement is made during $(t, t + dt]$. Now if a replacement is made during this interval, then the individual being replaced is either the initial one itself, or else the one which entered the population at some prior time $t - s$ $(0 < s \leq t)$. These considerations lead to the relation

$$u(t) = k(t) + \int_0^t u(t - s)f(s)ds. \tag{2.26}$$

This is called the integral equation of renewal theory, and is the starting point in earlier discussion of the theory.

If K and F do not have densities, then the above equation can be written as

$$U(t) = K(t) + \int_0^t U(t - s)F\{ds\} \tag{2.27}$$

as can be seen from (2.26).

For the discrete renewal process with span $d, U\{I\}$ is an atomic measure with atoms at n with weights u_n, where

$$u_n = \sum_{k=0}^{\infty} P\{S_k = nd\} \quad (n \geq 0). \tag{2.28}$$

The integral equation (2.27) reduces in this case to

$$u_n = k_n + \sum_{m=0}^{n} u_{n-m f m} \quad (n \geq 0), \tag{2.29}$$

where for $n \geq 0$

$$k_n = P\{X_1 = nd\}, \quad f_n = P\{X_k = nd\} \quad (k \geq 2). \tag{2.30}$$

It is analytically more convenient to consider the renewal measure associated with the process $\{S_n\}$ induced by F and add on to it an atom at the origin with unit weight. We denote it by

$$U_0\{I\} = \sum_{n=0}^{\infty} P\{S_n \in I\}. \tag{2.31}$$

Since

$$U(t) = \int_0^t K(t-s) U_0\{ds\} \quad (t \geq 0) \tag{2.32}$$

it suffices to consider the more general integral equation below.

Theorem 2.6 (The Integral Equation of Renewal Theory) *Let z be a bounded function on $[0, \infty)$. The unique bounded solution of the integral equation*

$$Z(t) = z(t) + \int_0^t Z(t-s) F\{ds\} \quad (t \geq 0) \tag{2.33}$$

such that Z is bounded over all finite intervals is given by

$$Z(t) = \int_0^t z(t-s) U_0\{ds\} \quad (t \geq 0). \tag{2.34}$$

Proof. (i) Clearly, Z given by (2.34) is bounded over all finite intervals. We have

$$\begin{aligned} Z(t) &= E\sum_{0}^{\infty} z(t-S_n) = z(t) + \sum_{0}^{\infty} Ez(t-S_{n+1}) \\ &= z(t) + E\sum_{0}^{\infty} E[z(t-S_n-X_{n+1}) \mid X_{n+1}] \\ &= z(t) + EZ(t-X_{n+1}) = z(t) + \int_0^t Z(t-s)F\{ds\}. \end{aligned}$$

This shows that (2.34) satisfies Eq.(2.33).

(ii) Suppose Z is a solution of (2.33) that is bounded over all finite intervals. We have

$$\begin{aligned} Z(t) &= z(t) + EZ(t-X_1) \\ &= z(t) + E[z(t-X_1) + EZ(t-X_1-X_2)] \\ &= z(t) + Ez(t-X_1) + EZ(t-X_1-X_2) \end{aligned}$$

and by induction

$$Z(t) = \sum_{m=0}^{n-1} Ez(t-S_m) + EZ(t-S_n).$$

Here

$$\begin{aligned} |EZ(t-S_n)| &= \left| \int_0^t Z(t-s)F_n\{ds\} \right| \\ &\leq \underset{0\leq s\leq t}{\text{L.U.B}} |Z(s)||F_n(t)| \to 0 \end{aligned}$$

as $n \to \infty$, since $\sum_0^\infty F_n(t) < \infty$. Therefore

$$Z(t) = \sum_{m=0}^{\infty} Ez(t-S_m) = \int_0^t z(t-s)U_0\{ds\}$$

which agrees with (2.34). □

Remark 2.2 (Laplace Transforms) For $z = K$, Eq. (2.33) yields the unique solution

$$\int_0^t K(t-s)U_0\{ds\} = U(t) = \sum_{1}^{\infty} F_n(t).$$

This expression for $U(t)$ in terms of an infinite series is not very helpful in evaluating it for finite t. Accordingly, we use transforms in (2.27). For $\theta > 0$

$$K^*(\theta) = \int_0^\infty e^{-\theta x} K\{dx\}, \quad F^*(\theta) = \int_0^\infty e^{-\theta x} F\{dx\}, \tag{2.35}$$

$$U^*(\theta) = \int_0^\infty e^{-\theta t} U\{dt\} \tag{2.36}$$

be the Laplace transforms of K, F, and U, respectively. Then from (2.27) we obtain

$$U^*(\theta) = \frac{K^*(\theta)}{1 - F^*(\theta)} \quad (\theta > 0) \tag{2.37}$$

since $F^*(\theta) < 1$. By inverting the right side of (2.37) we can find the unique solution of (2.27).

Example 2.-5 (The Gamma Process) Consider the renewal process induced by the gamma density

$$f(x) = e^{-\lambda x} \lambda^p x^{p-1}/(p-1)!, \tag{2.38}$$

where $0 < \lambda < \infty$ and p is a positive integer. The mean and the variance of this are given by

$$\mu = E(X_k) = \frac{p}{\lambda}, \quad \sigma^2 = \mathrm{Var}(X_k) = \frac{p}{\lambda^2}. \tag{2.39}$$

Here F_n has the gamma density

$$f_n(x) = e^{-\lambda x} \lambda^{np} x^{np-1}/(np-1)!$$

so that

$$F_n(t) = 1 - \sum_0^{np-1} e^{\lambda t} \frac{(\lambda t)^r}{r!} \quad (n \geq 1).$$

The distribution of $N(t)$ is given by

$$\begin{aligned} P_n(t) &= F_n(t) - F_{n+1}(t) \\ &= \sum_0^{np+p-1} e^{-\lambda t} \frac{(\lambda t)^r}{r!} - \sum_0^{np-1} e^{-\lambda t} \frac{(\lambda t)^r}{r!} \\ &= \sum_{np}^{np+p-1} e^{-\lambda t} \frac{(\lambda t)^r}{r!} \quad (n = 0\ 1, 2, \ldots). \end{aligned} \tag{2.40}$$

When $p = 1$, this process reduces to the one considered in Example 1.2, and we find that

$$P_n(t) = e^{-\lambda t}\frac{(\lambda t)^n}{n!} \quad (n = 0, 1, 2, \dots).$$

To find the mean and the variance of $N(t)$ we use transforms as in (2.37) but for complex θ with $\mathrm{Re}(\theta) > 0$. Then

$$F^*(\theta) = \int_0^\infty e^{-\theta x} f(x)dx = \frac{\lambda^p}{(\theta+\lambda)^p} \tag{2.41}$$

and

$$U^*(\theta) = \frac{\lambda^p}{(\theta+\lambda)^p - \lambda^p}. \tag{2.42}$$

The denominator on the right side of (2.42) has p zeros θ_0 (=0), $\theta_1, \theta_2, \dots, \theta_{p-1}$ which are such that $\mathrm{Re}(\theta_m) < 0$ $(m = 1, 2, \dots, p-1)$. In fact $\theta_m = -\lambda + \lambda\omega_m$, where $\omega_0 = 1$, $\omega_1, \omega_2, \dots, \omega_{p-1}$ are the pth roots of unity. We can therefore write

$$U^*(\theta) = \frac{A_0}{\theta} + \sum_{m=1}^{p-1} \frac{A_m}{\theta - \theta_m}, \tag{2.43}$$

where the coefficients $A_0, A_1, \dots, A_{p-1}$ are given by

$$A_m = \lim_{\theta\to\theta_m} \frac{(\theta-\theta_m)\lambda^p}{(\theta+\lambda)^p - \lambda^p} = \frac{\lambda+\theta_m}{p} \tag{2.44}$$

for $m = 0, 1, 2, \dots, p-1$. We also find that

$$-\sum_1^{p-1}\frac{A_m}{\theta_m} = \lim_{\theta\to 0}\left[\frac{\lambda^p}{(\theta+\lambda)^p - \lambda^p} - \frac{\lambda}{p\theta}\right] = \frac{1-p}{2p}.$$

Inverting (2.43) and using the fact that $U(0) = 0$ we obtain

$$\begin{aligned} U(t) &= A_0 t + \sum_1^{p-1} \frac{A_m}{\theta_m}(e^{\theta_m t} - 1) \\ &= \frac{\lambda t}{p} + \frac{1-p}{2p} + \sum_1^{p-1}\frac{A_m}{\theta_m}e^{\theta_m t}. \end{aligned} \tag{2.45}$$

This shows that the renewal density is given by

$$u(t) = \frac{\lambda}{p} + \sum_1^{p-1} A_m e^{\theta_m t}. \tag{2.46}$$

For $p = 1$ this reduces to $u(t) = \lambda$ as was expected. For $p > 1$ we find from (2.45) and (2.46) that as $t \to \infty$

$$U(t) = \frac{\lambda t}{p} + \frac{1-p}{2p} + o(1), \quad u(t) = \frac{\lambda}{p} + o(1). \tag{2.47}$$

Using (2.5) we also find that as $t \to \infty$

$$\mathrm{Var} N(t) = \frac{\lambda t}{p^2} + o(1). \tag{2.48}$$

6.3 Renewal Theorems

We first refer to Theorem 2.4 (The Elementary Renewal Theorem) and provide a second proof of it using a Tauberian theorem. This approach also leads to a limit theorem concerning the renewal density when it exists. These results are stated as follows:

Theorem 3.1 (i) *As $t \to \infty$, $t^{-1}U(t) \to c$ $(0 < c < \infty)$ iff $0 < \mu < \infty$. In this case $c = \mu^{-1}$.*

(ii) *Let $0 < \mu < \infty$. If the renewal density $u(t)$ exists and is monotone, then $u(t) \to \mu^{-1}$ as $t \to \infty$.*

Proof. (i) For $\theta > 0$ we denote the transforms as in Remark 2.2. Then

$$U^*(\theta) = \frac{K^*(\theta)}{1 - F^*(\theta)} \quad (\theta > 0).$$

Here $K^*(\theta) \to 1$ as $\theta \to 0+$ and so

$$\lim_{\theta \to 0+} \theta U^*(\theta) = \lim_{\theta \to 0+} \frac{\theta K^*(\theta)}{1 - F^*(\theta)} = \mu^{-1} \ (0 < \mu \leq \infty).$$

The desired result now follows from Theorem A4.1 which states that $t^{-1}U(t) \to c$ as $t \to \infty$ iff $\theta U^*(\theta) \to c$ as $\theta \to 0+$.

(ii) If the renewal density exists and is monotone, then since

$$U^*(\theta) = \int_0^\infty e^{-\theta t} u(t) dt \ \sim \ \theta^{-1}\mu^{-1}$$

it follows from Theorem A4.3 that $u(t) \sim \mu^{-1}$. □

Less elementary renewal theorems are also available in the literature. Their proofs use a combination of probabilistic and purely analytical tools. The following is a brief account of some of these theorems, stated only for continuous renewal processes, and moreover, the limits are interpreted as zero if $\mu = \infty$.

(i) D. Blackwell proved that for $h > 0$

$$U(t) - U(t-h) \to \frac{h}{\mu} \quad \text{as } t \to \infty. \tag{3.1}$$

(ii) The Key Renewal Theorem due to W. L. Smith states that

$$\int_0^t z(t-s)U\{ds\} \to \frac{1}{\mu}\int_0^\infty z(s)ds, \tag{3.2}$$

where z is a non-negative, nonincreasing function on $[0,\infty)$, which is integrable over $[0,\infty)$.

(iii) Later, Smith extended (3.2) to a larger class of functions which he called functions of class K. Apparently unaware of Smith's result, W. Feller introduced his class of directly Riemann-integrable functions and established (3.2) for them. These two classes are equivalent.

Perhaps the most important result for the purpose of applications is the Key Renewal Theorem, and we proceed to prove it.

Theorem 3.2 (The Key Renewal Theorem) *Let $\{S_n\}$ be continuous renewal process and Q a non-negative, nonincreasing function on $[0,\infty)$ for which*

$$\int_0^\infty Q(t)dt < \infty. \tag{3.3}$$

Then as $t \to \infty$

$$\int_0^t Q(t-s)U\{ds\} \to \frac{1}{\mu}\int_0^\infty Q(s)ds. \tag{3.4}$$

Proof. We can write

$$\int_0^t Q(t-s)U\{ds\} = I_1 + I_2,$$

where

$$I_1 = \int_0^{t/2} Q(t-s)U\{ds\}, \quad I_2 = \int_{t/2}^t Q(t-s)U\{ds\}.$$

Here

$$I_1 \leq Q(t/2)U(t/2) = \frac{t}{2}Q\left(\frac{t}{2}\right)U(t/2)[t/2]^{-1} \to 0$$

since $U(t/2)/(t/2) \to \mu^{-1}$ and $\frac{t}{2}Q(t/2) \to 0$ as $t \to \infty$. Therefore it remains to consider I_2. We proceed in two main steps.

(i) For $h > 0$ we have

$$Q = \int_0^\infty Q(t)dt = \sum_{n=0}^{\infty} \int_{nh}^{nh+h} Q(t)dt.$$

Since Q is nonincreasing we have

$$Q(nh + h) \leq Q(t) \leq Q(nh)$$

for $nh \leq t \leq nh + h$ $(n \geq 0)$. This leads to

$$h\sum_{n=0}^{\infty} Q(nh + h) \leq Q \leq h\sum_{0}^{\infty} Q(nh)$$

or

$$0 \leq Q - h\sum_{1}^{\infty} Q(nh) \leq hQ(0) < \varepsilon \tag{3.5}$$

if we choose $0 < h < \varepsilon/Q(0)$.

(ii) We can write

$$\begin{aligned} I_2 &= \int_0^{t/2} Q(s)U\{t - ds\} \\ &= \sum_{n=0}^{N-1} \int_{nh}^{nh+h} Q(s)U\{t - ds\} + \int_{Nh}^{t/2} Q(s)U\{t - ds\}, \end{aligned}$$

where $N = [t/2h]$. It is easily seen that

$$\begin{aligned} &\sum_{n=0}^{N-1} Q(nh + h)[U(t - nh) - U(t - nh - h)] \\ &\qquad \leq I_2 \leq \sum_{n=0}^{N} Q(nh)[U(t - nh) - U(t - nh - h)]. \end{aligned} \tag{3.6}$$

Now we can choose t so large that

$$\left| \frac{U(t-nh) - U(t-nh-h)}{h} - \frac{1}{\mu} \right| < \varepsilon \tag{3.7}$$

and

$$h \sum_{N+1}^{\infty} Q(nh) < \varepsilon. \tag{3.8}$$

From (3.5)–(3.8) we therefore find that

$$\left(\frac{1}{\mu} - \varepsilon \right) (Q - 2\varepsilon) < I_2 < \left(\frac{1}{\mu} + \varepsilon \right) (Q + \varepsilon). \tag{3.9}$$

Since ε is arbitrary, this implies that

$$I_2 \to Q/\mu \quad \text{as } t \to \infty, \tag{3.10}$$

which completes the proof. □

6.4 The Age and the Remaining Lifetime

The age $X(t)$ of the item in use at time t and its remaining lifetime $Y(t)$ are defined by (1.10) and (1.11). Of related interest is the increment

$$N(s+t) - N(s) = \max\{n - N(s) \colon Y(s) + X_{\nu(s)+1} + \cdots + X_n \ \leq s + t\} \tag{4.1}$$

with $\nu(s) = N(s) + 1$. Here the random variables $Y(s), X_{\nu(s)+k}$ $(k \geq 1)$ are mutually independent as in the Poisson case (Example 2.1). However, in the general case $Y(s)$ depends on $X_1, X_2, \ldots, X_{N(s)}, N(s)$. We have

$$\begin{aligned} X(t) &= X(s) - X(0) + (t-s) - \sum_{1}^{\nu(t)-1} X_k \quad (0 \leq s < X_1 \leq t) \\ &= X(s) + (t-s) - \sum_{\nu(s)}^{\nu(t)-1} X_k \quad (X_1 \leq s \leq t), \end{aligned} \tag{4.2}$$

$$Y(t) = Y(s) - (t-s) + \sum_{\nu(s)+1}^{\nu(t)} X_k \quad (0 \leq s \leq t), \tag{4.3}$$

the sum in each case being interpreted as zero if $\nu(t) = \nu(s)$. Since $\nu(s)$ is a stopping time relative to $\{S_n\}$ it follows that $\{X(t), Y(t)\}$ is a time-homogeneous Markov process on the state space $[0,\infty)\times[0,\infty]$. (The marginal processes $\{X(t)\}$ and $\{Y(t)\}$ are also Markovian, although this is not true in general for two-dimensional Markov processes.) Its transition d.f. is given below, where we note that $Y(0) = X_1$ and conditional on $X(0) = x_0$, the distribution of X_1 is

$$K(x) = \frac{F(x+x_0) - F(x_0)}{1 - F(x_0)} \ (x \geq 0). \tag{4.4}$$

Theorem 4.1 *We have*

$$\begin{aligned} &P\{X(t) < x, Y(t) > y \mid X(0) = x_0\} \\ &= \int_{(t-x)^t}^{t} U\{ds\}[1 - F(t-s+y)] \quad (0 \leq x < t + x_0, y \geq 0) \qquad (4.5) \\ &= \frac{1 - F(t+y+x_0)}{1 - F(x_0)} + \int_0^t U\{ds\}[1 - F(t-s+y)] \quad (x \geq t + x_0, y \geq 0). \qquad (4.6) \end{aligned}$$

Proof. For $0 \leq x < t$ we have

$$\begin{aligned} &P\{X(t) < x, Y(t) > y \mid X(0) = x_0\} \\ &= P\{0 \leq t - S_{N(t)} < x, S_{N(t)+1} - t > y, N(t) > 0\} \\ &= \sum_{n=1}^{\infty} \int_{t-x}^{t} P\{S_n \,\epsilon\, ds\} P\{S_{n+1} > t + y \mid S_n = s\} \\ &= \sum_{n=1}^{\infty} \int_{t-x}^{t} P\{S_n \,\epsilon\, ds\}[1 - F(t-s+y)] = \int_{t-x}^{t} U\{ds\}[1 - F(t-s+y)]. \end{aligned}$$

For $t \leq x < t + x_0$ the same result holds with the lower limit in the integral being replaced by 0. Finally

$$\begin{aligned} P\{X(t) = t + x_0, Y(t) > y \mid X(0) = x_0\} \\ = P\{X_1 > t + y\} = 1 - K(t+y). \end{aligned}$$

This leads to the desired result for $x \geq t + x_0, y \geq 0$. □

Example 4.1 (The Poisson Process) Let $F(x) = 1 - e^{-\lambda x}$. Then, as we have shown in Example 2.1, $K(x) = F(x)$. From Theorem 4.1 we find that

$$\begin{aligned} &P\{X(t) < x, Y(t) > y \mid X(0) = x_0\} \\ &= e^{-\lambda y}[1 - e^{-\lambda \min(x,t)}] \quad (0 \le x < t + x_0, y \ge 0) \\ &= e^{-\lambda y} \quad (x \ge t + x_0, y \ge 0). \end{aligned}$$

For the marginal processes $\{X(t)\}$ and $\{Y(t)\}$ we have

$$\begin{aligned} P\{X(t) < x \mid X(0) = x_0\} &= 1 - e^{-\lambda \min(x,t)} \quad (0 \le x < t + x_0) \\ &= 1 \quad (x \ge t + x_0) \end{aligned}$$

and

$$P\{Y(t) > y \mid X(0) = x_0\} = e^{-\lambda y} \quad (y \ge 0).$$

It follows that $X(t)$ and $Y(t)$ are independent, $Y(t)$ has density $\lambda e^{-\lambda y}$, $X(t)$ has density $\lambda e^{-\lambda x}$ in $0 \le x < t$, and there is an atom at $t + x_0$ with weight $e^{-\lambda t}$.

Theorem 4.2 *We have*

$$\lim_{t \to \infty} P\{X(t) < x, Y(t) > y \mid X(0) = x_0\} = \frac{1}{\mu} \int_0^x [1 - F(s + y)] ds, \quad (4.7)$$

the limit being interpreted as zero if $\mu = \infty$.

Proof. It suffices to consider the case $0 \le x < t$ in Theorem 4.1. We can write

$$P\{X(t) < x, Y(t) > y \mid X(0) = x_0\} = \int_0^t z(t - s) U\{ds\},$$

where

$$z(s) = 1 - F(s + y) \quad (0 \le s \le x).$$

Function z is non-negative, nonincreasing, and

$$\int_0^\infty z(s) ds = \int_0^x [1 - F(s + y)] ds < \infty.$$

By the Key Renewal Theorem we therefore obtain

$$\int_0^t z(t - s) U\{ds\} \to \frac{1}{\mu} \int_0^x [1 - F(s + y)] ds,$$

the limit being interpreted as zero if $\mu = \infty$. □

Remark 4.1 Suppose we choose the distribution of $X(0)$ to be the same as the limit distribution of $X(t)$, as given by (4.7), namely $\mu < \infty$ and

$$\Phi(x) = \int_0^x \frac{1 - F(s)}{\mu} ds \quad (x \geq 0). \tag{4.8}$$

Then as shown in Example 2.2, $U(t)$ has density μ^{-1}, so Theorem 4.1 gives

$$\begin{aligned} &P\{X(t) < x, Y(t) > y \mid X(0) = x_0\} \\ &= \int_0^{\min(x,t)} \frac{1 - F(s+y)}{\mu} ds \quad (0 \leq x < t + x_0, y \geq 0) \\ &= \frac{1 - F(t + y + x_0)}{1 - F(x_0)} + \int_0^t \frac{1 - F(s+y)}{\mu} ds \quad (x \geq t + x_0, y \geq 0). \end{aligned}$$

For the absolute (unconditional) distribution of $\{X(t), Y(t)\}$ we find after some easy calculations that

$$\begin{aligned} P\{X(t) < x, Y(t) > y\} &= \int_0^\infty \frac{1 - F(x_0)}{\mu} dx_0 P\{X(t) < x, Y(t) > y \mid X(0) = x_0\} \\ &= \int_0^x \frac{1 - F(s+y)}{\mu} ds \quad (x \geq 0, y \geq 0). \end{aligned} \tag{4.9}$$

This means that $\{X(t), Y(t)\} \stackrel{d}{=} \{X(0), Y(0)\}$, and the Markov process $\{X(t), Y(t)\}$ is stationary. It also follows that (4.8) with $\mu < \infty$ is the only distribution of $X(0)$ with property (4.9).

Remark 4.2 (The Recurrence Paradox) This concerns the interval between successive renewal epochs. If it is argued that the length of this interval is identical in distribution to X_k, it would lead to a paradox. To see this we note that this interval is $(S_{N(t)}, S_{N(t)+1}]$ and its length is

$$S_{N(t)+1} - S_{N(t)} = X_{N(t)+1} \quad (t \geq 0). \tag{4.10}$$

Because of the presence of $N(t) + 1$ in (4.10) the distribution of $X_{N(t)+1}$ is different from that of X_k and it turns out in fact that in steady state $X_{N(t)+1}$ is longer (in distribution) than X_k $(k \geq 2)$. These results are proved as follows. We assume that $X(0) = 0$ and $K = F$.

Theorem 4.3 *We have the following:*

$$\text{(i)}\ P\{X_{N(t)+1} \leq x\} = \int_{(t-x)^+}^t [F(x) - F(t-s)] U\{ds\}, \tag{4.11}$$

(ii) $$\lim_{t\to\infty} P\{X_{N(t)+1} \le x\} = \frac{1}{\mu}\int_0^x [F(x) - F(s)]ds, \tag{4.12}$$

the limit being interpreted as zero if $\mu = \infty$.

(iii) *If* $\mu < \infty$, *then*

$$\frac{1}{\mu}\int_0^x [F(x) - F(s)]ds \le F(x) \quad (x \ge 0) \tag{4.13}$$

so that in steady state $X_{N(t)+1}$ *is stochastically longer than* X_1.

Proof. Denote

$$Z_x(t) = P\{X_{N(t)+1} \le x\} \quad (x \ge 0).$$

A renewal argument yields the integral equation

$$Z_x(t) = z_x(t) + \int_0^t F\{ds\}Z_x(t-s), \tag{4.14}$$

where

$$\begin{aligned} z_x(t) &= P\{X_{N(t)+1} \le x, N(t) = 0\} \\ &= P\{t < X_1 \le x\} = [F(x) - F(t)]1_{\{t\le x\}}. \end{aligned}$$

We note that for fixed $x \ge 0, z_x(t)$ is a bounded function of $t \ge 0$. By Theorem 2.6 Eq.(4.14) has the unique bounded solution

$$Z_x(t) = \int_0^t z_x(t-s)U\{ds\},$$

which reduces to (4.11). Also, since $z_x(t)$ is non-negative, nonincreasing, and

$$\int_0^\infty z_x(t)dt = \int_0^x [F(x) - F(t)]dt < \infty,$$

the Key Renewal Theorem yields (4.12). To prove (4.13) we note that

$$\mu F(x) \equiv F(x)\int_0^x [1 - F(s)]ds \ge \int_0^x [F(x) - F(s)]ds.$$

□

Remark 4.3 We can re-state (4.12) as follows: $X_{N(t)+1} \xrightarrow{d} X_\infty$, where X_∞ has the d.f. given by

$$P\{X_\infty \le x\} = \frac{1}{\mu}\int_0^x yF\{dy\} \quad (x \ge 0). \tag{4.15}$$

It follows that

$$E\left(X_\infty^r\right) = \frac{1}{\mu}\int_0^\infty x^{r+1}F\{dx\} = \frac{1}{\mu}E\left(X_1^{r+1}\right)$$

if the corresponding moments exist. In particular, if $\sigma^2 = \mathrm{Var}(X_1) < \infty$, then

$$E(X_\infty) = \frac{1}{\mu}(\sigma^2 + \mu^2) \geq \mu. \tag{4.16}$$

6.5 The Stationary Renewal Process

We say that a renewal process $\{S_n\}$ is stationary if its renewal-counting process $\{N(t)\}$ has stationary independent increments; that is, for $0 \leq s < s + t$

$$N(s+t) - N(s) \stackrel{d}{=} N(t) - N(0). \tag{5.1}$$

Theorem 5.1 *A renewal process $\{S_n\}$ is stationary iff $E(X_k) = \mu < \infty$ $(k \geq 2)$ and the initial age distribution is given by*

$$\Phi(x) = \int_0^x \frac{1 - F(s)}{\mu} ds \quad (x \geq 0). \tag{5.2}$$

Proof. From (4.1) we have

$$N(s+t) - N(s) \stackrel{d}{=} N_{Y(s)}(t) - N_{Y(s)}(0), \tag{5.3}$$

where $\{N_{Y(s)}(t), t \geq 0\}$ is the renewal-counting process induced by (K_Y, F), where K_Y is the d.f. of $Y(s)$. However, by Remark 4.1, $Y(s) \stackrel{d}{=} Y(0)$ iff the distribution of $X(0)$ is given by (5.2) with $\mu < \infty$. □

Example 5.1 Suppose that $X_k \equiv d$ $(k \geq 2)$. Then

$$\frac{1 - F(x)}{\mu} = \frac{1}{d} \quad \text{for } x < d, \text{ and} = 1 \quad \text{for } x \geq d. \tag{5.4}$$

Thus Φ in (5.2) has uniform density in $(0, d)$. This shows that the renewal processes with constant lifetimes d is stationary iff $X(0)$ has uniform density in $(0, d)$.

6.6 The Case of the Infinite Mean

In this section we shall consider the renewal process $\{S_n\}$ induced by F and denote $U(t) = \sum_0^\infty F_n(t)$ $(t \geq 0)$. We assume that

$$1 - F(t) \sim \frac{t^{-\alpha} L(t)}{\Gamma(1-\alpha)} \quad (t \to \infty) \tag{6.1}$$

with $0 < \alpha < 1$ and L a slowly varying function on $[0, \infty)$. From Theorem A4.3 we find that

$$I - F^*(\theta) \sim \theta^\alpha L\left(\frac{1}{\theta}\right) \quad (\theta \to 0+). \tag{6.2}$$

This implies that the mean lifetime $\mu = \infty$. The Elementary Renewal Theorem gives $N(t)t^{-1} \to 0$ a.s. as $t \to \infty$, but the relation (6.1) yields more information concerning N and the process $\{X, Y\}$. We need two preliminary results. We first show (Lemma 6.1) that under condition (6.1) the sequence of partial sums $\{S_n\}$ belongs to the domain of attraction of the stable distribution with exponent α. Denoting this distribution as G_α, we know that

$$\int_0^\infty e^{-\theta x} G_\alpha(dx) = e^{-c\theta^\alpha} \quad (\theta > 0) \tag{6.3}$$

with $0 < c < \infty$. The next result (Lemma 6.2) shows that under (6.1) the renewal function U of the process is of regular variation with exponent α.

Lemma 6.1 *If* (6.1) *holds, then for a given c in* $(0, \theta)$ *there exists a sequence $\{a_n\}$ of scaling constants such that $a_n \to \infty$ and*

$$P\left\{\frac{S_n}{a_n} \leq x\right\} \to G_\alpha(x) \quad \text{as } n \to \infty. \tag{6.4}$$

Proof. By Theorem A2.2 we can choose a_n so that

$$n[1 - F(a_n)] \to \frac{c}{\Gamma(1-\alpha)}. \tag{6.5}$$

Clearly $a_n \to \infty$. Then as $n \to \infty$

$$na_n^{-\alpha} L(a_n) = \frac{a_n^{-\alpha} L(a_n)}{[1 - F(a_n)]\Gamma(1-\alpha)} \cdot n[1 - F(a_n)]\Gamma(1-\alpha) \to c$$

on account of (6.1). Also

$$na_n^{-\alpha} L\left(\frac{a_n}{\theta}\right) = na_n^{-\alpha} L(a_n) \cdot \frac{L\left(\frac{a_n}{\theta}\right)}{L(a_n)} \to c$$

since L is slowly varying. Now (6.2) gives

$$1 - F^*\left(\frac{\theta}{a_n}\right) \sim \theta^\alpha a_n^{-\alpha} L\left(\frac{a_n}{\theta}\right) \sim \frac{c}{n}\theta^\alpha. \tag{6.6}$$

Therefore

$$E(e^{-\theta \frac{S_n}{a_n}}) = F^*\left(\frac{\theta}{a_n}\right)^n = \left[1 - \frac{c}{n}\theta^\alpha + o\left(\frac{1}{n}\right)\right]^n \to e^{-c\theta^\alpha}, \tag{6.7}$$

which implies the desired result (6.4). □

Lemma 6.2 *If* (6.1) *holds, then*

$$U(t) \sim \frac{\sin \pi\alpha}{\pi\alpha} \cdot \frac{t^\alpha}{L(t)} \cdot \Gamma(1-\alpha) \quad (t \to \infty). \tag{6.8}$$

Proof. We have

$$U^*(\theta) = \frac{1}{1 - F^*(\theta)} \sim \frac{\theta^{-\alpha}}{L\left(\frac{1}{\theta}\right)} \quad (\theta \to 0+) \tag{6.9}$$

using (6.2). By Theorem A4.1 we therefore obtain

$$U(t) \sim \frac{t^\alpha}{\Gamma(1+\alpha)} \cdot \frac{1}{L(t)} \tag{6.10}$$

which leads to (6.8) since $\Gamma(1-\alpha)\Gamma(1+\alpha) = \pi\alpha/\sin \pi\alpha$. □

Theorem 6.1 *If* (6.1) *holds, then*

$$P\left\{\frac{L(t)N(t)}{t^\alpha} \geq cx^{-\alpha}\right\} \to G_\alpha(x) \quad \text{as } t \to \infty. \tag{6.11}$$

Proof. We have

$$P\{N(t) \geq n\} = P\{S_n \leq t\}.$$

If we choose a_n as in Lemma 6.1 we have (6.4). Now let $t \to \infty$, $n \to \infty$ in such a way that $t/a_n \to x > 0$ (fixed). Then

$$P\left\{\frac{L(t)N(t)}{t^\alpha} \geq \frac{nL(t)}{t^\alpha}\right\} = P\left\{\frac{S_n}{a_n} \leq \frac{t}{a_n}\right\} \tag{6.12}$$

and we find that

$$\frac{nL(t)}{t^\alpha} = \frac{L(t)}{L(a_n x)} \cdot \frac{L(a_n x)}{L(a_n)} \cdot \frac{n a_n^{-\alpha} L(a_n)}{(t/a_n)^\alpha} \to c x^{-\alpha}. \tag{6.13}$$

Therefore we get the desired result from (6.12) and (6.13). □

Theorem 6.2 *If* (6.1) *holds, then as* $t \to \infty$, $\{X(t)t^{-1}, Y(t)t^{-1}\}$ *has the limit density*

$$\frac{\alpha}{\beta(1-\alpha,\alpha)} \quad (1-x)^{\alpha-1}(x+y)^{-\alpha-1} \quad (0 < x < 1, y > 0). \tag{6.14}$$

Proof. In our renewal process $X(0) \equiv 0$ and $Y(0)$ has the d.f. F, so proceeding as in Theorem 4.1 we find that

$$P\{X(t) \le x, Y(t) \le y\} = \int_{t-x}^{t} U\{d\tau\}[F(t+y-\tau) - F(t-\tau)] \\ + F(t+y) - F(t)$$

for $x < t$, which is the only case we need to consider. Therefore

$$P\left\{\frac{X(t)}{t} \le x, \frac{Y(t)}{t} \le y\right\} = \int_{t-tx}^{t} U\{d\tau\}[F(t+ty-\tau) - F(t-\tau)] \\ + F(t+ty) - F(t).$$

Here $F(t+ty) - F(t) \to 0$ as $t \to \infty$. The substitution $\tau = tu$ reduces the integral to

$$\begin{aligned} I &= \int_{1-x}^{1} U\{tdu\}[F(t+ty-tu) - F(t-tu)] \\ &= \frac{\sin \pi\alpha}{\pi\alpha} \int_{1-x}^{1} \frac{U\{tdu\}}{U(t)} \cdot \left[\frac{1-F(t-tu)}{1-F(t)} - \frac{1-F(t+ty-tu)}{1-F(t)}\right] \\ &\cdot \frac{[1-F(t)]\Gamma(1-\alpha)}{t^{-\alpha}L(t)} \cdot \frac{U(t)\pi\alpha L(t)}{t^\alpha \sin \pi\alpha \Gamma(1-\alpha)}. \end{aligned}$$

By Lemma 6.2, $U\{tdu\}/U(t)$ converges as $t \to \infty$ to the measure with density $\alpha u^{\alpha-1}$. Using this lemma and (6.1) we find that

$$\begin{aligned} I &\to \frac{\sin \pi\alpha}{\pi\alpha} \int_{1-x}^{1} \alpha u^{\alpha-1}[(1-u)^{-\alpha} - (1+y-u)^{-\alpha}]du \\ &= \frac{1}{\beta(1-\alpha,\alpha)} \int_0^x (1-u)^{\alpha-1}[u^{-\alpha} - (y+u)^{-\alpha}]du \end{aligned}$$

$$= \frac{\alpha}{\beta(1-\alpha,\alpha)} \int_0^x (1-u)^{\alpha-1} \int_0^y (u+v)^{-\alpha-1} dv\, du$$

$$= \frac{\alpha}{\beta(1-\alpha,\alpha)} \int_0^x \int_0^y (1-u)^{\alpha-1} (u+v)^{-\alpha-1} dv\, du.$$

This proves the desired result. □

6.7 The Random Walk on the Real Line: Introduction

Let $\{X_k, k \geq 1\}$ be a sequence of IID random variables with the d.f. F, $S_0 = 0$, $S_n = X_1 + X_2 + \cdots + X_n$ $(n \geq 1)$ their partial sums, and F_n the d.f. of S_n. We have

$$F_0(x) = 0 \quad \text{for } x < 0, \text{ and } = 1 \quad \text{for } x \geq 0,\ F_1(x) = F(x) \tag{7.1}$$

and

$$F_{n+1}(x) = \int_{-\infty}^{\infty} F_n(x-y) F\{dy\} \quad (-\infty < x < \infty, n \geq 0). \tag{7.2}$$

We call $\{S_n, n \geq 0\}$ the random walk induced by F . If F is concentrated on $[0,\infty)$, $\{S_n\}$ becomes the renewal process considered in Secs. 6.1–6.6. We now consider the general case where F is not necessarily concentrated either on $[0,\infty)$ or on $(-\infty, 0]$. If F is concentrated on a set of the form $\{0, \pm d, \pm 2d, \ldots\}$ and is the largest number with this property, we say that the random walk is discrete with span d; otherwise it is continuous. We shall ignore the case where $X_k \equiv 0$.

The Bernoulli random walk is induced by F having atoms at $+1, -1$ with weights p, q $(0 < p < 1, q = 1-p)$. This is treated in Sec. 9.5. The classical gambler's ruin problem is actually a hitting time problem for the Bernoulli random walk. This problem is treated by martingale methods in Example 6.1 of Chap. 7. (The simple random walk in d dimensions is discussed in Sec. 9.4.) A more general problem deals with the exit time of a random walk from a finite subinterval of the real line. This problem arose in the statistical theory of sequential analysis, and the results derived by A. Wald for it have far more importance than originally expected. The results are derived by martingale methods in Sec. 7.7.

Since $S_{n+1} = S_n + X_{n+1}$ $(n \geq 0)$, the random walk $\{S_n\}$ is a Markov process on the state space $(-\infty, \infty)$. Earlier analysis of $\{S_n\}$ was based on point recurrence and interval recurrence, in analogy with the recurrence properties of Markov chains (see Chap. 10). We shall not treat these topics here, but only the fluctuation theory of random walks, which is concerned with the functionals

$$M_n = \max(0, S_1, S_2, \ldots, S_n), \quad m_n = \min(0, S_1, S_2, \ldots, S_n). \tag{7.3}$$

Our approach to this theory is based on two renewal processes imbedded in $\{S_n\}$. These are the ascending and descending ladder processes. (Another approach is based on ladder sets; see Sec. 9.6.) As a preliminary we prove the following theorem:

Theorem 7.1 *Let $0 < s < 1$, ω being real, $i = \sqrt{-1}$. Then*

$$(1-s)\sum_0^\infty s^n E(e^{i\omega S_n}) = e^{-\int_{-\infty}^{\infty}(1-e^{i\omega x})\nu_s\{dx\}}, \tag{7.4}$$

where ν_s is the Lévy measure given by

$$\nu_s\{dx\} = \sum_1^\infty \frac{s^n}{n} P\{S_n \in dx\} \quad (-\infty < x < \infty) \tag{7.5}$$

and is totally finite, with

$$\nu_s\{(-\infty,\infty)\} = \sum_1^\infty \frac{s^n}{n} < \infty. \tag{7.6}$$

Proof. Denote by $\phi(\omega)$ the c.f. of X_k. Then the left side of (7.4) is

$$\begin{aligned} &= \frac{1-s}{1-s\phi(\omega)} = e^{-\log[1-s\phi(\omega)]+\log(1-s)} \\ &\qquad = e^{-\sum_1^\infty \frac{s^n}{n}\int_{-\infty}^{\infty}(1-e^{i\omega x})P\{S_n \in dx\}} \\ &\qquad = e^{-\int_{-\infty}^{\infty}(1-e^{i\omega x})\nu_s\{dx\}}. \end{aligned}$$

It is easily seen that ν_s is a Lévy measure, in fact a probability measure (except for a normalizing constant). □

6.8 The Maximum and Minimum Functionals

We first prove the following result which simplifies many detailed calculations.

Lemma 8.1 *For $n \geq 0$*

$$(M_n, M_n - S_n) \stackrel{d}{=} (S_n - m_n, -m_n). \tag{8.1}$$

Proof. The permutation of the random variables

$$(X_1, X_2, \ldots, X_n) \to (X_n, X_{n-1}, \ldots, X_1)$$

results in the permutation of their partial sums

$$(S_1, S_2, \ldots, S_n) \to (S_1', S_2', \ldots, S_n'),$$

where

$$S_k' = X_n + X_{n-1} + \cdots + X_{n-k+1} = S_n - S_{n-k} \quad (1 \le k \le n).$$

Our assumptions imply that $S_k' \stackrel{d}{=} S_k$. Consequently,

$$M_n \stackrel{d}{=} \max_{0 \le k \le n} S_k' = S_n + \max_{0 \le k \le n} (-S_{n-k}) = S_n - m_n$$

and

$$M_n - S_n \stackrel{d}{=} \max_{0 \le k \le n} (S_k') - S_n' = S_n - m_n - S_n = -m_n.$$

□

This leads to the following lemma.

Lemma 8.2 *For $0 < s < 1$, ω being real, $i = \sqrt{-1}$ denote*

$$u^*(s, \omega) = \sum_0^\infty s^n E[e^{i\omega S_n}; \pi_n = n], \tag{8.2}$$

$$v^*(s, \omega) = \sum_0^\infty s^n E[e^{i\omega S_n}; \bar{\pi}_n = n], \tag{8.3}$$

where for $n \ge 0$

$$\pi_n = \min\{k \le n\colon S_k = M_n\}, \quad \bar{\pi}_n = \max\{k \le n\colon S_k = m_n\}.$$

Then for ω_1, ω_2 real we have

$$\sum_0^\infty s^n E[e^{i\omega_1 M_n + i\omega_2(M_n - S_n)}] = \sum_0^\infty s^n E[e^{i\omega_1(S_n - m_n) + i\omega_2(-m_n)}]$$
$$= u^*(s, \omega_1) v^*(s, -\omega_2). \tag{8.4}$$

Proof. We have

$$E[e^{i\omega_1 M_n + i\omega_2(M_n - S_n)}] = \sum_{k=0}^n E[e^{i\omega_1 M_n + i\omega_2(M_n - S_n)}; \pi_n = k]$$

$$= \sum_{k=0}^n E[e^{i\omega_1 S_k - i\omega_2(S_n - S_k)}; S_k > S_\ell(0 \le l \le k-1), S_k \ge S_\ell(k \le l \le n)]$$

$$= \sum_{k=0}^{n} E[e^{i\omega_1 S_k}; S_k > S_\ell \ (0 \le e \le k-1)]$$
$$\cdot E[e^{-i\omega_2(S_n - S_k)}; S_k \ge S_\ell \ (k \le e \le n) | S_k].$$

Here the first factor is

$$= E[e^{i\omega_1 S_k}; \pi_k = k],$$

and the second factor is

$$= E[e^{-i\omega_2 S_{n-k}}; S_{n-k} \le S_{\ell'} (0 \le \ell' \le n-k)]$$
$$= E[e^{-i\omega_2 S_{n-k}}; \bar{\pi}_{n-k} = n-k]$$

since by Lemma 1, $S_n - S_\ell \stackrel{d}{=} S_{n-\ell} (k \le \ell \le n)$. Thus

$$E[e^{i\omega_1 M_n + i\omega_2(M_n - S_n)}] = \sum_{k=0}^{n} E[e^{i\omega_1 S_k}; \pi_k = k] E[e^{-i\omega_2 S_{n-k}}; \bar{\pi}_{n-k} = n-k]$$

and

$$\sum_0^\infty s^n E[e^{i\omega_1 M_n + i\omega_2(M_n - S_n)}] = u^*(s, \omega_1) v^*(s, -\omega_2).$$

The proof is completed by the application of Lemma 8.1. □

Theorem 8.1 *For $0 < s < 1$, ω being real, we have*

$$(1-s)\sum_0^\infty s^n E(e^{i\omega S_n}) = (1-s)\left[\sum_0^\infty s^n E(e^{i\omega M_n})\right] \cdot (1-s)\sum_0^\infty s^n E(e^{i\omega m_n}). \tag{8.5}$$

Proof. Choosing $\omega_1 = \omega, \omega_2 = 0$ and $\omega_1 = 0, \omega_2 = -\omega$ in Lemma 8.2, we obtain, respectively,

$$\sum_0^\infty s^n E[e^{i\omega M_n}] = u^*(s, \omega) v^*(s, 0), \tag{8.6}$$

$$\sum_0^\infty s^n E[e^{i\omega m_n}] = u^*(s, 0) v^*(s, \omega). \tag{8.7}$$

Also, choosing $-\omega_2 = \omega_1 = \omega$ in Lemma 8.2 we find that

$$\sum_0^\infty s^n E[e^{i\omega S_n}] = u^*(s, \omega) v^*(s, \omega), \tag{8.8}$$

and for $\omega = 0$ this yields

$$(1-s)^{-1} = u^*(s,0)v^*(s,0). \tag{8.9}$$

Collecting all these results we obtain the desired result (8.5). □

Theorem 8.2 (Wiener–Hopf Factorization) *For $0 < s < 1$ let T_s be a random variable independent of $\{S_n\}$, having the geometric distribution $(1-s)s^n$ $(n \geq 0)$. Then*

$$E(e^{i\omega S_T}) = E(e^{i\omega M_T})E(e^{i\omega m_T}), \tag{8.10}$$

this factorization being unique among such factors (c.f.s of distributions concentrated on $(0, \infty)$ and $(-\infty, 0)$, respectively) except for a translation, namely M_T and m_T are replaced by $M_T + c$ and $m_T - c$, where $c = c(s) > 0$.

Proof. The identity (8.10) is a re-statement of Theorem 8.1. By Theorem 7.1 the left side of (8.10) is an infinitely divisible c.f. Now it is known that an infinitely divisible c.f. can be factorized uniquely upto a factor $e^{i\omega c}$ in terms of c.f.s of distributions concentrated on $(0, \infty)$ and $(-\infty, 0)$. □

Theorem 8.3 (i) *We have*

$$(1-s)\sum_0^\infty s^n E(e^{i\omega M_n}) = e^{-\int_0^\infty (1-e^{i\omega x})\nu_s\{dx\}}, \tag{8.11}$$

$$(1-s)\sum_0^\infty s^n E(e^{i\omega m_n}) = e^{-\int_{-\infty}^0 (1-e^{i\omega x})\nu_s\{dx\}}, \tag{8.12}$$

where ν_s is the Lévy measure given by (7.5).

(ii) *Denote*

$$A = \sum_1^\infty \frac{1}{n} P\{S_n \leq 0\}, \quad B = \sum_1^\infty \frac{1}{n} P\{S_n > 0\}. \tag{8.13}$$

Then as $n \to \infty$, $M_n \to M \leq \infty$ and $m_n \to m > -\infty$ with probability 1, where

$$E(e^{i\omega M}) = e^{-\int_0^\infty (1-e^{i\omega x})\nu_1\{dx\}}, \tag{8.14}$$

$$E(e^{i\omega m}) = e^{-\int_{-\infty}^0 (1-e^{i\omega x})\nu_1\{dx\}} \tag{8.15}$$

and $M < \infty$ iff $B < \infty$ and $m > -\infty$ iff $A < \infty$. Here $\nu_1 = \lim_{s\to 1} \nu_s$.

Proof. The results (i) follow from Theorems 7.1 and 8.1 on account of the uniqueness of the factorization as indicated in the proof of Theorem 8.2. (ii) Since the right side of (8.11) represents the c.f. of a compound Poisson distribution we can write

$$(1-s)\sum_0^\infty s^n E(e^{i\omega M_n}) = e^{-B_s+\int_0^\infty e^{i\omega x}\nu_s\{dx\}},$$

where

$$B_s = \sum_1^\infty \frac{s^n}{n} P\{S_n > 0\}.$$

Letting $s \to 1-$ in this and applying a Tauberian theorem we obtain by monotonicity

$$E(e^{i\omega M}) = e^{-B+\int_0^\infty e^{i\omega x}\nu_1\{dx\}}.$$

This shows that $M < \infty$ iff $B < \infty$. Similarly $m > -\infty$ iff $A < \infty$. □

Since, $A + B = \sum_1^\infty n^{-1} = \infty$ it follows that it is impossible for both M and m to be finite. This leads to the following corollary.

Corollary 8.1 *With probability* 1

(i) $M = \infty,\ m > -\infty$ *if* $A < \infty,\ B = \infty$
(ii) $M < \infty,\ m = -\infty$ *if* $A = \infty,\ B < \infty$
(iii) $M = \infty,\ m = -\infty$ *if* $A = B = \infty$.

6.9 Ladder Processes

Given the random walk $\{S_n\}$ we define the two sequences of random variables $\{T_k, k \geq 0\}$ and $\{\bar{T}_k, k \geq 0\}$ as follows: $T_0 = \bar{T}_0 = 0$ and

$$T_1 = \min\{n\colon S_n > 0\}, \quad T_k = \min\{n > T_{k-1}\colon S_n > S_{T_{k-1}}\}, \tag{9.1}$$

$$\bar{T}_1 = \min\{n > 0\colon S_n \leq 0\}, \quad \bar{T}_k = \min\{n > \bar{T}_{k-1}\colon S_n \leq S_{\bar{T}_{k-1}}\} \tag{9.2}$$

for $k \geq 1$. Here T_1 is the hitting time of set $(0, \infty)$ and by the strong Markov property it is a stopping time relative to $\{S_n\}$. It follows that $\{(T_k - T_{k-1}, S_{T_k} - S_{T_{k-1}}), k \geq 1\}$ is a sequence of IID random vectors. A similar remark applies to the sequence $\{(\bar{T}_k - \bar{T}_{k-1}, S_{\bar{T}_k} - S_{\bar{T}_{k-1}}), k \geq 1\}$.

Therefore $\{(T_k, S_{T_k}), k \geq 0\}$ is a renewal process on the state space $\{0,1,2,\ldots\}\times[0,\infty)$, and similarly $\{(\bar{T}_k, S_{\bar{T}_k}), k \geq 0\}$ is a renewal process on the state space $\{0,1,2,\ldots\}\times(-\infty,0]$. The first of these is called the (strong) ascending ladder process and the second is called the (weak) descending ladder process. For convenience we write $T, \bar{T}$ for $T_1, \bar{T}_1$. We have the following theorem.

Theorem 9.1 *For $0 < s < 1$, ω being real and $i\sqrt{-1}$ denote*

$$\chi(s,\omega) = E(s^T e^{i\omega S_T}), \quad \bar{\chi}(s,\omega) = E(s^{\bar{T}} e^{i\omega S_{\bar{T}}}). \tag{9.3}$$

Then

$$\chi(s,\omega) = 1 - e^{-\int_0^\infty e^{i\omega x}\nu_s\{dx\}} \tag{9.4}$$

and

$$\bar{\chi}(s,\omega) = 1 - e^{-\int_{-\infty}^0 e^{i\omega x}\nu_s\{dx\}}. \tag{9.5}$$

Also, the distribution of (T, S_T) is proper iff $B = \infty$, while that of $(\bar{T}, S_{\bar{T}})$ is proper iff $A = \infty$.

Proof. We have

$$\begin{aligned}\sum_{k=0}^\infty E(s^{T_k} e^{i\omega S_{T_k}}) &= \sum_{k=0}^\infty \sum_{n=k}^\infty s^n E[e^{i\omega S_n}; T_k = n] \\ &= \sum_{n=0}^\infty s^n E[e^{i\omega S_n}; \pi_n = n] = u^*(s,\omega)^{-1}.\end{aligned}$$

Here the first sum is

$$= \sum_{k=0}^\infty [\chi(s,\omega)]^k = [1 - \chi(s,\omega)]^{-1}.$$

Also, from Theorem 7.1 and Eq. (8.8), $u^*(s,\omega)$ is given by

$$u^*(s,\omega) = e^{\int_0^\infty e^{i\omega x}\nu_s\{dx\}}.$$

Thus we arrive at (9.4), and the proof of (9.5) is similar. By letting $s \to 1-$ and $\omega = 0$ in these two results we find that

$$P\{T < \infty, S_T < \infty\} = 1 - e^{-B} \tag{9.6}$$

and

$$P\{\bar{T} < \infty, S_{\bar{T}} > -\infty\} = 1 - e^{-A}. \tag{9.7}$$

This completes the proof. $\square$

Remark 9.1 From (9.4) and (9.5) and (8.8) we find that

$$1 - s\phi(\omega) = [1 - \chi(s, \omega)][1 - \bar{\chi}(s, \omega)]. \tag{9.8}$$

This identity can be viewed as a Wiener–Hopf factorization. However, it is probabilistically less meaningful than Theorem 8.2.

The results (9.6) and (9.7) show that the ascending ladder process is nonterminating iff $B = \infty$, while the descending ladder process $\{\bar{T}_R, S_{\bar{T}_R}\}$ is nonterminating iff $A = \infty$. Along with Corollary 8.1, this completes the fluctuation theory of $\{S_n\}$. The three cases (i)–(iii) below are as indicated in Corollary 8.1.

Theorem 9.2 *For the random walk $\{S_n\}$ there are three possibilities: (i) $S_n \to \infty$ with probability 1, (ii) $S_n \to -\infty$ with probability 1 or (iii) with probability 1*

$$-\infty = \lim_{n\to\infty} \inf S_n < \lim_{n\to\infty} \sup S_n = +\infty. \tag{9.9}$$

Proof. (i) Here with probability 1, $M_n \to \infty, m_n \to m > -\infty$. Also, the ascending ladder process $\{T_k, S_{T_k}\}$ is nonterminating. Thus as $k \to \infty$, $T_k \to \infty$ and $S_{T_k} = M_{T_k} \to \infty$ with probability 1. We have

$$\min_{n \geq T_k} S_n - M_{T_k} = \min_{n \geq T_k} (S_n - S_{T_k}).$$

Since the random variables $S_n - S_{T_k}$ are independent of S_{T_K} and have the same distribution as S_{n-T_k} $(n \geq T_k)$ we find that

$$\min_{n \geq T_k} (S_n) - M_{T_k} \stackrel{d}{=} m > -\infty.$$

For $A' > 0, B' > 0$ we therefore have

$$P\left\{\min_{n \geq T_k} S_n > A'\right\} \geq P\{M_{T_k} > A'\} P\{m > -B'\}.$$

Given $\varepsilon > 0$ we can choose A, B so large that

$$P\{M_{T_k} > A'\} > 1 - \varepsilon, \quad P\{m > -B'\} > 1 - \varepsilon$$

which is possible by the hypothesis. Thus

$$P\left\{\min_{n\geq T_k}(S_n) > A\right\} \geq (1-\varepsilon)^2 > 1-2\varepsilon. \tag{9.10}$$

This shows that all S_n $(n \geq T_k)$ lie above $A > 0$, and so $\lim S_n = +\infty$ with probability 1. This proves (i), and the proof of (ii) is similar. To prove (iii) we note that the subsequences $\{S_{T_n}\}$ and $\{S_{\bar{T}_n}\}$ both diverge, and so (9.9) must hold for the upper and lower limits. □

Corollary 8.1 and Theorem 9.2 provide us with a criterion for the classification of random walks on the real line.

Definition 9.1 (i) The random walk *drifts* to $+\infty$ if as $n \to \infty$

$$M_n \to \infty, m_n \to m > -\infty, \quad \text{and} \quad S_n \to +\infty \text{ with probability 1.} \tag{9.11}$$

(ii) The random walk *drifts* to $-\infty$ if as $n \to \infty$

$$M_n \to M < \infty, m_n \to -\infty, \text{ and } S_n \to -\infty \text{ with probability 1.} \tag{9.12}$$

(iii) The random walk *oscillates* if as $n \to \infty$

$$M_n \to \infty \text{ and } m_n \to -\infty \text{ with probability 1.} \tag{9.13}$$

In this case (9.9) holds.

Example 9.1 Let $\{S_n\}$ be the random walk induced by the normal density with mean μ and unit variance. For the standard normal d.f. N we have

$$1 - N(t) \sim \frac{1}{\sqrt{2\pi}} t^{-1} e^{-\frac{1}{2}t^2} \quad (t \to \infty).$$

Using this we find that for $\mu < 0$

$$\begin{aligned} P\{S_n > 0\} &= P\left\{\frac{S_n - n\mu}{\sqrt{n}} > -\mu\sqrt{n}\right\} = 1 - N(-\mu\sqrt{n}) \\ &\sim \frac{1}{\sqrt{2\pi}} \cdot \frac{1}{(-\mu)} n^{-1/2} (e^{-\frac{1}{2}\mu^2})^n \quad (n \to \infty) \end{aligned}$$

and

$$-\frac{\mu}{n} P\{S_n > 0\} \sim \frac{1}{\sqrt{2\pi}} \cdot \frac{1}{\sqrt{n}} (e^{-\frac{1}{2}\mu^2})^n \quad (n \to \infty).$$

Therefore, series B converges faster than the geometric series with the common ratio $e^{-\frac{1}{2}\mu^2} < 1$, so $B < \infty$. We conclude that

$$A = \infty \text{ and } B < \infty \quad \text{if } \mu < 0.$$

Similarly

$$A < \infty \text{ and } B = \infty \quad \text{if } \mu > 0.$$

If $\mu = 0$ we have

$$P\{S_n > 0\} = P\{S_n \le 0\} = \frac{1}{2}$$

so

$$A = B = \frac{1}{2}\sum_1^\infty \frac{1}{n} = \infty \quad \text{if } \mu = 0.$$

We conclude that this random walk drifts to $+\infty =$ if $\mu > 0$, drifts to $-\infty$ if $\mu < 0$, and oscillates if $\mu = 0$.

In the case where F does not have a finite mean it may happen that

$$E(X_k^+) = \infty, \quad \text{but } E(X_k^-) < \infty \tag{9.14}$$

or else

$$E(X_k^+) < \infty, \quad \text{but } E(X_k^-) = \infty. \tag{9.15}$$

We shall then say that mean μ of F exists and write $\mu = \infty$ in case (9.14) and $\mu = -\infty$ in case (9.15).

Theorem 9.3 *Suppose that mean μ of F exists.*

(i) *If $0 < \mu \le \infty$, then the random walk drifts to $+\infty$.*
(ii) *If $-\infty \le \mu < 0$, then the random walk drifts to $-\infty$.*
(iii) *If $\mu = 0$, then the random walk oscillates.*

Proof. If μ is finite, then by the strong law of large numbers

$$\frac{S_n}{n} \to \mu \text{ as } n \to \infty \tag{9.16}$$

with probability 1. This shows that if $\mu > 0, S_n \to +\infty$ a.s. and by Definition 9.1 the random walk drifts to $+\infty$. The same conclusion is reached if $\mu = \infty$ since the strong law of large numbers also holds in this case. This proves (i), and the proof of (ii) is similar.

If $\mu = 0$, then by the theory of interval recurrence the random walk visits every finite interval infinitely often with probability 1. This rules out the drift of the random walk to $+\infty$ or $-\infty$. $\square$

Example 9.2 Let $\{S_n\}$ be the random walk induced by a stable distribution. Its characteristic function is given by $e^{-\psi(\omega)}$, where

$$\psi(\omega) = c|\omega|^\alpha \left\{1 + i\beta\frac{\omega}{|\omega|}\Omega(|\omega|, \alpha)\right\}, \tag{9.17}$$

where $c > 0, 0 < \alpha \leq 2,\ -1 \leq \beta \leq 1$, and

$$\begin{aligned} \Omega(|\omega|, \alpha) &= \tan\frac{\pi\alpha}{2} \quad \text{for } \alpha \neq 1 \\ &= \frac{2}{\pi}\log|\omega| \quad \text{for } \alpha = 1. \end{aligned} \tag{9.18}$$

If $0 < \alpha < 1,\ \beta = \mp 1$, then F is concentrated on either $(0, \infty)$ or on $(-\infty, 0)$ (see Sec. 1.10). Except when $\alpha = 1,\ \beta \neq 0$, F is strictly, stable, namely $S_n/n^{1/\alpha} \stackrel{d}{=} X_1$, as can be verified from (9.17). Thus, in particular,

$$P\{S_n > 0\} = P\left\{\frac{S_n}{n^{1/\alpha}} > 0\right\} = P\{X_1 > 0\} = \gamma \text{ (say)}. \tag{9.19}$$

For $\alpha \neq 1$, C. Chung–Che found the value of γ to be

$$\gamma = \frac{1}{2} + \frac{1}{\pi\alpha}\arctan\left(-\beta\tan\frac{\pi\alpha}{2}\right). \tag{9.20}$$

We have

$$A = \sum_1^\infty \frac{1}{n}P\{S_n \leq 0\} = (1-\gamma)\sum_1^\infty \frac{1}{n} = \infty,$$

$$B = \sum_1^\infty \frac{1}{n}P\{S_n > 0\} = \gamma\sum_1^\infty \frac{1}{n} = \infty.$$

Therefore, random walks induced by strictly stable distributions oscillate.

We continue our discussion of ladder processes. For convenience we write $Z = S_T, \bar{Z} = S_{\bar{T}}$.

Theorem 9.4 (i) *The distribution of $(T.Z)$ is proper iff $B = \infty$, in which case*

$$E(T) = e^A \leq \infty. \tag{9.21}$$

(ii) *If* $EX_1^+ = \infty$ *then* $E(Z) = \infty$. *If* F *has a finite mean* $\mu > 0$, *then*

$$E(Z) = e^A < \infty. \tag{9.22}$$

(iii) *If* F *has mean zero and finite variance* σ^2, *then*

$$E(Z) = \frac{\sigma}{\sqrt{2}c}, \quad E(\bar{Z}) = -\frac{\sigma c}{\sqrt{2}}, \tag{9.23}$$

where

$$c = e^{\sum_1^\infty \frac{1}{n}\left\{P\{S_n>0\}-\frac{1}{2}\right\}} \quad (0 < c < \infty). \tag{9.24}$$

Conversely, if Z *and* $\bar{Z}$ *have finite means, then* F *has mean zero and finite variance* σ^2, *and further*

$$E(Z)E(\bar{Z}) = -\frac{1}{2}\sigma^2. \tag{9.25}$$

Proof. (i) From Theorem 9.1 we find that

$$P\{T < \infty,\ Z < \infty\} = \lim_{s\to 1-,\ \omega\to 0} \chi(s,\omega) = 1 - e^{-B}$$

which reduces to 1 iff $B = \infty$. For the marginal distribution of T we have

$$E(s^T) = 1 - e^{-\sum_1^\infty \frac{s^n}{n} P\{S_n>0\}}$$

so that

$$\frac{1 - E(s^T)}{1-s} = e^{\sum_1^\infty \frac{s^n}{n} - \sum_1^\infty \frac{s^n}{n} P\{S_n>0\}} = e^{\sum_1^\infty \frac{s^n}{n} P\{S_n\le 0\}}.$$

If $B = \infty$, then

$$E(T) = \lim_{s\to 1-} \frac{1 - E(s^T)}{1-s} = e^A \le \infty.$$

(ii) We have $Z \ge X_1^+$ so that $E(Z) = \infty$ if $EX_1^+ = \infty$. If F has a finite mean $\mu > 0$, then $B = \infty$, $A < \infty$ by Theorem 9.3 and the Wald Equation (Theorem 7.7.1) gives

$$E(S_T) = \mu E(T) = \mu e^A < \infty.$$

(iii) Suppose that F has mean 0 and finite variance σ^2. From Theorem 9.1 we obtain

$$\frac{-\chi'(s,0)}{1-\chi(s,0)} = (-i)\sum_1^\infty \frac{s^n}{n} E[S_n; S_n > 0]. \tag{9.26}$$

Using Lemma 9.1 (below) and Theorem A4.4 we find that

$$\sum_1^\infty \frac{s^n}{n} E[S_n; S_n > 0] \sim \frac{\sigma}{\sqrt{2}(1-s)^{1/2}} \ (s \to 1-),$$

so that

$$-i\chi'(s,0)\frac{\sqrt{1-s}}{1-\chi(s,0)} \to \frac{\alpha}{\sqrt{2}}. \tag{9.27}$$

Now, as $s \to 1-$, $-i\chi'(s,0) \to E(Z) \leq \infty$. Therefore

$$\lim_{s\to 1-} \frac{\sqrt{1-s}}{1-\chi(s,0)} \tag{9.28}$$

exists, but may be zero. Similarly

$$\lim_{s\to 1-} \frac{\sqrt{1-s}}{1-\bar{\chi}(s,0)} \tag{9.29}$$

exists, but may be zero. However, letting $\omega \to 0$ in the Wiener–Hopf factorization (9.8) we obtain

$$\frac{\sqrt{1-s}}{1-\chi(s,0)} \frac{\sqrt{1-s}}{1-\bar{\chi}(s,0)} = 1$$

which shows that neither of the limits (9.28), (9.29) can be zero. To evaluate these limits we observe that

$$\begin{aligned}\frac{\sqrt{1-s}}{1-\chi(s,0)} &= e^{-\frac{1}{2}\sum_1^\infty \frac{s^n}{n} + \sum_1^\infty \frac{s^n}{n} P\{S_n > 0\}} \\ &= e^{\sum_1^\infty \frac{s^n}{n}\left\{P\{S_n>0\} - \frac{1}{2}\right\}} = e^{\sum_1^\infty c_n s^n}\end{aligned}$$

(say). Here $nc_n \to 0$ and $\lim_{s\to 1-}\sum_1^\infty c_n s^n$ exists. By the Tauberian theorem of Sec. A.1 it follows that $\sum c_n$ converges (to $\log c$, say), and

$$\lim_{s\to 1-} \frac{\sqrt{1-s}}{1-\chi(s,0)} = c. \tag{9.30}$$

From, (9.27) we obtain $E(Z) = \frac{\sigma}{\sqrt{2}c}$ and similarly $E(\bar{Z}) = -\frac{\sigma}{\sqrt{2}}c$.

To prove the converse, suppose that Z and $\bar{Z}$ have finite means. Letting $s \to 1-$ in the Wiener–Hopf factorization (9.8) we obtain

$$\frac{1-\chi(1,\omega)}{\omega}\frac{1-\bar{\chi}(1,\omega)}{\omega} = \frac{1-\phi(\omega)}{\omega^2}. \tag{9.31}$$

Here $\chi'(1,\omega) \to iE(Z)$ and $\bar{\chi}(1,\omega) \to iE(\bar{Z})$ as $\omega \to 0$. Therefore the right side of (9.31) has a limit as $\omega \to 0$, which implies that $\phi'(0) = 0$ and $\phi''(0)$ is finite. Thus F has mean 0 and variance $\sigma^2 < \infty$ and (9.25) follows. □

Lemma 9.1 *If F has mean 0 and finite variance σ^2, then*

$$E[S_n; S_n > 0] \sim \frac{\sigma}{\sqrt{2\pi}} n^{1/2} \quad (n \to \infty). \tag{9.32}$$

Proof. We can write (9.32) as

$$E\left(\frac{S_n}{\sigma\sqrt{n}}\right)^+ \to \frac{1}{\sqrt{2\pi}}. \tag{9.33}$$

By the central limit theorem we have

$$\begin{aligned}\lim_{n\to\infty} P\left\{\left(\frac{S_n}{\sigma\sqrt{n}}\right)^+ \le x\right\} &= 0 \qquad \text{for } x < 0\\ &= N(x) \quad \text{for } x \ge 0,\end{aligned} \tag{9.34}$$

where we note that the limit distribution has mean $(2\pi)^{-1/2}$. The desired result follows from the result

$$E\left[\left(\frac{S_n}{\sigma\sqrt{n}}\right)^+\right]^2 \le E\left(\frac{S_n}{\sigma\sqrt{n}}\right)^2 = 1$$

so that $\left\{\left(\frac{S_n}{\sigma\sqrt{n}}\right)^+\right\}$ is uniformly integrable. □

6.10 Limit Theorems for M_n

In this section we establish some limit theorems for M_n. Similar results hold for m_n.

Theorem 10.1 (Law of Large Number for the Maximum) *If mean μ of F exists, then with probability* 1

$$\frac{M_n}{n} \to \mu^+ \quad \text{as } n \to \infty. \tag{10.1}$$

Proof. If $-\infty \le \mu < 0$ then by Theorem 9.3, $M_n \to M < \infty$ with probability 1, and the result, (10.1) follows trivially. Let $0 \le \mu \le \infty$. By the strong law of large numbers we have with probability 1

$$\frac{S_n}{n} \to \mu \quad \text{as } n \to \infty. \tag{10.2}$$

This gives

$$\liminf_{x\to\infty} \frac{M_n}{n} \ge \lim_{n\to\infty} \frac{S_n}{n} = \mu, \tag{10.3}$$

with probability 1. If $\mu = \infty$, result (10.1) follows immediately. For $0 \le \mu < \infty$, we shall prove that

$$\limsup_{n\to\infty} \frac{M_n}{n} \le \mu \tag{10.4}$$

with probability 1, and then (10.1) will follow immediately. Suppose that with probability 1

$$\limsup_{n\to\infty} \frac{M_n}{n} > \mu. \tag{10.5}$$

Then for some $\varepsilon > 0$ and a subsequence $\{n_k\}$ with $n_k \to \infty$ we have

$$\frac{M_{n_k}}{n_k} > \mu + \varepsilon.$$

Now $\Pi_{n_k} = \min\{n\colon S_n = M_{n_k}\} \le n_k$ and consequently with probability 1

$$\frac{S_{\pi_{n_k}}}{\pi_{n_k}} \ge \frac{M_{n_k}}{n_k} > \mu + \varepsilon \quad \text{a.s.},$$

which is clearly impossible, in view of (10.2). □

Theorem 10.2 (Central Limit Theorem for the Maximum) *If F has mean $\mu > 0$ and finite variance α^2, then as $n \to \infty$*

$$P\left\{\frac{M_n - n\mu}{\alpha\sqrt{n}} \le x\right\} \to N(x) \ (-\infty < x < \infty). \tag{10.6}$$

Proof. By Lemma 8.1, we have $M_n \stackrel{d}{=} S_n - m_n$. Let $W_n = S_n - m_n$ $(n \ge 0)$. Since $\mu > 0$, the random walk drifts to $+\infty$ by Theorem 9.3 and therefore $m_n \to m > -\infty$ a.s. Now

$$\frac{W_n - n\mu}{\alpha\sqrt{n}} = \frac{S_n - n\mu}{\alpha\sqrt{n}} - \frac{m_n}{\alpha\sqrt{n}}. \tag{10.7}$$

Here on the right side $m_n/\sqrt{n} \to 0$ a.s., and the distribution of $(S_n - n\mu)/\alpha\sqrt{n}$ converges to the standard normal. □

Theorem 10.3 *If F has mean 0 and finite variance σ^2, then as $n \to \infty$,*

$$P\left\{\frac{M_n}{\sigma\sqrt{n}} \leq x\right\} \to N_+(x) \quad (x \geq 0). \tag{10.8}$$

Proof. Since $\mu = 0, \sigma^2 < \infty$ we have $E(Z) = \sigma/\sqrt{2}c$ by Theorem 9.4 (iii). We have

$$M_n = Z_1 + Z_2 + \cdots + Z_{N(n)}, \tag{10.9}$$

where $N(n) = \max\{k\colon T_k \leq n\}$ (which is the number of ascending ladder epochs up to n). Therefore

$$\frac{M_n}{\sigma\sqrt{n}} = \frac{Z_1 + Z_2 + \cdots + Z_{N(n)}}{N(n)E(Z)} \cdot \frac{N(n)}{c\sqrt{2n}}. \tag{10.10}$$

From (9.30) we find that

$$1 - E(s^T) \sim c^{-1}(1-s)^{1/2} \ (s \to 1-). \tag{10.11}$$

By Theorem 6.1 we therefore obtain

$$P\left\{\frac{N(n)}{c\sqrt{2n}} \geq x^{-1/2}\right\} \to G_{1/2}(x),$$

where $G_{1/2}$ is the stable distribution with the Laplace transform $e^{-\sqrt{2\theta}}$ $(\theta > 0)$. We can write this as

$$P\left\{\frac{N(n)}{c\sqrt{2n}} < x\right\} \to 1 - G_{1/2}\left(\frac{1}{x^2}\right) = N_+(x). \tag{10.12}$$

On the right side of (10.10) the first factor tends to 1 by the strong law of large numbers, while the second factor has the limit distribution N_+. This leads to the desired result. □

Our proof of Theorem 10.2 suggest that M_n has a limit distribution whenever S_n has a limit distribution. Suppose now that F belongs to the domain of attraction (necessarily) of a stable distribution G with exponent

$\alpha (0 < \alpha \leq 2)$. Thus there exist constants $a_n > 0$ and b_n such that as $n \to \infty$

$$P\left\{\frac{S_n - b_n}{a_n} \geq x\right\} \to G_a(x) \tag{10.13}$$

We are interested in the case of a random walk drifting to $+\infty$. If $1 < \alpha \leq 2$, F has a finite mean μ and by Theorem 9.3, the random walk drifts to $+\infty$ iff $\mu > 0$. If $0 < \alpha < 1$, then for a random walk drifting to $+\infty$ we must have $\lim S_n = +\infty$ with probability 1 by Definition 9.1, and the limit distribution G_α is therefore concentrated on $(0, \infty)$. Conversely, by a result of M. Rosenblatt if G_α is concentrated on $(0, \infty)$ then the random walk drifts to $+\infty$ for $0 < \alpha < 1$. We shall not consider the case $\alpha = 1$. In (10.13), we can choose the centering constant $b_n = nc_\alpha$, where $c_\alpha = 0$ for $0 < \alpha < 1$ and $c_\alpha = \mu$ for $1 < \alpha \leq 2$. We have then the following result. The proof is exactly the same as for Theorem 10.2.

Theorem 10.4 *Suppose F belongs to the domain of attraction of either* (i) *a stable distribution G_α with exponent $1 < \alpha \leq 2$ and mean $\mu > 0$ or* (ii) *a stable distribution G_α with exponent $0 < \alpha < 1$, concentrated on $(0, \infty)$. If* (10.13) *holds, then as $n \to \infty$*

$$P\left\{\frac{M_n - b_n}{a_n} \leq x\right\} \to G_\alpha(x). \tag{10.14}$$

6.11 Problems for Solution

1. For the function U defined by

$$U(t) = \sum_{r=0}^{[t]} \frac{(-1^r)}{r!}(t-r)^r e^{(t-r)} - 1 \quad (t > 0)$$

show that

$$U(t) = 2t - 1/3 + o(1) \quad \text{as } t \to \infty.$$

(Hint: consider the renewal process induced by the uniform distribution in (0, 1).)

2. For the renewal process of Example 2.3 prove the follow results:

   a. The age $X(t)$ has the density

$$g(x,t) = [1 - F(x)]\left\{\frac{\lambda}{p} + \sum_{m=1}^{p-1} A_m e^{\theta_m(t-x)}\right\} \quad (0 \le x < t).$$

   b. The residual lifetime $Y(t)$ has the density

$$h(y,t) = \frac{\lambda}{p}[1 - F(y)] + \sum_{m=1}^{p-1} A_m e^{\theta_m(t+y)}\left[1 - \int_0^y e^{-\theta_m u} F(du)\right] \ (y \ge 0).$$

3. If Y_T is the residual lifetime at an epoch selected at random from an interval $(0, T)$, independently of the renewal process induced by F, show that

$$\lim_{T\to\infty} P\{Y_T \le y\} = \int_0^y \frac{1 - F(x)}{\mu} dx.$$

4. Show that the limit distribution of $N(t+s) - N(t)$ as $t \to \infty$ is identical with the distribution of the number of renewals in $(0, s]$ for the stationary renewal process.
5. Denoting by G_α the stable distribution of Sec. 6, show that $F_\alpha(x) = 1 - G_\alpha(x^{-1/\alpha})$ is a d.f with density

$$f_\alpha(x) = \frac{c}{\pi\alpha}\sum_1^\infty \frac{\Gamma(k\alpha+1)}{k!}\sin k\pi\alpha(-cx)^{k-1}.$$

   It is seen that $f_0(x) = ce^{-cx}$, $f_{1/2}(x) = \sqrt{\frac{2}{\pi}}e^{-\frac{1}{2}x^2}\,(c = \sqrt{2})$.
6. If F satisfies condition (6.1) show that (i) $X(t)/t$ has the limit distribution with the beta density

$$\beta(1-\alpha,\alpha)^{-1}x^{-\alpha}(1-x)^{\alpha-1} \quad (0 < x < 1)$$

   (ii) $Y(t)/t$ has the limit distribution with the beta density

$$\beta(1-\alpha,\alpha)^{-1}y^{-\alpha}(1+y)^{-1} \ (y \ge 0),$$

   and (iii) $X_{N(t)+1}/t$ has the limit distribution with the density

$$\beta(1-\alpha,\alpha)^{-1}z^{-\alpha-1}f(z),$$

   where $f(z) = 1 - (1-z)^\alpha$ for $0 \le z < 1$ and $= 1$ for $z \ge 1$.

7. Consider items (such as components of industrial equipment) with a lifetime distribution F. Suppose that the replacement policy is such that each item is replaced whenever it fails or has been in use for c units of time. Replacements take no time.

   (i) Show that the successive replacement epochs $\{S_n\}$ form a renewal process and find the distribution of the time between successive replacements.

   (ii) If $N(t)$ is the number of replacements during a time-interval $(0, t]$ show that as $t \to \infty$

   $$\frac{N(t)}{t} \to \left[\int_0^c [1 - F(x)]dx\right]^{-1} \text{ with probability 1.}$$

   (iii) Show that the successive failure epochs form a renewal process with the distribution G of the time between successive failure being given by

   $$1 - G(t) = [1 - F(c)]^k [F(c) - F(t - kc)] \quad \text{for } kc < t \leq kc + c, k \geq 0.$$

   If $N_1(t)$ is the number of failures during $(0, t]$, show that as $t \to \infty$,

   $$\frac{N_1(t)}{t} \to F(c) \left[\int_0^c [F(c) - F(x)]dx\right]^{-1} \text{ with probability 1.}$$

8. If F is an unsymmetric stable distribution with exponent $\alpha = 1$, show that the random walk induced by it is drifting.
9. Let $\{S_n\}$ be the random walk induced by the symmetric Cauchy density and d a nonzero constant. Show that the random walk $\{S_n + nd, n > 0\}$ oscillates.
10. *Walter Bartky's Multiple Sampling Scheme.* In this scheme, an initial sample of M_0 units is examined; if the number X_0 of defectives in this sample is such that $X_0 \leq c$, the lot is accepted, and if $X_0 \geq c + a$, the lot is rejected. But if $c < X_0 < c + a$, additional samples of size M are examined according to the following procedure. Let X_n be the number of defectives in the nth additional sample, and $S_n = X_1 + X_2 + \cdots + X_n$. Then, if $X_0 + S_n = c + n$ the lot is accepted, and if $X_0 + S_n \geq c + n + a$ it is rejected while if $c + n < X_0 + S_n < c + n + a$, an additional sample is examined $(n \geq 1)$. Clearly, additional samples will be required iff $X_0 = c + i (1 \leq i \leq a - 1)$, and the number of additional samples

required is given by

$$N = \min\{n \geq 1; X_0 + S_n = c + n \text{ or } X_0 + S_n \geq c + n + a\}$$
$$= \min\{n \geq 1; S_n = -i \text{ or } S'_n \geq a - i\}, \qquad (11.1)$$

where $S'_n = S_n - n$. If p is the fraction defective in an infinite lot and $q = 1 - p$, show that the probability of eventuality accepting the lot is given by

$$\prod = \sum_{m \leq c} \binom{M_0}{m} p^m q^{M_0 - m} + \frac{h_a}{g_a}, \qquad (11.2)$$

and the expected number of additional samples is given by

$$E = \frac{h_a}{g_a}(g_1 + g_2 + \cdots + g_{a-1}) - (h_1 + h_2 + \cdots + h_{a-1}), \qquad (11.3)$$

where g_i and h_i are functions of p, and the defined as follows:

$$\frac{1}{K(x) - x} = g_1 + g_2 x + g_3 x^2 + \cdots$$
$$\frac{\sum_{i=1}^{a-1} \binom{M_0}{c+i} p^{c+i} q^{M_0 - c - i}}{K(x) - x} = h_1 + h_2 x + h_3 x^2 + \cdots$$
$$K(x) = (q + px)^M.$$

11. Let $P\{X_k = j\} = k_j > 0$ $(-r \leq j \leq s)$, $M(z) = \sum_{-r}^{s} k_j z^j$. For the random walk $\{S_n\}$ induced by this distribution, let

$$N = \min\{n \geq 1; S_n \geq a\}. \qquad (11.4)$$

Show that

$$P\{N < \infty\} = \sum_{k=1}^{s} \zeta_k^{-a} \prod_{j \neq k} \left(\frac{1 - \zeta_j}{\zeta_k - \zeta_j} \right), \qquad (11.5)$$

where $\zeta_1, \zeta_2, \ldots, \zeta_s$ are the roots of equation $M(z) = 1$ such that $|\zeta_k| \geq |\zeta_1|$ $(k = 1, 2, \ldots, s)$.

12. Suppose that the random walk $\{S_n\}$ is induced by a discrete distribution $\{k_j, j \geq 0\}$ with the generating function $K(s) = \sum_0^\infty k_j s^j$ $(0 < s < 1)$. Let

$$N = \min\{n \geq 1; S_n - n \leq -i\}. \qquad (11.6)$$

Show that $E(s^N) = \xi^i$, where $\xi \equiv \xi(s)$ is the unique continuous root of the equation $\xi = sK(\xi)$, with $\xi(0+) = 0$.

13. Show that for $0 < s < 1$,

$$1+\sum_{1}^{\infty} s^n P\{S_1 \leq 0, S_2 \leq 0, \ldots, S_{n-1} \leq 0, S_n = 0\}$$
$$= \exp\left\{\sum_{1}^{\infty} \frac{s^n}{n} P\{S_n = 0\}\right\}. \tag{11.7}$$

Hence deduce that except when $X_k = 0$ a.s., the series $\sum_1^\infty n^{-1}P\{S_n = 0\}$ converges.

14. Let F be a stable distribution (see Example 9.2.). Ignore the cases where $0 < \alpha < 1, \beta = \pm 1$ or $\alpha = 1, \beta \neq 0$. With γ defined by (9.19) prove
    (i) $E(s^T) = 1 - (1-s)^\gamma$;
    (ii) the random variables $\{T_k, k \geq 0\}$ belong to the domain of attraction of a stable distribution with exponent γ; and
    (iii) the random variables $\{S_{T_k}, k \geq 0\}$ belong to the domain of attraction of a stable distribution with exponent $\alpha\gamma$.

15. Show that

$$E(M_n) = \sum_{m=1}^{n} \frac{1}{m} E[S_m; S_m > 0]. \tag{11.8}$$

16. If F has mean 0 and finite variance σ^2, show that

$$E(M_n) \sim \sqrt{\frac{2}{\pi}}\sigma n^{1/2} \quad (n \to \infty). \tag{11.9}$$

Further Reading

Gut, A (1988). *Stopped Random Walks: Limit Theorems and Applications.* New York: Springer Verlag.

CHAPTER 7

Martingales in Discrete Time

7.1 Introduction and Examples

A martingale is a *fair process.* To motivate the discussion we consider the following example.

Example 1.1 (Classical Gambling) In the classical gambler's ruin problem, a player places bets at time $1, 2, 3, \ldots$, the stake (amount of bet) on each play being \$1. He continues to play until his accumulated gain (capital) increases from the initial value i to a or drops to 0. The game is fair iff $p = q$. Denote by Y_k the player's gain on the kth play ($k \geq 1$). Then

$$P\{Y_k = +1\} = p, \quad p\{Y_k = -1\} = q, \tag{1.1}$$

and $E(Y_k) = p - q = 0$ if $p = q$. Since $Y_1, Y_2, \ldots$ are independent and identically distributed, we have

$$E(Y_{n+1}|Y_1, Y_2, \ldots, Y_n) = 0 \quad (n \geq 1). \tag{1.2}$$

The player's capital at the end of the nth play is then $i + X_n$, where

$$X_n = Y_1 + Y_2 + \cdots + Y_n \quad (n \geq 1). \tag{1.3}$$

We have then

$$\begin{aligned} E(X_{n+1}|X_1, X_2, \ldots, X_n) &= E(X_n + Y_{n+1}|Y_1, Y_2, \ldots, Y_n) \\ &= X_n + E(Y_{n+1}|Y_1, Y_2, \ldots, Y_n) = X_n. \end{aligned} \tag{1.4}$$

We have thus an example $\{X_n\}$ of a martingale.

Definition 1.1 A sequence $\{Y_n, n \geq 1\}$ of random variables with $E|Y_n| < \infty$ is called *absolutely fair* if

$$E(Y_1) = 0, \quad E(Y_{n+1}|Y_1, Y_2, \ldots, Y_n) = 0. \tag{1.5}$$

Definition 1.2 A sequence $\{X_n, n \geq 1\}$ of random variables with $E|X_n| < \infty$ is called a *martingale* if

$$E(X_{n+1}|X_1, X_2, \ldots, X_n) = X_n \quad (n \geq 1). \tag{1.6}$$

The connection between the two ideas is explained in the following theorem.

Theorem 1.1 *A sequence $\{X_n\}$ of random variables with $E|X_n| < \infty$ is a martingale iff it can be represented in the form*

$$X_n = Y_1 + Y_2 + \cdots + Y_n + c \quad (n \geq 1), \tag{1.7}$$

where $\{Y_n\}$ is absolutely fair and c is a constant.

Proof. (i) Let $\{Y_n\}$ be absolutely fair and c a constant. If (1.7) holds, then $E|X_n| < \infty$ and

$$\begin{aligned} E(X_{n+1}|Y_1, Y_2, \ldots, Y_n) &= E(X_n + Y_{n+1}|Y_1, Y_2, \ldots, Y_n) \\ &= X_n + E(Y_{n+1}|Y_1, Y_2, \ldots, Y_n) \\ &= X_n + 0 = X_n \end{aligned}$$

since $\{Y_n\}$ is absolutely fair. We can express this equivalently as

$$E(X_{n+1}|X_1, X_2, \ldots, X_n) = X_n,$$

which shows that $\{X_n\}$ is a martingale.

(ii) Conversely, suppose $\{X_n\}$ is a martingale. Put

$$Y_1 = X_1 - E(X_1), Y_k = X_k - X_{k-1} \quad (k \geq 2).$$

Then

$$Y_1 + Y_2 + \cdots + Y_n + E(X_1) = X_n$$

so that (1.7) holds with $c = E(X_1), c$ finite. Now $E|Y_n| < \infty$ and

$$\begin{aligned} &E(Y_1) = 0, \\ &\quad E(Y_{n+1}|Y_1, Y_2, \ldots, Y_n) \\ &\qquad = E(X_{n+1} - X_n|X_1 - E(X_1), X_2 - X_1, \ldots, X_n - X_{n-1}) \\ &\qquad = E(X_{n+1} - X_n|X_1, X_2, \ldots, X_n) \end{aligned}$$

$$= E(X_{n+1}|X_1, X_2, \ldots, X_n) - X_n$$
$$= X_n - X_n = 0.$$

This shows that sequence $\{Y_n\}$ is absolutely fair. □

In the gambling context, let X_n represent the player's capital at the end of the nth play. Then the *martingale increment* $Y_n = X_n - X_{n-1}$ represents the player's gain in the nth play ($n \geq 2$). Theorem 1.1 states that $\{X_n\}$ is a martingale iff $\{Y_n\}$ is absolutely fair.

Example 1.2 (Polya's Urn Model for the Spread of Infection) An urn contains r red and g green balls. A series of drawings is made from this urn in the following manner. A ball is taken at random. It is then returned to the urn; in addition, c balls of the same color are put in the urn. Let X_n denote the proportion of red balls at the nth drawing. We shall show that the sequence $\{X_n\}$ is a martingale.

Denote by Y_n the number of red balls at the nth drawing. Then $Y_1 = r$ and for $n \geq 1$,

$$\begin{aligned} Y_{n+1} &= Y_n + c \quad \text{if a red ball is drawn on the } n\text{th drawing} \\ &= Y_n \qquad\quad \text{otherwise.} \end{aligned}$$

This shows that the conditional distribution of Y_{n+1} given $Y_1, Y_2, \ldots, Y_n$ depends only on Y_n. Since at the nth drawing there will be $r + g + nc - c$ balls in the urn find that

$$\begin{aligned} E(Y_{n+1}|Y_1, Y_2, \ldots, Y_n) &= E(Y_{n+1}|Y_n) \\ &= (Y_n + c)\frac{Y_n}{r+g+nc-c} + Y_n\left(1 - \frac{Y_n}{r+g+nc-c}\right) \\ &= Y_n \frac{r+g+nc}{r+g+nc-c}. \end{aligned}$$

By definition $X_n = Y_n/(r + g + nc - c)$, so we can write this last result as

$$E(X_{n+1}|X_1, X_2, \ldots, X_n) = X_n \quad (n \geq 1),$$

which shows that $\{X_n\}$ is a martingale.

Our objective is to extend the idea of fairness to a more general setting. In particular we wish to convey the following possibilities.

(1) The stake on each play may be an arbitrary amount (instead of being a unit) and may depend on the outcomes of previous plays. The amount

gained on each play will depend on the stake and the rules of the game. Thus in the general case the random variables Y_k have an arbitrary distribution rather then (1.1) and are not necessarily independent.

(2) The player may decide not to bet on some plays (that is, he may skip some plays).

(3) The player may decide to quit the game while he is still ahead.

These possibilities extend the game of pure chance to a game of strategy and chance.

7.2 Some Terminology

Given a stochastic process $\{X_n, n \geq 1\}$ we denote by $\mathcal{F}_n^X$ the sub-sigma-fields generated by events of the type

$$\{a_1 \leq X_1 \leq b_1, a_2 \leq X_2 \leq b_2, \ldots, a_n \leq X_n \leq b_n\}.$$

The family $\{\mathcal{F}_n^X, n \geq 1\}$ is called the history of the process $\{X_n\}$.

For many purposes we need to enlarge the scope of the history of the process. A family $\{\mathcal{F}_n, n \geq 1\}$ of sub-sigma-fields will be called a history if it is increasing, namely

$$\mathcal{F}_1 \subset \mathcal{F}_2 \subset \mathcal{F}_3 \subset \cdots.$$

Such a family is also called a filtration. If it also happens that this family is rich enough to include $\{\mathcal{F}_n^X\}$, that is, if

$$\mathcal{F}_n^X \subset \mathcal{F}_n \quad (n \geq 1),$$

then $\{\mathcal{F}_n\}$ is called a history of the process $\{X_n\}$. We shall then say that $\{X_n\}$ is adapted to $\{\mathcal{F}_n\}$. In practical applications $\mathcal{F}_n$ will be of the type $\mathcal{F}_n^Z$, with $Z_n = (X_n, Y_n)$, where $\{Y_n\}$ is a second stochastic process that we need to observe along with the process of interest to $\{X_n\}$. Thus $\{\mathcal{F}_n^Z\}$ is larger then $\{\mathcal{F}_n^X\}$, and therefore $\{X_n\}$ is adapted to $\{\mathcal{F}_n^Z\}$.

We may view a history $\{\mathcal{F}_n\}$ as an increasing information pattern. In the above example, $\{\mathcal{F}_n^Z\}$ summarizes the information arising from both processes $\{X_n\}$ and $\{Y_n\}$, while $\{\mathcal{F}_n^X\}$ corresponds to incomplete information on $\{Z_n\}$ through its component $\{X_n\}$.

In the following sections we shall assume that the process $\{X_n\}$ under consideration is adapted to some filtration $\{\mathcal{F}_n\}$.

We shall say that a random variable Z is $\mathcal{F}_n$-measurable if events of the type $\{a \leq Z \leq b\}$ belong to the sigma-field $\mathcal{F}_n$.

7.3 Martingales Relative to a Sigma-Field

With the terminology of Sec. 7.2 we are now ready to introduce martingales in a more general framework.

Definition 3.1 Let $\{X_n\}$ be a sequence of random variables with $E|X_n| < \infty$ and adapted to the filtration $\{\mathcal{F}_n\}$. We say $\{X_n\}$ is a martingale relative to $\{\mathcal{F}_n\}$ if

$$E(X_{n+1}|\mathcal{F}_n) = X_n \tag{3.1}$$

with probability 1.

Since $\{\mathcal{F}_n\}$ includes, in particular, $\{\mathcal{F}_n^X\}$, Definition 3.1 implies Definition 1.2, which states that $\{X_n\}$ is a martingale relative to its own history. When $\{\mathcal{F}_n\}$ is not specified, we shall take it to be the history of the given process $\{X_n\}$.

In Example 1.1 we have in effect proved that $\{X_n\}$ is a martingale relative to the sigma-field generated by $\{Y_n\}$.

Example 3.1 (Double or Nothing) In classical gambling (Example 1.1) with $p = q$ suppose that the stake on the kth play (instead of being a dollar) is given by B_k $(k \geq 1)$, where $B_1 = 1$ and for $k \geq 2$,

$$\begin{aligned} B_k &= 2^{k-1} \quad \text{if } Y_1 = Y_2 = \cdots = Y_{k-1} = 1 \\ &= 0 \qquad\quad \text{otherwise.} \end{aligned}$$

The player's gain on the kth play is then $B_k Y_k$, where Y_k has the distribution (1.1). His capital at the end of the nth play is given by X_n, where

$$X_n = \sum_{k=1}^{n} B_k Y_k.$$

Let $\{\mathcal{F}_n\}$ be the sigma-field generated by $\{Y_n\}$. Then B_n is $\mathcal{F}_{n-1}$-measurable. We have

$$\begin{aligned} E(X_{n+1}|\mathcal{F}_n) &= E(X_n + B_{n+1}Y_{n+1}|\mathcal{F}_n) = X_n + E(B_{n+1}Y_{n+1}|\mathcal{F}_n) \\ &= X_n + B_{n+1}E(Y_{n+1}|\mathcal{F}_n) = X_n + B_{n+1} \cdot 0 = X_n. \end{aligned}$$

This shows that $\{X_n\}$ is a martingale relative to $\{\mathcal{F}_n\}$.

Example 3.2 (Likelihood Ratios) Let $\{X_n\}$ be a stochastic process such that the joint distribution of $(X_1, X_2, \ldots, X_n)$ has a density that is strictly positive and continuous. Let $(X_1, X_2, \ldots, X_n)$ be an observed sample. Suppose we wish to test the hypothesis

$$H_0: \text{ density is } f_n(X_1, X_2, \ldots, X_n)$$

against the alternative

$$H_1: \text{ density is } g_n(X_1, X_2, \ldots, X_n).$$

The test procedure is based on the likelihood ratio

$$L_n = \frac{g_n(X_1, X_2, \ldots, X_n)}{f_n(X_1, X_2, \ldots, X_n)}.$$

We shall prove that under H_0, $\{L_n, n \geq 1\}$ is a martingale relative to $\{\mathcal{F}_n^X\}$. We have

$$\begin{aligned} &E(L_{n+1}|X_1, X_2, \ldots, X_n) \\ &\quad = \int_{-\infty}^{\infty} \frac{g_{n+1}(X_1, X_2, \ldots, X_n, x)}{f_{n+1}(X_1, X_2, \ldots, X_{n,x})} f(x \mid X_1, X_2, \ldots, X_n)dx, \end{aligned}$$

where $f(x|X_1, X_2, \ldots, X_n)$ is the conditional density of X_{n+1}, given $X_1, X_2, \ldots, X_n$. Since under H_0

$$f(x|X_1, X_2, \ldots, X_n) = \frac{f_{n+1}(X_1, X_2, \ldots, X_n, x)}{f_n(X_1, X_2, \ldots, X_n)},$$

it follows that under H_0

$$\begin{aligned} &E(L_{n+1} \mid X_1, X_2, \ldots, X_n) \\ &\quad = \frac{1}{f_n(X_1, X_2, \ldots, X_n)} \int_{-\infty}^{\infty} g_{n+1}(X_1, X_2, \ldots, X_n, x)dx \\ &\quad = \frac{g_n(X_1, X_2, \ldots, X_n)}{f_n(X_1, X_2, \ldots, X_n)} = L_n. \end{aligned}$$

This proves our assertion. □

Theorem 3.1 *Suppose $\{X_n\}$ is a martingale relative to $\{\mathcal{F}_n\}$. Then we have*

(i) $E(X_{n+1}|\mathcal{F}_k) = X_k$ *for* $1 \leq k \leq n$.
(ii) $E(X_n) = E(X_1)$ $(n \geq 1)$.
(iii) *If Z is $\mathcal{F}_n$-measurable, then $E(ZX_{n+1}|\mathcal{F}_n) = ZX_n$ if the expectation on the left side exists.*

Proof. (i) For $k = n$, this is merely Definition 3.1. For $k = n-1$ we have

$$E(X_{n+1}|\mathcal{F}_{n-1}) = E[E(X_{n+1}|\mathcal{F}_n)|\mathcal{F}_{n-1}]$$
$$= E(X_n|\mathcal{F}_{n-1}) = X_{n-1},$$

and the proof is completed by induction.

(ii) From Definition 3.1 it follows that

$$E(X_n) = EE(X_{n+1}|\mathcal{F}_n) = E(X_{n+1}).$$

(iii) This follows from the definition.

□

Example 3.3 In the urn model of Example 1.2, we saw that $\{X_n\}$ is a martingale, X_n being the proportion of red balls at the nth drawing. Since $X_1 = r/(r+g)$ we find from Theorem 3.1(ii) that $E(X_n) = E(X_1) = r/(r+g)$ for $n \geq 1$. Since the probability of a red ball at the nth drawing is given by $E(X_n)$, it follows that this probability is the same as on the first drawing.

7.4 Decision Functions; Optional Stopping

Definition 4.1 A random variable δ_n is a decision function if it takes only two values 0 and 1, and is $\mathcal{F}_{n-1}$-measurable ($n \geq 2$). We take $\delta_1 = 1$.

In the gambling context let

$$\begin{aligned} \delta_n &= 1 \quad \text{if the player bets on the } n\text{th play} \\ &= 0 \quad \text{if he skips the } n\text{th play.} \end{aligned}$$

Here the decision to bet or skip the nth play is based only on the history of the process up to the $(n-1)$th trail; that is, δ_n is $\mathcal{F}_{n-1}$-measurable. A choice of decision functions (δ_n) and stakes constitutes a system. Theorem 4.2 below states that a fair game remains fair under any system.

Definition 4.2 By optional stopping we mean a choice of decision functions $\{\delta_n\}$ such that

$$\delta_1 = \delta_2 = \cdots = \delta_T = 1, \quad \delta_{T+1} = \delta_{T+2} = \cdots = 0, \tag{4.1}$$

where T is a random variable taking values $1, 2, \ldots,$ and $+\infty$.

We have

$$\{T < n\} = \{\delta_n = 0\} \in \mathcal{F}_{n-1} \tag{4.2}$$

so that

$$\{T = n\} = \{T < n+1\} - \{T < n\} \in \mathscr{F}_n. \tag{4.3}$$

This shows that T is a stopping time relative to $\{\mathscr{F}_n\}$.

Theorem 4.1 (The System Theorem) *Let $\{X_n\}$ be a martingale relative to $\{\mathscr{F}_n\}$ and $\{\delta_n\}$ a sequence of decision functions. Put*

$$\tilde{X}_1 = X_1,$$
$$\tilde{X}_n = X_1 + \delta_2(X_2 - X_1) + \delta_3(X_3 - X_2) + \cdots + \delta_n(X_n - X_{n-1}) \quad (n \geq 2). \tag{4.4}$$

Then $\{\tilde{X}_n\}$ is also a martingale relative to $\{\mathscr{F}_n\}$ and $E(\tilde{X}_n) = E(X_1)$.

Proof. Clearly, $\tilde{X}_n$ is $\mathscr{F}_n$-measurable and $E|\tilde{X}_n| < \infty$. We have

$$\begin{aligned} E(\tilde{X}_{n+1}|\mathscr{F}_n) &= E[\tilde{X}_n + \delta_{n+1}(X_{n+1} - X_n)|\mathscr{F}_n] \\ &= \tilde{X}_n + \delta_{n+1}[E(X_{n+1}|\mathscr{F}_n) - X_n] \\ &= \tilde{X}_n + \delta_{n+1} \cdot 0 = \tilde{X}_n \end{aligned}$$

so that $\{\tilde{X}_n\}$ is a martingale relative to $\{\mathscr{F}_n\}$. By Theorem 3.1(ii) we have $E(\tilde{X}_n) = E(\tilde{X}_1) = E(X_1)$. □

Theorem 4.2 (Optional Stopping Theorem) *Let $\{X_n\}$ be a martingale relative to $\{\mathscr{F}_n\}$ and T a stopping time relative to $\{\mathscr{F}_n\}$. Let*

$$\begin{aligned} \tilde{X}_n &= X_n \quad \text{for } n \leq T \\ &= X_T \quad \text{for } n > T. \end{aligned} \tag{4.5}$$

Then $\{\tilde{X}_n\}$ is also a martingale relative to $\{\mathscr{F}_n\}$ and $E(\tilde{X}_n) = E(X_1)$.

Proof. In view of (4.1)–(4.3), Theorem 4.1 applies and we obtain the desired results. □

7.5 Submartingales and Supermartingales

We now extend our results to cover games that are either favorable or unfavorable to the player. In classical gambling this happens when $p > q$

or $p < q$. Since $E(Y_k) = p - q \gtrless 0$ according as $p \gtrless q$, we see from (1.3) that

$$E(X_{n+1}|Y_1, Y_2, \ldots, Y_n) \gtrless X_n.$$

This motivates the following definition.

Definition 5.1 Let $\{X_n, n \geq 1\}$ be a sequence of random variables with $E|X_n| < \infty$, and adapted to a filtration $\{\mathcal{F}_n\}$.

(i) $\{X_n\}$ is a submartingale relative to $\{\mathcal{F}_n\}$ if

$$E(X_{n+1}|\mathcal{F}_n) \ \geq X_n. \tag{5.1}$$

with probability 1.

(ii) $\{X_n\}$ is a supermartingale relative to $\{\mathcal{F}_n\}$ if

$$E(X_{n+1}|\mathcal{F}_n) \leq X_n. \tag{5.2}$$

with probability 1.

Thus submartingales correspond to favorable games and supermartingales to unfavorable games (there is a reversal of terminology here). In Example 3.1 (double or nothing) $\{X_n\}$ is a submartingale or a supermartingale according as $p > q$ or $p < q$.

Clearly, $\{X_n\}$ is a supermartingale relative to $\{\mathcal{F}_n\}$ iff $\{-X_n\}$ is a submartingale relative to $\{\mathcal{F}_n\}$. Therefore we need to consider only (say) submartingales.

For a submartingale we have the following results analogous to Theorems 3.1 and 4.1. Proofs are omitted.

Theorem 3.1s. *Suppose $\{X_n\}$ is a submartingale relative to $\{\mathcal{F}_n\}$. Then we have*

(i) $E(X_{n+1}|\mathcal{F}_k) \geq X_k$ *for* $1 \leq k \leq n$.

(ii) *The sequence* $\{E(X_n), n \geq 1\}$ *is nondecreasing:*

$$E(X_1) \leq E(X_2) \leq \cdots .$$

(iii) *If $Z \geq 0$ is $\mathcal{F}_n$-measurable, then*

$$E(ZX_{n+1}|\mathcal{F}_n) \geq ZX_n$$

if the expectation on the left side exists.

Theorem 4.1s *Let $\{X_n\}$ be a submartingale relative to $\{\mathcal{F}_n\}$ and $\{\tilde{X}_n\}$ as in* (4.4). *Then $\{\tilde{X}_n\}$ is a submartingale relative to $\{\mathcal{F}_n\}$ and $E(\tilde{X}_n) \geq E(X_1)$.*

Theorem 3.1s tells us that in classical gambling (Example 1.1) the expected gain remains positive (if $p > q$) or negative (if $p < q$) at the end of every play, since

$$E(i + X_n) \geq i + E(X_1) = i + p - q \gtrless i$$

according as $p \gtrless q$.

From Theorem 4.1s we conclude, in particular, that a favorable game remains favorable under any system and similarly for an unfavorable game.

Theorem 4.2s (Optional Stopping Theorem for Submartingales) *Let $\{X_n\}$ be a submartingale relative to $\{\mathcal{F}_n\}$ and T a stopping time relative to $\{\mathcal{F}_n\}$. Let*

$$\begin{aligned} \tilde{X}_n &= X_n \quad \textit{for } n \leq T \\ &= X_T \quad \textit{for } n > T. \end{aligned} \tag{5.3}$$

Then $\{\tilde{X}_n\}$ is a submartingale relative to $\{\mathcal{F}_n\}$ and

$$E(X_1) \leq E(\tilde{X}_n) \leq E(X_n) \quad (n \geq 1). \tag{5.4}$$

Proof. Proceeding as in Theorem 4.2 we find that $\{\tilde{X}_n\}$ is a submartingale and $E(\tilde{X}_n) \geq E(X_1)$. To prove $E(\tilde{X}_n) \leq E(X_n)$ we proceed as follows. We have

$$\begin{aligned} \tilde{X}_n &= X_1 + \sum_{k=2}^{n} \delta_k (X_k - X_{k-1}) \\ &= \sum_{k=1}^{n} \delta_k X_k - \sum_{1}^{n-1} \delta_{k+1} X_k \quad (\delta_1 = 1) \\ &= \sum_{k=1}^{n-1} (\delta_k - \delta_{k+1}) X_k + \delta_n X_n \\ &= \sum_{k=1}^{n} I_k X_k + \delta_n X_n, \end{aligned} \tag{5.5}$$

where $I_k = \delta_k - \delta_{k+1}$ $(1 \leq k \leq n-1)$. We find that

$$I_k = 1 \quad \text{if } k = T, \quad I_k = 0 \quad \text{if } k \neq T$$

so that I_k is $\mathcal{F}_k$-measurable. Since $I_k \geq 0$ we have by Theorem 3.1s(i) and (iii)

$$E(I_k X_n) = EE(I_k X_n | \mathcal{F}_k) \geq E(I_k X_k) \quad (1 \leq k \leq n-1)$$

and therefore (5.5) gives

$$\begin{aligned} E(\tilde{X}_n) &\leq \sum_{k=1}^{n-1} E(I_k X_n) + E(\delta_n X_n) \\ &= E(I_1 + I_2 + \cdots + I_{n-1} + \delta_n) X_n = E(X_n) \end{aligned}$$

as was required to be proved. □

We next explore the connection between martingales and submartingales. A simple result is as follows.

Theorem 5.1 *If $\{X_n\}$ is a martingale relative to $\{\mathcal{F}_n\}$, then $\{|X_n|\}$ is a submartingale relative to $\{\mathcal{F}_n\}$.*

Proof. From

$$E(X_{n+1} | \mathcal{F}_n) = X_n$$

we obtain

$$|X_n| = |E(X_{n+1} | \mathcal{F}_n)| \leq E(|X_{n+1}| | \mathcal{F}_n),$$

which proves the submartingale property of $\{|X_n|\}$. □

The following theorem establishes a more general result.

Theorem 5.2 *Let u be a convex function for which $E|u(X_n)| < \infty$.*

(i) *If $\{X_n\}$ is a martingale relative to $\{\mathcal{F}_n\}$, then $\{u(X_n)\}$ is a submartingale relative to $\{\mathcal{F}_n\}$.*
(ii) *If $\{X_n\}$ is a submartingale relative to $\{F_n\}$ and u is nondecreasing, then $\{u(X_n)\}$ is a submartingale relative to $\{\mathcal{F}_n\}$.*

Proof. Jensen's inequality gives

$$Eu(X_{n+1}) \geq u\left[E(X_{n+1})\right]. \tag{5.6}$$

This holds also for conditional expectations, so

$$E\left[u(X_{n+1}) | \mathcal{F}_n\right] \geq u\left[E(X_{n+1} | \mathcal{F}_n)\right] = u(X_n). \tag{5.7}$$

This proves the desired result for martingales. If u is nondecreasing, then since $E(X_{n+1}|\mathcal{F}_n) \geq X_n$ for submartingales we have using (5.6)

$$E[u(X_{n+1})|\mathcal{F}_n] \geq u\left[E(X_{n+1}|\mathcal{F}_n)\right] \geq u(X_n) \tag{5.8}$$

which proves the submartingale property. □

Corollary 5.1 *If $\{X_n\}$ is a submartingale relative to $\{\mathcal{F}_n\}$, then so is $\{X_n^+\}$.*

Theorem 5.3 *A sequence $\{X_n\}$ is a submartingale relative to $\{\mathcal{F}_n\}$ iff there exists a martingale $\{M_n\}$ relative to $\{\mathcal{F}_n\}$ and a positive increasing sequence $\{A_n\}$ adapted to $\{\mathcal{F}_n\}$ such that*

$$X_n = M_n + A_n \ (n \geq 1). \tag{5.9}$$

Proof. (i) Suppose (5.9) holds. Then

$$E(X_{n+1}|\mathcal{F}_n) = E(M_{n+1}|\mathcal{F}_n) + E(A_{n+1}|\mathcal{F}_n),$$

where $E(M_{n+1}|\mathcal{F}_n) = M_n$ since $\{M_n\}$ is a martingale and $E(A_{n+1}|\mathcal{F}_n) \geq E(A_n|\mathcal{F}_n)$ since $\{A_n\}$ is increasing. Therefore

$$E(X_{n+1}|\mathcal{F}_n) \geq M_n + E(A_n|\mathcal{F}_n) = M_n + A_n = X_n$$

which shows that $\{X_n\}$ is a submartingale.

(ii) Conversely, suppose that $\{X_n\}$ is a submartingale. Let

$$Y_1 = X_1 - E(X_1), \quad Y_n = X_n - E(X_n|\mathcal{F}_{n-1}) \quad (n \geq 2). \tag{5.10}$$

Then $E(Y_1) = 0$ and

$$E(Y_{n+1}|\mathcal{F}_n) = E(X_{n+1}|\mathcal{F}_n) - E(X_{n+1}|\mathcal{F}_n) = 0 \quad (n \geq 1). \tag{5.11}$$

Thus $\{Y_n\}$ is absolutely fair and by Theorem 1.1 the sequence $\{M_n\}$, where

$$M_n = \sum_1^n Y_k + E(X_1) \quad (n \geq 1), \tag{5.12}$$

is a martingale. Further, let

$$Z_1 = 0, \quad Z_n = E(X_n|\mathcal{F}_{n-1}) - X_{n-1} \quad (n \geq 2). \tag{5.13}$$

We have $Z_n \geq 0$, since $\{X_n\}$ is a submartingale. Let

$$A_n = \sum_{1}^{n} Z_k \quad (n \geq 1). \tag{5.14}$$

Then $\{A_n\}$ is a positive increasing sequence, and

$$M_n + A_n = \sum_{1}^{n}(Y_k + Z_k) + E(X_1) = X_1 + \sum_{2}^{n}(X_k - X_{k-1}) = X_n,$$

as was required to be proved. □

Theorem 5.4 *Suppose that $\{X_n\}$ is a supermartingale relative to a filtration $\{\mathcal{F}_n\}$ such that*

$$\sup_{n\geq 1} E|X_n| < \infty. \tag{5.15}$$

Then we can find two non-negative supermartingales $\{Y_n\}, \{Z_n\}$ relative to $\{\mathcal{F}_n\}$ such that

$$X_n = Y_n - Z_n \quad (n \geq 1). \tag{5.16}$$

Proof. We have

$$\begin{aligned} E(X_{n+1+k}|\mathcal{F}_k) &= E[E(X_{n+1+k}|\mathcal{F}_{n+k})|\mathcal{F}_k] \\ &\leq E(X_{n+k}|\mathcal{F}_k). \end{aligned}$$

Therefore the sequence $\{E(X_{n+k}|\mathcal{F}_k), n \geq 1\}$ is nonincreasing and converges to a limit ξ_k, where ξ_k is a $\mathcal{F}_k$-measurable random variable. From the supermartingale property we find that

$$X_k \geq E(X_{k+1}|\mathcal{F}_k) \geq \xi_k \quad (k \geq 1). \tag{5.17}$$

It turns out that the sequence $\{\xi_k, k \geq 1\}$ is a martingale. To see this we need to prove first that the ξ_k have finite expectations. By Fatou's lemma we have

$$\begin{aligned} E|\xi_k| &= E \lim_{n\to\infty} |E(X_{n+k}|\mathcal{F}_k)| \leq \liminf_{n\to\infty} E|E(X_{n+k}|\mathcal{F}_k)| \\ &\leq \liminf_{n\to\infty} EE(|X_{n+k}||\mathcal{F}_k|)| = \liminf_{n\to\infty} E|X_n| < \infty \end{aligned} \tag{5.18}$$

by our assumption (5.15). Also because of monotone convergence

$$\begin{aligned} E(\xi_{k+1}|\mathcal{F}_k) &= \lim_{n\to\infty} E[E(X_{n+k+1}|\mathcal{F}_{k+1})|\mathcal{F}_k] \\ &= \lim_{n\to\infty} E(X_{n+k+1}|\mathcal{F}_k) = \xi_k. \end{aligned} \tag{5.19}$$

From (5.18) and (5.19) it follows that $\{\xi_k\}$ is indeed a martingale with

$$\sup_{k\geq 1} E|\xi_k| < \infty. \tag{5.20}$$

Since $\{\xi_k\}$ is a martingale, $\{|\xi_k|\}$ is a submartingale by Theorem 5.1. By applying the above arguments to $\{|\xi_k|\}$ we see that the sequence $E(X_{n+k}|\mathcal{F}_k)$ is nondecreasing and converges to a random variable η_k such that $\eta_k \geq |\xi_k|$ and the sequence $\{\eta_k\}$ is a martingale. Now denote

$$Y_n = X_n + \eta_n - \xi_n, \quad Z_n = \eta_n - \xi_n \quad (n \geq 1). \tag{5.21}$$

Then $\{Y_n\}$ and $\{Z_n\}$ are both non-negative supermartingales and $X_n = Y_n - Z_n$ $(n \geq 1)$, as desired. □

7.6 Optional Skipping and Sampling Theorems

The decision functions δ_n introduced in Sec. 7.4 give rise to the possibility that the gambler may skip some plays. Let N_k be the play on which the gambler places his kth bet. Thus $N_1 = 1$, while for $k \geq 2$

$$N_k = \min\{n\colon \delta_1 + \delta_2 + \cdots + \delta_n = k\}$$

and

$$\{N_k = n\} = \{\delta_1 + \delta_2 + \cdots + \delta_{n-1} = k-1\} \cap \{\delta_n = 1\}.$$

Therefore

$$\{N_k = n\} \in \mathcal{F}_{n-1} \quad (n \geq 2). \tag{6.1}$$

If we assume that $\delta_1 + \delta_2 + \cdots = \infty$ with probability 1, then the game will continue indefinitely and

$$N_1 < N_2 < \cdots < \infty. \tag{6.2}$$

Let $Y_1 = X_1, Y_n = X_n - X_{n-1}$ $(n \geq 2)$. Then Theorem 1.1 tells us that $\{Y_n\}$ is an absolutely fair process. The gambler's gain on the kth play on which he does bet is given by Y_{N_k}, and the system theorem tells us that the sequence $\{Y_{N_k}\}$ is absolutely fair.

A generalization of the above considerations leads to the optional skipping theorem, where $\{N_k\}$ is a sequence of random variables satisfying conditions (6.1) and (6.2). Since we are aiming at a degree of generality, an additinal condition needs to be imposed on the subsequence $\{Y_{N_k}\}$ generated. We have then the following result, stated for submartingales.

Theorem 6.1 (Optional Skipping Theorem) *Let $\{Y_n, n \geq 1\}$ be a stochastic process adapted to a filtration $\{\mathscr{F}_n\}$ such that $E|Y_n| < \infty (n \geq 1)$ and*

$$E(Y_{n+1}|\mathscr{F}_n) \geq 0 \tag{6.3}$$

with probability 1. *Let $\{N_k, k \geq 1\}$ be a sequence of random variables taking integer values and satisfying conditions* (6.1) *and* (6.2). *Denote*

$$\tilde{Y}_k = Y_{N_k} \quad (k \geq 1). \tag{6.4}$$

If $E|\tilde{Y}_k| < \infty$ $(k \geq 1)$, then for the process $\{\tilde{Y}_k, k \geq 1\}$ we have

$$E(\tilde{Y}_{k+1}|\tilde{Y}_1, \tilde{Y}_2, \cdots, \tilde{Y}_k) \geq 0 \tag{6.5}$$

with probability 1. *If the equality holds in* (6.3), *then it also holds in* (6.5).

Proof. Let A be an event determined by $\tilde{Y}_1, \tilde{Y}_2, \cdots, \tilde{Y}_k$ and

$$A_n = A \cap \{N_{k+1} = n\} \quad (n > k+1). \tag{6.6}$$

Denoting $Z_n = 1_{A_n}$ we note that $Z_n \geq 0$ and Z_n is $\mathscr{F}_{n-1}$-measurable since $A_n \in \mathscr{F}_{n-1}$. Now

$$\begin{aligned} E(\tilde{Y}_{k+1} Z_n) &= E(Y_n Z_n) = EE(Y_n Z_n|\mathscr{F}_{n-1}) \\ &= EZ_n E(Y_n|\mathscr{F}_{n-1}) \geq 0 \end{aligned}$$

on account of (6.3). Adding this over $n = k+2, k+3, \ldots$ and using the fact that $N_{k+1} < \infty$ with probability 1 we obtain

$$E(\tilde{Y}_{k+1}; A) \geq 0$$

from which the desired result (6.5) follows. □

Next we extend Theorem 4.2s to a sequence of stopping times $\{T_k, k \geq 1\}$. The process $\{X_{T_k}, k \geq 1\}$ is obtained from the original process if the gambler samples his capital only at certain epochs $\{T_k\}$. This leads to the optimal sampling theorem, which states that under certain conditions

$\{X_{T_k}\}$ is also a submartingale. To motivate the discussion we consider two simple cases first.

(i) Let T be a stopping time and for $k \geq 1$, denote $T_k = \min(k, T)$. The process $\{X_{T_k}\}$ then reduces to $\{\tilde{X}_k\}$ defined by (5.13) and Theorem 4.2s states that this process is a submartingale. We have

$$X_T = \sum_{k=1}^{\infty} 1_{\{T \geq k\}} Y_k, \tag{6.7}$$

where $Y_1 = X_1$ and $Y_k = X_k - X_{k-1} (k \geq 2)$. If T is a bounded random variable, that is, if there exists an integer $N < \infty$ such that $T \leq N$ with probability 1, then

$$E|X_T| \leq \sum_{k=1}^{N} E|Y_k| < \infty. \tag{6.8}$$

(ii) Let T_1, T_2 be two stopping times such that $T_1 \leq T_2 \leq N < \infty$ with probability 1. Then using (6.7) we find that

$$X_{T_2} - X_{T_1} = \sum_{k=1}^{N} Z_k Y_k, \tag{6.9}$$

where $Z_k = 1_{\{T_1 < k \leq T_2\}}$. We have $Z_k \geq 0$ and Z_k is $\mathcal{F}_{k-1}$-measurable, since

$$\{T_1 < k \leq T_2\} = \{T_1 \leq k-1\} \cap \{T_2 \leq k-1\}^c \in \mathcal{F}_{k-1}.$$

Therefore

$$\begin{aligned} E(Z_k Y_k) &= E\left[Z_k (X_k - X_{k-1})\right] = EE\left[Z_k (X_k - X_{k-1}) | \mathcal{F}_{k-1}\right] \\ &\geq EZ_k \left[E(X_k | \mathcal{F}_{k-1}) - X_{k-1}\right] \geq 0 \end{aligned}$$

and from (6.9) we obtain

$$E(X_{T_1}) \leq EX(T_2). \tag{6.10}$$

In the general case we need to prove that X_{T_k} has finite mean. Also, we need to guarantee that our arguments extend to the case of infinite sums in (6.9). We first prove the following lemma.

Lemma 6.1 *Suppose that $\{X_n\}$ is a submartingale relative to a filtration $\{\mathcal{F}_n\}$ with*

$$\sup_{n \geq 1} E(X_n^+) < \infty. \tag{6.11}$$

Then for any stopping time T we have

$$E|X_T| < \infty. \tag{6.12}$$

Proof. Denote

$$\tilde{X}_n = X_n \quad \text{for} \quad n \leq T, \text{ and} = X_T \quad \text{for} \quad n > T. \tag{6.13}$$

Then

$$|\tilde{X}_n| \to |X_T| \tag{6.14}$$

with probability 1. By Theorem 4.2s $\{\tilde{X}_n\}$ is a submartingale relative to $\{\mathcal{F}_n\}$ and

$$E(\tilde{X}_n) \geq E(X_1). \tag{6.15}$$

By Corollary 5.1, $\{X_n^+\}$ is also a submartingale relative to $\{\mathcal{F}_n\}$, so by Theorem 4.2s,

$$E(\tilde{X}_n^+) \leq E(X_n^+). \tag{6.16}$$

Using (6.15) and (6.16), we find that

$$\begin{aligned} E|\tilde{X}_n| &= 2E(\tilde{X}_n^+) - E(\tilde{X}_n) \leq 2E(X_n^+) - E(X_1) \\ &\leq 2M - E(X_1) < \infty. \end{aligned}$$

In view of (6.14), Fatou's lemma gives

$$E|X_T| \leq \liminf_{n \to \infty} E|\tilde{X}_n| < \infty.$$

□

Theorem 6.2 (Optional Sampling Theorem) *Suppose that $\{X_n\}$ is a submartingale relative to a filtration $\{\mathcal{F}_n\}$, and $\{T_k,\ k \geq 1\}$ a sequence of integer-valued random variables such that the inequalities*

$$1 \leq T_1 \leq T_2 \leq \cdots < \infty \tag{6.17}$$

hold with probability 1, and

$$\{T_k = n\} \in \mathcal{F}_n \quad (n \geq 1). \tag{6.18}$$

Denote

$$\tilde{X}_k = X_{T_k} \quad (k \geq 1). \tag{6.19}$$

Assume that

$$E|\tilde{X}_k| < \infty \quad (k \geq 1) \tag{6.20}$$

and

$$\liminf_{N\to\infty} E[X_N^+; T_k > N] = 0 \quad (k \geq 1). \tag{6.21}$$

Then $\{\tilde{X}_k\}$ *is also a submartingale, and*

$$E(X_1) \leq E(\tilde{X}_k) \quad (k \geq 1). \tag{6.22}$$

If (6.20) *holds and*

$$\liminf_{N\to\infty} E[\,|X_N|; T_k > N] = 0 \quad (k \geq 1), \tag{6.23}$$

then if $\{X_n\}$ *is a martingale,* $\{\tilde{X}_k\}$ *is also a martingale and*

$$E(X_1) = E(\tilde{X}_k) \quad (k \geq 1). \tag{6.24}$$

Proof. Let A be an event determined by $\tilde{X}_1, \tilde{X}_2, \dots, \tilde{X}_k$. Then for $m \leq n$ we have

$$\begin{aligned} &E(X_n; A, T_k = m, T_{k+1} \geq n) \\ &\quad = E(X_n; A, T_k = m, T_{k+1} = n) + E(X_n; A, T_k = m, T_{k+1} > n). \end{aligned}$$

Since $A \cap \{T_k = m,\ T_{k+1} > n\} \in \mathcal{F}_n$ and $\{X_n\}$ is a submartingale, we have

$$E(X_n; A, T_k = m, T_{k+1} > n) \leq E(X_{n+1}; A, T_k = m, T_{k+1} > n). \tag{6.25}$$

Therefore

$$\begin{aligned} E(X_n; A, T_k = m,\ T_{k+1} \geq n) \leq\ & E(\tilde{X}_{k+1}; A, T_k = m, T_{k+1} = n) \\ & + E(X_{n+1};\ A,\ T_k = m,\ T_{k+1} \geq n+1). \end{aligned}$$

Adding this over $n = m, m+1, \dots, N-1$ and simplifying we obtain

$$\begin{aligned} E(X_m; A, T_k = m,\ T_{k+1} \geq m) \leq\ & E(\tilde{X}_{k+1}; A, T_k = m, m \leq T_{k+1} < N) \\ & + E(X_N; A, T_k = m, T_{k+1} \geq N). \end{aligned}$$

This can be written as

$$E(\tilde{X}_k; A, T_k = m) \le E(\tilde{X}_{k+1}; A, T_k = m, T_{k+1} \le N) \\ + E(X_N; A, T_k = m, T_{k+1} \ge N). \tag{6.26}$$

Letting $N \to \infty$ we obtain

$$E(\tilde{X}_k; A, T_k = m) \le E(\tilde{X}_{k+1}; A, T_k = m) \\ + \liminf_{N\to\infty} E(X_N; A, T_k = m, T_{k+1} \ge N)$$

since $T_{k+1} < \infty$ with probability 1. Now

$$\liminf_{N\to\infty} E(X_N; A, T_k = m, T_{k+1} \ge N) \\ \le \liminf_{N\to\infty} E(X_N^+; T_{k+1} \ge N) = 0$$

by an assumption (6.21). Thus

$$E(\tilde{X}_k; A, T_k = m) \le E(\tilde{X}_{k+1}; A, T_k = m). \tag{6.27}$$

Adding these over $m = 1, 2, \ldots$ we obtain

$$E(\tilde{X}_k; A) \le E(\tilde{X}_{k+1}; A) \tag{6.28}$$

which proves the submartingale property of $\{\tilde{X}_k\}$.

If $\{X_n\}$ is a martingale, we have equality in (6.25) and so (6.26) becomes

$$E(\tilde{X}_k; A, T_k = m) = E(\tilde{X}_{k+1}; A, T_k = m, T_{k+1} \le N) \\ + E(X_N; A, T_k = m, T_{k+1} \ge N).$$

Since

$$\liminf_{N\to\infty} |E(X_N; A, T_k = m, T_{k+1} \ge N)| \\ \le \liminf_{N\to\infty} E(|X_N|; A, T_{k+1} \ge N) = 0$$

by our assumption (6.23), we obtain equality in (6.27) and so in (6.28). It follows that $\{\tilde{X}_k\}$ is a martingale.

Taking $T_1 = 1$ without loss of generality, we obtain (6.22) in the submartingale case and (6.24) in the martingale case, in view of Theorem 3.1(ii) and Theorem 3.1s(ii). □

Several sufficient conditions are known for assumptions (6.20) and (6.23) of Theorem 6.2 to hold. In particular, we have already shown that if each stopping time T_k is a bounded random variable, then (6.20) holds; moreover, since $P\{T_k > N\} \Rightarrow 0$ as $N \to \infty$, then (6.23) also holds. In the examples discussed in this chapter we shall use the conditions of the following lemma.

Lemma 6.2 *For each k if $E(T_k) < \infty$, and with probability* 1

$$E[|X_{n+1} - X_n|\,|\mathcal{F}_n] \leq c, \quad T_k > n \tag{6.29}$$

for some constant c, then assumptions (6.20) *and* (6.23) *of Theorem* 6.2 *hold.*

Proof. Denote $Z_1 = |X_1|, Z_n = |X_n - X_{n-1}|\ (n \geq 2), W_n = Z_1 + Z_2 + \cdots + Z_n (n \geq 1)$ and $\tilde{W}_k = W_{T_k} (k \geq 1)$. From $|X_n| \leq W_n$ we see that

$$E|\tilde{X}_k| \leq E(\tilde{W}_k) \tag{6.30}$$

since

$$E|\tilde{X}_k| = \sum_{n=1}^{\infty} E[|X_n|;\ T_k = n] \leq \sum_{n=1}^{\infty} E[W_n; T_k = n] = E(\tilde{W}_k).$$

On the other hand,

$$\begin{aligned} E(\tilde{W}_k) &= \sum_{n=1}^{\infty}\sum_{m=1}^{n} E[Z_m; T_k = n] = \sum_{m=1}^{\infty}\sum_{n=m}^{\infty} E[Z_m; T_k = n] \\ &= \sum_{m=1}^{\infty} E[Z_m; T_k \geq m]. \end{aligned}$$

Now

$$\begin{aligned} E[Z_m; T_k \geq m] &= EE[Z_m; T_k \geq m|\mathcal{F}_{m-1}] \\ &\leq cP\{T_k \geq m\} \end{aligned}$$

by condition (6.29). Therefore

$$E(\tilde{W}_k) \leq c\sum_{m=1}^{\infty} P\{T_k \geq m\} = cE(T_k) < \infty. \tag{6.31}$$

We have thus shown that $E(\tilde{X}_k)$ and $E(\tilde{W}_k)$ exist and satisfy the inequalities

$$E|\tilde{X}_k| \leq E(\tilde{W}_k) \leq cE(T_k) < \infty. \tag{6.32}$$

Also

$$E[|X_N|; T_k > N] \le E[\tilde{W}_k; T_k > N] \to 0 \tag{6.33}$$

as $N \to \infty$, since $E(\tilde{W}_k) < \infty$. □

Example 6.1 (Classical Gambling) While classical gambling provided much of the motivation for the study of martingales, it is possible to apply the martingale theory to rederive the results of classical gambling. As in Example 1.1, let $i + X_n$ be the player's capital at the end of the nth play. Let $\{\mathcal{F}_n\}$ be the history of the process $\{Y_n\}$. The duration of the game is given by

$$D_i = \min\{n\colon X_n = -i \quad \text{or} \quad X_n = a - i\}. \tag{6.34}$$

since

$$\{D_i = n\} = \{0 < X_m < a \ (1 \le m \le n-1),\ X_n = -i \text{ or } X_n = a\} \tag{6.35}$$

it follows that D_i is a stopping time relative to $\{\mathcal{F}_n\}$. It can be proved by elementary methods that $D_i < \infty$ with probability 1 and $E(D_i) < \infty$. To apply Theorem 6.2 it remains to verify condition (6.29) of Lemma 6.2 for an appropriate martingale. Our objective is to calculate $E(D_i)$ and the probability u_i that the game ends with the player's ruin ($1 - u_i$ is then the probability that the game ends with the player winning).

(i) *The case* $p=q$. We recall from Example 1.1 that $\{X_n, n \ge 1\}$ is a martingale relative to $\{\mathcal{F}_n\}$, which $E(X_n) = E(Y_1 + Y_2 + \cdots + Y_n) = 0$ $(n \ge 1)$. We have

$$E[|X_{n+1} - X_n|\,|\mathcal{F}_n] = E[|Y_{n+1}|\,|\mathcal{F}_n] = E|Y_{n+1}| = 1 \quad \text{for } n < D_i.$$

Therefore result (6.24) under Theorem 6.2 gives

$$E(X_{D_i}) = 0.$$

Here X_{D_i} takes only two values $-i$ and $a-i$ with probabilities u_i and $1-u_i$, respectively. Therefore

$$(-i)u_i + (a-i)(1-u_i) = 0$$

which gives

$$u_i = 1 - \frac{i}{a} \tag{6.36}$$

as the probability of the player's ruin.

To derive the expected duration of the game in this case we consider $\{M_n,\ n \geq 1\}$, where

$$M_n = X_n^2 - n. \tag{6.37}$$

We have

$$M_{n+1} = (X_n + Y_{n+1})^2 - n - 1 = M_n + 2X_nY_{n+1} + Y_{n+1}^2 - 1$$

so that

$$E(M_{n+1}|\mathscr{F}_n) = M_n + 2X_nE(Y_{n+1}) + E(Y_{n+1}^2 - 1) = M_n.$$

Thus $\{M_n\}$ is a martingale relative to $\{\mathscr{F}_n\}$ with $E(M_n) = E(M_1) = E(Y_1^2) - 1 = 0$. We have

$$\begin{aligned} E[|M_{n+1} - M_n||\mathscr{F}_n] &\leq 2|X_n|E|Y_{n+1}| + E(Y_{n+1}^2) + 1 \\ &= 2|X_n| + 2 < 2(a-i) + 2 \quad \text{for } n < D_i \end{aligned}$$

with probability 1. Therefore (6.24) gives $E(M_{D_i}) = 0$ or $E(X_{D_i}^2) - E(D_i) = 0$. Thus

$$(-i)^2u_i + (a-i)^2(1-u_i) - E(D_i) = 0.$$

Using (6.36) in this we obtain

$$E(D_i) = i(a-i). \tag{6.38}$$

(ii) *The case* $p \neq q$. Denote

$$M_n = \left(\frac{q}{p}\right)^{X_n} \quad (n \geq 1). \tag{6.39}$$

We have

$$M_{n+1} = \left(\frac{q}{p}\right)^{X_n + Y_{n+1}} = M_n\left(\frac{q}{p}\right)^{Y_{n+1}}$$

so that

$$E(M_{n+1}|\mathscr{F}_n) = M_n\left[\left(\frac{q}{p}\right)^{+1}p + \left(\frac{q}{p}\right)^{-1}q\right] = M_n.$$

Thus $\{M_n\}$ is a martingale relative to $\{\mathscr{F}_n\}$ with $E(M_n) = E(M_1) = 1$. To verify condition (6.29) we note that

$$E[|M_{n+1} - M_n|\,|\mathscr{F}_n] \leq M_nE\left[\left(\frac{q}{p}\right)^{Y_{n+1}} + 1\right] = 2M_n.$$

Also, for $n < D_i$ we have $-i < X_n < a - i$ so that

$$M_n = \left(\frac{q}{p}\right)^{X_n} \leq \min\left\{\left(\frac{q}{p}\right)^{-i}, \left(\frac{q}{p}\right)^{a-i}\right\} = c_1 \text{ (say)}.$$

Therefore

$$E[|M_{n+1} - M_n||\mathcal{F}_n] \leq 2c_1 \text{ with probability 1 and } E(M_{D_i}) = 1 \text{ or}$$

$$E\left(\frac{q}{p}\right)^{D_i} = 1.$$

This gives

$$\left(\frac{q}{p}\right)^{-i} u_i + \left(\frac{q}{p}\right)^{a-i} (1 - u_i) = 1$$

or

$$u_i = \frac{\left(\frac{q}{p}\right)^i - \left(\frac{q}{p}\right)^a}{1 - \left(\frac{q}{p}\right)^a} \tag{6.40}$$

as the probability of the player's ruin.

To obtain the expected duration of the game we consider $\{M_n\}$, where

$$M_n = X_n - n(p - q). \tag{6.41}$$

As in the case $p = q$, $\{M_n\}$ is a martingale relative to $\{\mathcal{F}_n\}$, and Theorem 6.2 again applies. Thus $E(X_{D_i}) - (p - q)E(D_i) = 0$ or

$$(-i)u_i + (a - i)(1 - u_i) + (q - p)E(D_i) = 0.$$

Using (6.40) in this we obtain

$$E(D_i) = \frac{i}{q - p} - \frac{a}{q - p} \cdot \frac{1 - (\frac{q}{p})^i}{1 - (\frac{q}{p})^a}. \tag{6.42}$$

□

7.7 Application to Random Walks

Let $\{X_k, k \geq 1\}$ be a sequence of IID random variables, $S_0 = 0$ and $S_n = X_1 + X_2 + \cdots + X_n (n \geq 1)$ so that $\{S_n, n \geq 0\}$ is a random walk on the real line. We denote by $\{\mathcal{F}_n\}$ the history of the process, and consider a random variable N which is a stopping time relative to $\{\mathcal{F}_n\}$. In classical gambling (Example 6.1) we have already noted that the duration of game D_i is a stopping time for the Bernoulli random walk $\{X_n\}$. In A. Wald's sequential analysis the key role is played by the random variable

$$N = \min\{n\colon S_n \in I^c\} \tag{7.1}$$

where $\{S_n\}$ is a random walk on the real line and I is a finite interval $(-b, a)$. Since

$$\{N = n\} = \{S_m \in I(0 \le m \le n-1), S_n \in I^c\} \tag{7.2}$$

it follows that N is a stopping time relative to the history of $\{S_n\}$. It is known that N is a proper random variable with a finite mean. The results of Theorem 7.1 (below) were derived for this particular stopping time, but hold for more general stopping times under appropriate conditions. We have already used these results in Example 6.1; for the martingales defined by (6.37) and (6.41), it was easy to establish the desired results in a direct manner.

Theorem 7.1 (Wald Equations) *Suppose that N is a stopping time relative to $\{\mathcal{F}_n\}$ such that $E(N) < \infty$.*

(i) *If the X_k have a finite mean μ, then*

$$E(S_N) = \mu E(N). \tag{7.3}$$

(ii) *If the X_k have a finite variance σ^2 and*

$$|S_n - n\mu| < c \quad \textit{for } n < N \tag{7.4}$$

for some constant c, then

$$E(S_N - N\mu)^2 = \sigma^2 E(N). \tag{7.5}$$

Proof. (i) Denote $M_n = S_n - n\mu(n \ge 0)$. Then since $M_{n+1} = M_n + X_{n+1} - \mu$, we have

$$\begin{aligned} E(M_{n+1}|\mathcal{F}_n) &= M_n + E[(X_{n+1} - \mu)|\mathcal{F}_n] \\ &= M_n + E(X_{n+1} - \mu) = M_n. \end{aligned}$$

This shows that $\{M_n, n \ge 0\}$ is a martingale relative to $\{\mathcal{F}_n\}$ with $E(M_n) = 0(n \ge 0)$. Also,

$$E(|M_{n+1} - M_n||\mathcal{F}_n) = E|X_{n+1} - \mu|$$

so that condition (6.29) of Lemma 6.2 is satisfied with $c = E|X_{n+1} - \mu|$. By Theorem (6.2) we therefore obtain $E(M_N) = 0$, which reduces to $E(S_N - N\mu) = 0$, as desired.

(ii) Let $M_n = (S_n - n\mu)^2 - n\sigma^2 (n \geq 0)$. We have

$$\begin{aligned} M_{n+1} &= (S_n - n\mu + X_{n+1} - \mu)^2 - (n+1)\sigma^2 \\ &= M_n + 2(S_n - n\mu)(X_{n+1} - \mu) + (X_{n+1} - \mu)^2 - \sigma^2 \end{aligned}$$

so that

$$\begin{aligned} E(M_{n+1}|\mathcal{F}_n) &= M_n + 2(S_n - n\mu)E(X_{n+1} - \mu) + E(X_{n+1} - \mu)^2 - \sigma^2 \\ &= M_n. \end{aligned}$$

Therefore $\{M_n\}$ is a martingale relative to $\{\mathcal{F}_n\}$, with $E(M_n) = 0$ $(n \geq 0)$. We have

$$E(|M_{n+1} - M_n|\,|\mathcal{F}_n) \leq 2|S_n - n\mu|E|X_{n+1} - \mu| + E|X_{n+1} - \mu|^2 + \sigma^2$$

and the right side of this inequality is bounded for $n < N$ in view of condition (7.4). Applying Theorem 6.2 we thus obtain $E(M_N) = 0$ or $E(S_N - N\mu)^2 - \sigma^2 E(N) = 0$, as desired. □

We now introduce the moment generating function (m.g.f.)

$$\phi(\theta) = E(e^{\theta X_k}) \tag{7.6}$$

of the random variables X_k, assumed to exist for $\theta \in (\theta_1, \theta_2)$, where $-\infty \leq \theta_1 < 0 < \theta_2 \leq \infty$. Also, let $0 < s < 1$ and denote $D = \{(s, \theta) : 0 < s < 1, \theta_1 < \theta < \theta_2\}$.

Theorem 7.2 *Let the random variable N be as in Theorem 7.1, and $(s, \theta) \in D$. If for some constant c*

$$s^n e^{\theta S_n} < c \quad \text{for } n < N \tag{7.7}$$

with probability 1, then

$$E(s^N e^{\theta S_N}) = 1 + [s\phi(\theta) - 1] \sum_{n=0}^{\infty} E[s^n e^{\theta S_n}; N > n]. \tag{7.8}$$

Proof. For $(s, \theta) \in D$, denote

$$M_n = s^n e^{\theta S_n} - 1 - [s\phi(\theta) - 1] \sum_{m=0}^{n-1} s^m e^{\theta S_m} \quad (n \geq 0). \tag{7.9}$$

Then M_n is $\mathcal{F}_n$-measurable and

$$E(M_n) = s^n \phi(\theta)^n - 1 - [s\phi(\theta) - 1] \sum_{m=0}^{n-1} s^m \phi(\theta)^m = 0. \tag{7.10}$$

We have

$$\begin{aligned} M_{n+1} &= s^{n+1} e^{\theta S_{n+1}} - 1 - [s\phi(\theta) - 1] \sum_{m=0}^{n} s^m e^{\theta S_m} \\ &= M_n + s^{n+1} e^{\theta S_n} [e^{\theta X_{n+1}} - \phi(\theta)]; \end{aligned}$$

so

$$\begin{aligned} E(M_{n+1}|\mathcal{F}_n) &= M_n + s^{n+1} e^{\theta S_n} E[e^{\theta X_{n+1}} - \phi(\theta)] \\ &= M_n. \end{aligned}$$

Thus $\{M_n, n \geq 0]$ is a martingale relative to $\{\mathcal{F}_n\}$. Since

$$\begin{aligned} E(|M_{n+1} - M_n||\mathcal{F}_n &\leq s^{n+1} e^{\theta S_n} [E e^{\theta X_{n+1}} + \phi(\theta)] \\ &= 2\phi(\theta) s^{n+1} e^{\theta S_n} \end{aligned}$$

condition (6.29) of Lemma 6.2 is satisfied in view of (7.7). Theorem 6.2 now yields the result $E(M_N) = 0$, which can be written as

$$\begin{aligned} E(s^N e^{\theta S_N}) &= 1 + [s\phi(\theta) - 1] E \sum_{m=0}^{\infty} [s^m e^{\theta S_m}; m < N] \\ &= 1 + [s\phi(\theta) - 1] \sum_{m=0}^{\infty} E[s^m e^{\theta S_m}; m < N] \end{aligned}$$

since

$$\begin{aligned} \sum_{m=0}^{\infty} E[s^m e^{\theta S_m}; m < N] &\leq \sum_{m=0}^{\infty} c P\{N > m\} \\ &= cE(N) < \infty. \end{aligned}$$

□

Corollary 7.1 (The Wald Identity) *Let θ be such that the m.g.f. $\phi(\theta)$ exists and $\phi(\theta) > 1$. If for some constant c*

$$\phi(\theta)^{-n} e^{\theta S_n} < c \quad \text{for } n < N \tag{7.11}$$

with probability 1, then

$$E[\phi(\theta)^{-N} e^{\theta S_N}] = 1. \tag{7.12}$$

Proof. The desired result follows from Theorem (7.2) if we choose $s = \phi(\theta)^{-1}$. □

We remark here that for the stopping time N defined by (7.2) condition (7.11) is satisfied, since for $n < N$

$$\begin{aligned}\phi(\theta)^{-n}e^{\theta S_n} \le e^{\theta S_n} &\le e^{\theta a} \quad \text{for } \theta > 0 \\ &\le e^{-\theta b} \quad \text{for } \theta < 0.\end{aligned}$$

7.8 Convergence Properties

Perhaps the most important property of martingales from the point of view of applications is that under appropriate conditions they converge with probability 1. We shall prove this for non-negative supermartingales. From this the convergence property follows for general supermartingales, in view of Theorem 5.4.

We first need the following result, which is really a version of the optional sampling theorem; however, we do not need the heavy technicalities of the proof of this theorem, and moreover, we are considering stopping times that are not necessarily finite with probability 1.

Lemma 8.1 *Suppose that* $\{X_n\}$ *is a non-negative supermartingale relative to a filtration* $\{\mathcal{F}_n\}$*, and* T_1, T_2 *stopping times relative to* $\{\mathcal{F}_n\}$ *such that* $T_1 \le T_2$*. If* $X_\infty = 0$*, then*

$$E(X_{T_1}) \ge E(X_{T_2}). \tag{8.1}$$

Proof. For any stopping time T we can write in the notation of (5.3),

$$\begin{aligned}E(X_T; T \le n) + E(X_n; T > n) &= E(\tilde{X}n) \\ &= E\sum_{k=1}^{n} 1_{\{T\ge k\}} Y_k,\end{aligned}$$

where $Y_1 = X_1$ and $Y_k = X_k - X_{k-1}$ $(k \ge 2)$. Therefore

$$E(X_{T_1}; T_1 \le n) - E(X_{T_2}; T_2 \le n) = E\sum_{k=1}^{n} Z_k Y_k - EZ_{n+1}X_n,$$

where $Z_k = (-1)1_{\{T_1 < k \le T_2\}} \le 0$. Proceeding as in (6.9) and (6.10) and remembering that $\{X_n\}$ is a supermartingale we find that

$$E(X_{T_1}; T_1 \le n) \ge E(X_{T_2}; T_2 \le n). \tag{8.2}$$

Since $\{X_n\}$ is non-negative, $E(X_{T_i}; T_i \le n)$ increases to $E(X_{T_i}; T_i < \infty) = E(X_{T_i})(i = 1, 2)$, since $X_\infty = 0$ by assumption. This gives (8.1), as desired. □

Theorem 8.1 *Let $\{X_n\}$ be a non-negative supermartingale relative to a filtration $\{\mathcal{F}_n\}$ such that $X_\infty = 0$. Then $\lim_{n\to\infty} X_n$ exists and is finite with probability* 1.

Proof. If $\lim X_n = \infty$ on a set of positive probability p, then for some n,

$$P\{X_n > a\} > \frac{1}{2}p \quad (a > 0).$$

This would mean that

$$E(X_n) \ge E(X_n; X_n > a) > aP\{X_n > a\} > a\frac{1}{2}p.$$

Choosing $a = 2p^{-1}E(X_1)$ we find that $E(X_n) > E(X_1)$, which is a contradiction, since $\{X_n\}$ is a supermartingale.

If $\{X_n\}$ does not converge with probability 1, there exist two numbers a, b $(a < b)$ such that the set

$$D = \left\{\liminf_{n\to\infty} X_n < a < b < \limsup_{n\to\infty} X_n\right\} \tag{8.3}$$

has positive probability. Now let $T_0 = 0$ and for $k = 0, 1, 2, \ldots$

$$T_{2k+1} = \min\{n > T_{2k} : X_n < a\}, \tag{8.4a}$$

$$T_{2k+2} = \min\{n > T_{2k+1} : X_n > b\}. \tag{8.4b}$$

Thus in sample sequence for which these random variables are finite, T_{2k+1} are the hitting times of the set $(0, a)$ and T_{2k+2} the hitting times of the set (b, ∞). They are stopping times relative to $\{\mathcal{F}_n\}$. We have $0 < T_1 < T_2 < T_3 < \cdots \le \infty$, and moreover, as $k \to \infty$

$$\{T_{2k+1} < \infty\} \to \bigcap_{\ell=0}^{\infty}\{T_{2\ell+1} < \infty\} = D, \tag{8.5a}$$

$$\{T_{2k+1} < \infty\} \to \bigcap_{\ell=0}^{\infty}\{T_{2\ell+2} < \infty\} = D. \tag{8.5b}$$

As regards $E(X_{T_{2k+1}})$ we find that

$$\begin{aligned} E(X_{T_{2k+1}}) &= E(X_{T_{2k+1}}; T_{2k+1} < \infty) + E(X_{T_{2k+1}}; T_{2k+1} = \infty) \\ &= E(X_{T_{2k+1}}; T_{2k+1} < \infty) \end{aligned}$$

because of our assumption $X_\infty = 0$. This gives

$$E(X_{T_{2k+1}}) < aP\{T_{2k+1} < \infty\}. \tag{8.6}$$

Similarly

$$E(X_{T_{2k+2}}) > bP\{T_{2k+2} < \infty\}. \tag{8.7}$$

By Lemma 8.1

$$bP\{T_{2k+2} < \infty\} < E(X_{T_{2k+2}}) < E(X_{T_{2k+1}}) < aP\{T_{2k+1} < \infty\}.$$

Letting $k \to \infty$ in this we obtain

$$bP(D) < aP(D) \quad (a < b) \tag{8.8}$$

which is a contradiction, unless $P(D) = 0$. Thus $\{X_n\}$ has a finite limit with probability 1. □

Corollary 8.1 *If $\{X_n\}$ is a supermartingale relative to a filtration $\{\mathcal{F}_n\}$ and*

$$\sup_{n \geq 1} E|X_n| < \infty, \tag{8.9}$$

then $\lim_{n\to\infty} X_n$ *exists and is finite with probability* 1.

Proof. The desired result follows from Theorems 5.4 and 8.1. □

Example 8.1 In Polya's urn model (Examples 1.2 and 3.3) we saw that if X_n is the proportion of red balls at the nth drawing, then $\{X_n\}$ is a martingale. By Theorem 8.1, $X_n \to X < \infty$ with probability 1, but this result does not tell us how to find the limit random variable X. Therefore we proceed directly as follows. It is clear that the probability $P_{k,n}$ of k red balls in the first n drawings does not depend on the precise drawings in which these red balls appeared, and is in fact given by

$$P_{k,n} = \binom{n}{k} \frac{r(r+c)(r+2c)\cdots(r+kc-c)g(g+c)\cdots(g+nc-kc-c)}{(r+g)(r+g+c)\cdots(r+g+n-c)}.$$

We can write this as

$$\begin{aligned} P_{k,n} &= \binom{n}{k} \frac{\Gamma(\frac{r}{c}+k)\Gamma(\frac{g}{c}+n-k)}{\Gamma\left(\frac{r+g}{c}+n\right)} \cdot \frac{\Gamma\left(\frac{r+g}{c}\right)}{\Gamma(\frac{r}{c})\Gamma(\frac{g}{c})} \\ &= \binom{n}{k} \frac{B(\frac{r}{c}+k, \frac{g}{c}+n-k)}{B(\frac{r}{c}, \frac{g}{c})}, \end{aligned} \tag{8.10}$$

where Γ and B are the gamma and beta functions, respectively. Now introducing the beta density

$$f(p) = \frac{1}{B(\frac{r}{c}, \frac{g}{c})} p^{\frac{r}{c}-1}(1-p)^{\frac{g}{c}-1} \tag{8.11}$$

$(0 \leq p \leq 1)$ we can verify that

$$P_{k,n} = \int_0^1 \binom{n}{k} p^k (1-p)^{n-k} f(p) dp. \tag{8.12}$$

This shows that $P_{k,n}$ is the mixture of the binomial distribution with the probability of success p having the beta density (8.11). Therefore

$$P\{X_n \leq x\} = \sum_{\{k:\, n^{-1}k \leq x\}} P_{k,n} = \int_0^1 F_n(nx; p) f(p) dp, \tag{8.13}$$

where

$$F_n(x; p) = \sum_{k=0}^{n} \binom{n}{k} p^k (1-p)^{n-k} \epsilon_0(x-k), \tag{8.14}$$

ϵ_0 being a distribution concentrated at 0. By the law of large numbers

$$F_n(nx; p) \to \epsilon_0(x-p)$$

as $n \to \infty$. This gives

$$\lim_{n\to\infty} P\{X_n \leq x\} = \int_0^1 \epsilon_0(x-p) f(p) dp = \int_0^x f(p) dp, \tag{8.15}$$

which shows that limit X has the beta density (8.11). □

7.9 The Concept of Fairness

We started the treatment of martingales by characterizing them as fair processes. The concept of fairness featured significantly in our discussion of classical gambling and motivated our definitions. Yet fairness is far from being a well defined notion, as is clear from the history of the St. Petersburg paradox. In this section we give a brief account of this paradox.

Consider a game in which a single play consists of tossing a fair coin until it falls heads. If this happens on the nth toss, the player receives 2^{n-1} dollars. We ask the following questions: (a) what is the player's expectation in each play? (b) What is a fair fee the player should pay in return for this expectation?

The player's expectation is given by

$$1\left(\frac{1}{2}\right)+2\left(\frac{1}{2}\right)^2+2^2\left(\frac{1}{2}\right)^3+\cdots=\frac{1}{2}+\frac{1}{2}+\cdots=\infty,$$

and therefore it could be argued that a fair fee would be ∞. However, if G denotes the player's gain in a single play, then $E(G)=\infty$ as we have already seen, but

$$P\{G>4\}=\sum_{n=4}^{\infty}\left(\frac{1}{2}\right)^n=0.125.$$

The question therefore arises (perhaps motivated by moral or social considerations) whether it is fair to pay an enormous fee $(=\infty)$ for a very small chance $\left(\frac{1}{8}\right)$ of winning a large fortune. This is the St. Petersburg paradox.

Several solutions were proposed to resolve this paradox, which are more or less satisfactory. Perhaps the most important one among these is a solution based on a utility function $u(x)$. Assume that the player's initial capital is i dollars and he pays a fee of z dollars. Then his expected final utility is $Eu(i+G-z)$. We say the fee z is fair if

$$Eu(i+G-z)=u(i).$$

In particular for $u(x)=c\ \log\ x+d$, this gives

$$\sum_{n=1}^{\infty}[c\ \log(i+2^{n-1}-z)+d]\left(\frac{1}{2}\right)^n=c\ \log i+d$$

or

$$\sum_{n=1}^{\infty} \log(i + 2^{n-1} - z) \left(\frac{1}{2}\right)^n = \log i.$$

For $i = 10$, this gives $z \approx 3$ and for $i = 1000$, $z \approx 6$.

From the point of view of modern probability, there is no paradox here, because random variables finite with probability 1, but having infinite mean, arise in many problems in applied probability.

7.10 Problems for Solution

1. Show that every subsequence $\{X_{nk},\ k \geq 1\}$ of a martingale $\{X_n\}$ is also a martingale relative to $\{\mathcal{F}_{nk}\}$.
2. Suppose that $\{X_n\}$ and $\{Y_n\}$ are submartingales relative to $\{\mathcal{F}_n\}$ and denote Z_n as in (i) and (ii) below. In each case show that $\{Z_n\}$ is a submartingale relative to $\{\mathcal{F}_n\}$.

   (i) $Z_n = X_n + Y_n$, (ii) $Z_n = \max(X_n, Y_n)$.

3. (a) Let $\{Y_n, n \geq 1\}$ be an absolutely fair sequence of random variables, $B_1 = 1$, and for each $n \geq 2$, let B_n be a bounded function of $Y_1, Y_2, \ldots, Y_{n-1}$. Let

   $$X_n = \sum_{k=1}^{n} B_k Y_k \quad (n \geq 1).$$

   Show that the sequence $\{X_n\}$ is a martingale relative to the sigma-field generated by $\{Y_n\}$.
   (b) A gambler bets on a sequence of independent tosses of a fair coin as follows. On each toss he bets some amount; if the outcome is heads he wins an amount equal to his bet, while if the outcome is tails he loses this amount. For the first toss he bets \$1. For each subsequent toss he bets either \$1 or \$2 depending on whether the previous toss ended in heads or tails. Let X_n be his capital at the end of the nth toss. Show that the sequence $\{X_n, n \geq 1\}$ is a martingale relative to its own history.
4. Suppose that $\{Y_n\}$ is a sequence of independent and identically distributed random variables such that

   $$P\{Y_k = +1\} = p \quad \text{and} \quad P\{Y_k = -1\} = q,$$

where $p+q=1$. Let the random variables $\{N_k\}$ be such that

$$\{N_k = n\} \in \mathscr{F}_{n-1}, \quad N_1 < N_2 < \cdots < \infty,$$

where $\{\mathscr{F}_n\}$ is the history of $\{Y_n\}$. Show that $\{Y_{N_k}, k \geq 1\}$ are independent and identically distributed random variables with the same distribution as $\{Y_n\}$.

5. Let $\{X_n, n \geq 0\}$ be a time-homogeneous Markov chain on the state space $x_0, x_1, x_2, \ldots, x_a$ such that $x_0 < x_1 < \ldots < x_a$. Suppose, in addition, that $\{X_n\}$ is a martingale relative to its history.

   (i) Prove that the states x_0 and x_a are absorbing.
   (ii) Assume that the chain does not contain any further closed sets. Show that the probabilities of ultimate absorption at x_0 and x_a, starting from state i $(1 \leq i \leq a-1)$ are given, respectively, by

$$\frac{x_i - x_a}{x_0 - x_a} \quad \text{and} \quad \frac{x_0 - x_i}{x_0 - x_a}.$$

6. Consider the classical gambling problem in the case where each play result in a tie with probability r, where $p+q+r=1$. Show that the probability of the gambler's ruin is given by

$$\begin{aligned} u_i &= \frac{\left(\frac{q}{p}\right)^i - \left(\frac{q}{p}\right)^a}{1 - \left(\frac{q}{p}\right)^a} \quad \text{if } p \neq q \\ &= 1 - \frac{i}{a} \quad \text{if } p = q \end{aligned}$$

   and the expected duration of the game is given by

$$\begin{aligned} E(D_i) &= \frac{i}{q-p} - \frac{a}{q-p} \cdot \frac{1 - \left(\frac{q}{p}\right)^i}{1 - \left(\frac{q}{p}\right)^a} \quad \text{if } p \neq q \\ &= \frac{i(a-i)}{1-r} \quad \text{if } p = q. \end{aligned}$$

7. For the Bernoulli random walk show that if

$$M_n = s^{n_\theta S_n} - 1 - [s\phi(\theta) - 1] \sum_0^{n-1} s^{m_\theta S_m} \quad (n \geq 1),$$

   then $\{M_n\}$ is a martingale.

Further Reading

Bremaud, P (1981). *Point Processes and Queues. Martingale Dynamics.* New York: Springer-Verlag.

Hall, P (1980). *Martingale Limit Theory and its Application.* San Diego: Academic Press.

Liptser, RS and AN Shiryaev (1989). *Theory of Martingales* (Translated by K. Dzjaparidze). Dordrecht: Kluwer Academic Publishers.

Meyer, PA (1972). *Martingales and Stochastic Integrals.* Berlin: Springer Verlag.

Neveu, J (1975). *Discrete Parameter Martingales.* Amsterdam: North Holland Publishing Company.

Rao KM (1972). Lecture notes for statistik 3. Matematisk Institut, Aarhus, Universitet.

Shafer, G (1988). The St. Petersburg paradox. In: *Encyclopedia of Statistical Sciences*, Vol. 8, 865–870. New York: John Wiley.

Williams, D (1991). *Probability with Martingales.* New York: Cambridge University Press.

CHAPTER 8
Branching Processes

8.1 Introduction

Branching processes arise from a particular model of population growth, dealing with problems in genetics, nuclear chain reactions, and demography (including survival of family names). Other models for population growth will be treated in Chap. 10.

In this chapter we describe the Bienayme–Watson–Galton branching process arising from the following model. Consider a population in which the individuals of the $(n+1)$th generation are produced by those of the nth generation in such a way that each individual can, independently of others, produce $0, 1, 2, \ldots$ successors (direct descendents or offspring) with probabilities $k_0, k_1, k_2, \ldots$. If X_n denotes the number of individuals in the nth generation $(n = 0, 1, \ldots)$, then we have the relation

$$X_{n+1} = Y_1^{(n+1)} + Y_2^{(n+1)} + \cdots + Y_{X_n}^{(n+1)}, \tag{1.1}$$

where $Y_i^{(n+1)}$ is the number of direct descendents of the ith individual of the nth generation. We assume that (a) the $Y_i^{(n+1)}$ $(i = 1, 2, \ldots)$ are IID with the distribution $\{k_j, j \geq 0\}$ and (b) $Y_i^{(n+1)}$ is independent of X_n. These assumptions are equivalent to the statement that there is no interaction among individuals and the birth rates do not change from generation to generation.

We assume $X_0 = 1$. For $X_0 = i$ (> 1) the i processes started by the i ancestors are i copies of the same process.

For $0 < s < 1$ denote

$$F_n(s) = E(s^{X_n}) \quad (n = 0, 1, \ldots). \tag{1.2}$$

We have

$$F_0(s) = s, \quad F_1(s) = E(s^{X_1}) = \sum_{j=0}^{0} k_j s^j = k(s). \tag{1.3}$$

Theorem 1.1 *For $m \geq 0, n \geq 0$ we have*

$$F_{m+n}(s) = F_m \circ F_n(s). \tag{1.4}$$

Proof. We note that X_{m+n} is the total number of individuals in the nth branch (line) descended from each of the X_m ancestors. Therefore,

$$\begin{aligned} F_{m+n}(s) &= E(s^{X_{m+n}}) = EE(s^{X_{m+n}}|X_m) \\ &= E[F_n(s)^{X^m}] = F_m \circ F_n(s). \end{aligned}$$

□

From (1.4) we obtain for $n \geq 0$

$$F_{n+1}(s) = F_n \circ F_1(s), \quad F_{n+1}(s) = F_1 \circ F_n(s). \tag{1.5}$$

Using (1.5) successively for $n = 1, 2, \ldots$ we can compute the *functional iterates* $F_2(s), F_3(s), \ldots$.

Using (1.5) we can also obtain the mean and variance of X_n. Let

$$\mu = \sum_0^\infty jk_j \leq \infty, \quad \sigma^2 = \sum_0^\infty j^2 k_j - \mu^2 \leq \infty \tag{1.6}$$

be the mean variance of the offspring distribution. We then have the following.

Theorem 1.2 *Let $X_0 = 1$. If μ and σ^2 are finite, then for $n \geq 1$ we have*

$$E(X_n) = \mu^n \tag{1.7}$$

$$\begin{aligned} \mathrm{Var}(X_n) &= \sigma^2 \mu^{n-1} \frac{1-\mu^n}{1-\mu} \quad \text{if } \mu \neq 1 \\ &= n\sigma^2 \quad \text{if } \mu = 1 \end{aligned} \tag{1.8}$$

and for $m < n$,

$$\begin{aligned} \mathrm{Covar}(X_m, X_n) &= \sigma^2 \mu^{n-1} \frac{1-\mu^m}{1-\mu} \quad \text{if } \mu \neq 1 \\ &= m\sigma^2 \quad \text{if } \mu = 1. \end{aligned} \tag{1.9}$$

Proof. The first relation in (1.5) gives

$$F'_{n+1}(s) = F'_n \circ K(s) \cdot K'(s)$$
$$F''_{n+1}(s) = F''_n \circ K(s) \cdot [K'(s)]^2 + F'_n \circ K(s) \cdot K''(s).$$

Letting $s \to 1-$ in these and nothing that $K(1-) = 1, K'(1-) = \mu$ and $K''(1-) = \sigma^2 + \mu^2 - \mu$, we obtain

$$\mu_{n+1} = \mu_n \mu \tag{1.10}$$

and

$$\sigma^2_{n+1} + \mu^2_{n+1} - \mu_{n+1} = (\sigma^2_n + \mu^2_n - \mu_n)\mu^2 + \mu_n(\sigma^2 + \mu^2 - \mu), \tag{1.11}$$

where $\mu_n = E(X_n)$ and $\sigma^2_n = \text{Var}(X_n)(n \geq 1)$. We note that $\mu_1 = \mu$ and $\sigma^2_1 = \sigma^2$. From (1.10) we obtain $\mu_2 = \mu_1 \cdot \mu = \mu^2$, and by induction (1.7) follows. Using this in (1.11) and simplifying we find that

$$\sigma^2_{n+1} = \mu^2 \sigma^2_n + \sigma^2 \mu^n.$$

This is gives

$$\sigma^2_2 = \mu^2 \sigma^2_1 + \sigma^2 \mu = \sigma^2(\mu + \mu^2)$$
$$\sigma^2_3 = \mu^2 \sigma^2_2 + \sigma^2 \mu^2 = \sigma^2(\mu^2 + \mu^3 + \mu^4)$$

and by induction

$$\begin{aligned} \sigma^2_n &= \sigma^2(\mu^{n-1} + \mu^n + \cdots + \mu^{2n-2}) \\ &= \sigma^2 \mu^{n-1} \frac{1 - \mu^n}{1 - \mu} \quad \text{if } \mu \neq 1 \\ &= n\sigma^2 \quad \text{if } \mu = 1 \end{aligned}$$

which is the result (1.8). Finally, from $F[s^{X_n}|X_m] = [F_{n-m}(s)]^{X_m} (m < n)$ we find that for $m < n$

$$E(X_m X_n) = EE(X_m X_n | X_m) = E\big(X_m \mu^{n-m} X_m\big) = \mu^{n-m} E\big(X^2_m\big).$$

This leads to the result (1.9) for the covariance. □

Example 1.1 (Binary Splitting) Consider particles that either split into two with probability p or remain in their current state with probability

$q = 1 - p$, during their lifetime. We have

$$k_0 = q, \quad k_2 = p \quad 0 \le p \le 1. \tag{1.12}$$

A direct argument gives

$$P\{X_1 = j | X_0 = i\} = P_{ij} = \binom{i}{j/2} p^{j/2} q^{i-j/2} \quad \text{if } j \text{ is even}$$
$$= 0 \quad \text{if } j \text{ is odd.} \tag{1.13}$$

To obtain P_{ij} by using Theorem 1.1 we note that $K(s) = q + ps^2$ and

$$K(s)^i = (q + ps^2)^i = \sum_{j=0}^{\infty} \binom{i}{j/2} p^{j/2} q^{i-j/2} s^j. \tag{1.14}$$

Since P_{ij} is the coefficient of s^j in $K(s)^i$ we get (1.13) as desired.

The mean and variance of the offspring distribution (1.12) are found to be $\mu = 2p$ and $\sigma^2 = 4pq$. Theorem 1.2 gives

$$E(X_n) = (2p)^n, \quad \operatorname{Var}(X_n) = 2^{n+1} p^n q \frac{1 - (2p)^n}{1 - 2p} \tag{1.15}$$

if $p \neq \frac{1}{2}$ and

$$E(X_n) = 1, \quad \operatorname{Var}(X_n) = n \tag{1.16}$$

if $p = \frac{1}{2}$.

Example 1.2 Suppose the offspring distribution is given by $\{pq^j, j \ge 0\}$ where $0 < p < 1$ and $q = 1 - p$. We have

$$F_1(s) = K(s) = \sum_0^{\infty} pq^j s^j = \frac{p}{1 - qs} \tag{1.17}$$

and

$$F_2(s) = F_1 \circ K(s) = \frac{p}{1 - qK(s)} = p \frac{1 - qs}{1 - qp - qs}. \tag{1.18}$$

We can write

$$F_1(s) = p \frac{q - p}{(q^2 - p^2) - (q - p)qs}.$$

This suggests the general expression

$$F_n(s) = p \frac{\alpha_n - \alpha_{n-1} qs}{\alpha_{n+1} - \alpha_n qs} (n \ge 1), \tag{1.19}$$

where $\alpha_n = q^n - p^n (n \geq 0)$. To verify this, assume (1.19) holds for $1, 2, \ldots, n$. Then

$$\begin{aligned} F_{n+1}(s) = F_n \circ K(s) &= p\frac{\alpha_n - \alpha_{n-1}qK(s)}{\alpha_{n+1} - \alpha_n qk(s)} \\ &= p\frac{\alpha_n(1-qs) - \alpha_{n-1}qp}{\alpha_{n+1}(1-qs) - \alpha_n qp} = p\frac{(\alpha_n - \alpha_{n-1}qp) - \alpha_n qs}{(\alpha_{n+1} - \alpha_n qp) - \alpha_{n+1}qs}. \end{aligned}$$

Now

$$\alpha_n - \alpha_{n-1}qp = (q^n - p^n) - (q^{n-1} - p^{n-1})qp = q^{n+1} - p^{n+1} = \alpha_{n+1}.$$

Therefore,

$$F_{n+1}(s) = p\frac{\alpha_{n+1} - \alpha_n qs}{\alpha_{n+2} - \alpha_{n+1}qs}.$$

This shows that (1.19) hold for $n+1$.

For $p = q$, the expression reduces to

$$F_n(s) = \lim_{q \to p} \frac{\alpha_n - \alpha_{n-1}qs}{\alpha_{n+1} - \alpha_n qs} = \frac{n - (n-1)s}{n + 1 - ns}.$$

Branching processes are sometimes called multiplicative processes, this term being suggested by the result (1.7). A more important feature of our model is that it gives rise to martingales. Let the family of sigma-fields $\{\mathcal{F}_n, n \geq 1\}$ denote the history of the process $\{X_n\}$ and denote

$$M_n = \frac{X_n}{\mu^n} \quad (n \geq 0). \tag{1.20}$$

We have the following.

Theorem 1.3 *Let the offspring distribution have a finite mean μ. Then the process $\{M_n,\ n \geq 0\}$ is a martingale relative to the family $\{\mathcal{F}_n\}$. Also, if the variance σ^2 is finite, then*

$$\begin{aligned} E(M_n) = 1, \quad \mathrm{Var}(M_n) &= \sigma^2 \frac{1 - \mu^{-n}}{\mu(\mu - 1)} \quad \textit{if } \mu \neq 1 \\ &= n\sigma^2 \qquad\qquad\quad \textit{if } \mu = 1. \end{aligned} \tag{1.21}$$

Proof. Using (1.1) we find that

$$E(M_{n+1}|\mathcal{F}_n) = E\left(\frac{X_{n+1}}{\mu^{n+1}}\middle|\mathcal{F}_n\right) = \frac{1}{\mu^{n+1}}\,\mu X_n = M_n,$$

which proves the martingale property. The results (1.21) follow from (1.7) and (1.8). □

8.2 The Problem of Extinction

If $k_0 = 1$ then from (1.1) we find that $X_1 = 0$ and by induction $X_n = 0$ $(n \geq 1)$. If $k_0 = 0$, then $X_1 \geq X_0$ and again by induction $X_{n+1} \geq X_n$ $(n \geq 0)$. These possibilities are not interesting from the present point of view, and therefore we assume that $0 < k_0 < 1$. We first need the following lemma.

Lemma 2.1 *For $0 < s < 1$ let $K(s) = k_0 + k_1 s + k_2 s^2 + \cdots$ be a p.g.f. and assume that $0 < k_0 < 1$ and $K'(1-) = \mu \leq \infty$.*

(*a*) *For $0 < s < 1$ the equation $\xi = sK(\xi)$ has a unique continuous solution $\xi = \xi(s)$ with $\xi(0+) = 0$.*
(*b*) *As $s \to 1, \xi(s) \to \zeta$, where ζ is the smallest positive root of the equation $\zeta = K(\zeta)$. Here $0 < \zeta < 1$ iff $\mu > 1$.*

Proof. Consider the equation $f(x) = s^{-1}$, where

$$f(x) = \frac{K(x)}{x} = \frac{k_0}{x} + k_1 + k_2 x + k_3 x^2 + \cdots (0 < x < 1).$$

We have $f(0+) = \infty$ and $f(1-) = 1$. Also

$$f'(x) = -\frac{k_0}{x^2} + k_2 + 2k_3 x + \cdots$$

$$f''(x) = \frac{2k_0}{x^3} + 2k_3 + \cdots > 0.$$

Therefore, $f'(x)$ is a monotone increasing function and vanishes at most once in $(0 < x < 1)$, which happens iff $f'(1-) = \mu - 1 > 0$. If $f'(x) = 0$ at $x = x_0$ $(0 < x_0 < 1)$, then $f(x)$ is monotone decreasing in $(0, x_0)$ and monotone increasing in $(x_0, 1)$. Since $f(1-) = 1$ and $f(x_0) < 1$, there exists exactly one ζ that $f(\zeta) = 1$, $0 < \zeta < 1$.

Now it is clear that the equation $f(x) = s^{-1}$ has a unique solution $\xi = \xi(s) \leq \zeta$, this solution being continuous in s. Hence, $\xi(1-) = \ell \leq \zeta$.

Since $f(\ell) = 1$ we must have $\ell = \zeta$. The lemma is therefore completely proved. □

Theorem 2.1 *Let $\mu \leq \infty$ be the mean number of direct descendents of an individual in a branching process, and assume that $0 < k_0 < 1$. Then if $\mu \leq 1$, the population becomes extinct with probability 1, while if $\mu > 1$, the population either becomes extinct with probability ζ^i, or explodes with probability $1 - \zeta^i$, where ζ is the smallest positive root of the equation $\zeta = K(\zeta)$, and $X_0 = i$ is the initial number in the population.*

Proof. It suffices to consider the case where $X_0 = 1$. We shall prove that for $0 \leq s < 1, F_n(s) \to \zeta$ as $n \to \infty$, where ζ is the smallest positive root of the equation $K(\zeta) = \zeta$. The result will then follow from Lemma 2.1.

For $0 \leq s \leq \zeta$ we have $s \leq K(s) \leq K(\zeta)$, which can be written as $F_0(s) \leq F_1(s) \leq \zeta$. Since $F_2(s) = F_1 \circ F_1(s)$ we obtain $F_1(s) \leq F_2(s) \leq K(\zeta) = \zeta$. Using induction we find from (1.5) that

$$F_0(s) \leq F_1(s) \leq \cdots \leq F_n(s) \leq \zeta.$$

Thus the sequence $\{F_n(s), n \geq 0\}$ is a bounded monotone non-decreasing sequence, and therefore $F_n(s) \to \ell = \ell(s)$ as $n \to \infty$, where $\ell \leq \zeta$. From $F_{n+1}(s) = K \circ F_n(s)$ we find that $\ell = K(\ell)$. However, since ζ is the smallest positive root of the equation we must have $\ell = \zeta$, and therefore $F_n(s) \to \zeta$ $(0 \leq s \leq \zeta)$.

If $\mu \leq 1$, then $\zeta = 1$, and so we have proved that $F_n(s) \to 1$ for $0 \leq s < 1$. If $\mu > 1$ we need a further argument for $\zeta \leq s < 1$. Here we have $\zeta = K(\zeta) \leq K(s) \leq 1$, which can be written as $\zeta \leq F_1(s) \leq 1$. Proceeding as before we find that $F_n(s) \to \ell = \ell(s)$ as $n \to \infty$, where $\ell \geq \zeta$, and $\ell = K(\ell)$. But, if $\ell > \zeta$, then $K(\ell) < \ell$, so $\ell = \zeta$. Therefore, $F_n(s) \to \zeta$ $(\zeta \leq s < 1)$. We have thus proved that as $n \to \infty, F_n(s) \to \zeta$ for $0 \leq s < 1$, as required. □

We call the branching process *subcritical*, *critical* or *supercritical* according to $\mu < 1, = 1$ or > 1. From Theorem 1.2 we see that in the critical case $\mu = 1$, although extinction is a sure event, large fluctuations around the mean are likely, since $\mathrm{Var}(X_n) = n\sigma^2$.

Example 2.1 Suppose that the offspring distribution is given by

$$k_0 = a, \quad k_j = (1-a)pq^{j-1} \ (j \geq 1), \tag{2.1}$$

where $0 < a < 1$. This is the geometric distribution with a modified initial term (or briefly, *modified geometric* distribution). Its generating function is given by

$$K(s) = a + (1-a)\sum_{j=1}^{\infty} pq^{j-1}s^j = a + \frac{(1-a)ps}{1-qs}$$

which simplifies to

$$K(s) = \frac{a+(p-a)s}{1-qs} \quad (0 < s < 1). \tag{2.2}$$

From $\mu = K'(1-)$ and $\sigma^2 = K''(1-) + K'(1-) - K'(1-)^2$ we find the mean and variance of this distribution as

$$\mu = \frac{1-a}{p}, \quad \sigma^2 = \frac{q}{p^2} + \frac{a}{p}\left(1 - \frac{a}{p}\right). \tag{2.3}$$

The equation $K(\zeta) = \zeta$ reduces to $qs^2 - (a+q)s + a = 0$, which has two roots 1, a/q. By Theorem 2.1 the probability of extinction is therefore

$$\zeta = \min\left(1, \frac{a}{q}\right). \tag{2.4}$$

A.J. Lotka fitted the distribution

$$k_0 = 0.4813, \quad k_j = (0.2290)(0.5586)^{j-1} \ (j \geq 1) \tag{2.5}$$

to the data contained in the 1920 US Census of white males. This is a modified geometric distribution (2.1) with $a = 0.4813, q = 0.5586$, and $p = 0.4414$. It has mean $\mu = 1.1746$ and variance $\sigma^2 = 2.7690$. The probability of extinction is $a/q = 0.8616$. This is certainly less than 1, but perhaps uncomfortably high.

Table 5.1 compares the observed data with fitted distribution according to (2.5). The total probability of up to 10 children (according to the observed date) is 0.9999, so we may ignore observations after $j = 10$ and solve the equation

$$x = k_0 + k_1 x + k_2 x^2 + \cdots + k_{10} x^{10}.$$

This gives $\zeta = 0.8790$ for the extinction probability, which is even higher than the value obtained from the fitted distribution.

Table 8.1. Frequency of families with i sons

No. of sons j	Observed	Fitted
0	0.4981	0.4813
1	0.2103	0.2290
2	0.1270	0.1279
3	0.0730	0.0715
4	0.0418	0.0399
5	0.0241	0.0223
6	0.0132	0.0125
7	0.0069	0.0070
8	0.0035	0.0039
9	0.0015	0.0022
10	0.0005	0.0012

8.3 The Extinction Time and the Total Progeny

The total number of individuals in the first n generations is given by

$$Z_n = X_0 + X_1 + X_2 + \cdots + X_n \quad (n \geq 0). \tag{3.1}$$

We shall call Z_n the progeny up to and including the nth generation. As $n \to \infty, Z_n \to Z$ with probability 1, where

$$Z = X_0 + X_1 + X_2 + \cdots \tag{3.2}$$

or equivalently,

$$Z = X_0 + X_1 + X_2 + \cdots + X_T, \tag{3.3}$$

where $T(\leq \infty)$ is the extinction time, namely

$$T = \min\{n \geq 1 \colon X_n = 0\}. \tag{3.4}$$

We shall call Z the total progeny. We have the following results for the random variables Z_n, Z, and T.

Theorem 3.1 *For the branching process with i (≥ 1) ancestors, the p.g.f. of Z_n is given by*

$$E[s^{Z_n} \mid Z_0 = i] = \xi_n(s)^i, \tag{3.5}$$

where

$$\xi_0(s) = s, \quad \xi_{n+1}(s) = sK \circ \xi_n(s) \ (n \geq 0). \tag{3.6}$$

Proof. From our assumptions it follows that

$$E[s^{Z_n}|Z_0] = [\xi_n(s)]^{Z_0},$$

where

$$\xi_n(s) = E[s^{Z_n}|Z_0 = 1].$$

Now

$$\begin{aligned} E(s^{Z_{n+1}} \mid Z_0) &= E[s^{X_0+X_1+X_2+\cdots+X_{n+1}} \mid X_0] \\ &= EE[s^{X_0+X_1+X_2+\cdots+X_{n+1}}|X_0, X_1] \\ &= EE[s^{X_0+Z_n}|X_0, X_1] \\ &= s^{X_0}E[\xi_n(s)^{X_1}|X_0] = s^{X_0}K_0\xi_n(s)^{X_0}, \end{aligned}$$

so that

$$\xi_{n+1}(s)^{Z_0} = s^{Z_0}[K_0\xi_n(s)]^{Z_0}.$$

This gives

$$\xi_{n+1}(s) = sK_0\xi_n(s).$$

□

For the extinction time T we have

$$P\{T \leq n|X_0 = i\} = P\{X_n = 0 \mid X_0 = i\} = \zeta_n^i, \tag{3.7}$$

where $\zeta_n = F_n(0)$. From $F_{n+1}(s) = F_1 \circ F_n(s) = K \circ F_n(s)$ we obtain

$$\zeta_1 = k_0, \quad \zeta_2 = K(\zeta_1), \quad \zeta_3 = K(\zeta_2), \ldots. \tag{3.8}$$

Also

$$P\{T < \infty|X_0 = i\} = \zeta^i, \tag{3.9}$$

where $\zeta = \lim_{n\to\infty} \zeta_n$ satisfies the equation $\zeta = K(\zeta)$.

Theorem 3.2 (i) *The p.g.f. of the total progeny of a branching process with i (≥ 1) ancestors is given by $\xi(s)^i$, where $\xi = \xi(s)$ is the unique continuous solution of $\xi = sK(\xi)$ with $\xi(0+) = 0$.*

(ii) *The distribution of Z is given by $\{g_n\}$, where*

$$g_n = \frac{i}{n}P_{n,n-i} \tag{3.10}$$

and P_{ij} is the coefficient of s^j in the expansion of $K(s)^i$ as a power series.

(iii) *If $\mu \leq 1$, then*

$$E(Z) = (1-\mu)^{-1} \text{ if } \mu < 1, \quad \text{and} \quad = \infty \text{ if } \mu = 1. \tag{3.11}$$

$$\operatorname{Var}(Z) = \sigma^2(1-\mu)^3 \text{ if } \mu < 1. \tag{3.12}$$

Proof. (i) Since $Z_n \to Z$ with probability 1, $\xi_n(s) \to \xi(s)$, where $\xi(s) = E(s^Z)$. Letting $n \to \infty$ in $\xi_{n+1}(s) = sK_0\xi_n(s)$ we obtain $\xi = sK(\xi)$.

(ii) The probability g_n is the coefficient of s^n in the p.g.f. $\xi(s)^i$. Now from Lagrange's expansion of $\xi(s)^i$ we find that

$$\begin{aligned}
\xi^i(s) &= \sum_{n=1}^{\infty} \frac{s^n}{n!}\left(\frac{d}{dx}\right)^{n-1} \left[ix^{i-1}K(x)^n\right]_{x=0} \\
&= \sum_{n=1}^{\infty} \frac{s^n}{n!}\left(\frac{d}{dx}\right)^{n-1} \left[i\sum_{j=0}^{\infty} P_{nj}x^{j+i-1}\right]_{x=0} \\
&= \sum_{n=i}^{\infty} \frac{i}{n} P_{n,n-i}s^n.
\end{aligned}$$

This gives the required expansion for g_n.

(iii) From $\xi = sK(\xi)$ we obtain

$$\begin{aligned}
\xi'(s) &= K(\xi) + sK'(\xi)\xi'(s) \\
\xi''(s) &= 2K'(\xi)\xi'(s) + sK''(\xi)\xi'(s)^2 + sK'(\xi)\xi''(s).
\end{aligned}$$

Letting $s \to 1-$ in these we obtain

$$\xi'(1-) = (1-\mu)^{-1}, \quad \xi''(1-) = \frac{\sigma^2}{(1-\mu)^3} + \frac{\mu}{(1-\mu)^2}$$

since $\xi(1-) = 1$ if $\mu \leq 1$. The desired results now follow from the fact that $\xi'(1-) = E(Z)$ and $\xi''(1-) = \operatorname{Var}(Z) + [E(Z)]^2 - E(Z)$. □

Example 3.1 In the binary splitting model of Example 1.1, assume that $0 < p < 1$. The equation $K(\zeta) = \zeta$ reduces to

$$p\zeta^2 - \zeta + q = 0,$$

which has two roots $1, q/p$. By Theorem 2.1 the probability of extinction is given by

$$\begin{aligned} \zeta &= 1 \quad \text{if } 0 < p \leq 1/2 \\ &= \frac{q}{p} \quad \text{if } \frac{1}{2} < p < 1. \end{aligned} \tag{3.13}$$

From Theorem 3.2(ii) we find that the distribution of the total progeny is given by

$$g_n = \frac{i}{n} P_{n,n-i} = \frac{i}{n} \binom{n}{\frac{n-i}{2}} p^{\frac{n-i}{2}} q^{\frac{n+i}{2}}.$$

This can be written as

$$g_n = \frac{i}{i+2m} \binom{i+2m}{m} p^m q^{i+m} \tag{3.14}$$

for $n = i + 2m$ $(m = 0, 1, 2, \ldots)$, and $g_n = 0$ otherwise.

8.4 The Supercritical Case

Theorem 2.1 describes the asymptotic behavior of the process $\{X_n\}$ for large n. To obtain a more precise idea as to the behavior of X_n (relative to its mean μ^n) we now consider the scaled random variable $M_n = X_n/\mu^n$. If $\mu \leq 1$,

$$P\{M_n = 0\} = P\left\{\frac{X_n}{\mu_n} = 0\right\} = P\{X_n = 0\} \to 1,$$

so that the limit of M_n exists and is 0 with probability 1. If $\mu > 1$ (the supercritical case), $P\{M_n = 0\} \to \zeta$ where $0 < \zeta < 1$. We shall prove in this case that $M_n \to M$ with probability 1, where M is a random variable. We assume that $X_0 = 1$ and the offspring distribution has a finite variance σ^2.

Theorem 4.1 *If $\mu > 1$, $\sigma^2 < \infty$, then as $n \to \infty$, $M_n \to M$ with probability 1, where M is a random such that*

$$E(M) = 1, \quad \mathrm{Var}(M) = \frac{\sigma^2}{\mu(\mu-1)} > 0. \tag{4.1}$$

Also, the Laplace transform $\phi(\theta) = E(e^{-\theta M})$ of the distribution of M satisfies the equation

$$\phi(\theta) = K \circ \phi\left(\frac{\theta}{\mu}\right) \tag{4.2}$$

and is in fact its unique solution among Laplace transforms of distributions on $[0, \infty)$ with mean 1.

Proof. By Theorem 1.3, $\{M_n\}$ is a martingale relative to its own sigma-field. Convergence follows from the fact that it is a non-negative martingale (Theorem 8.1 of Chap. 7). However, this does not tell us anything about the limit random variable M. We have therefore to consider another property of the process. From Theorem 1.3 we see that for $n_1 < n_2$,

$$\begin{aligned} E(M_{n_1} - M_{n_2})^2 &= \mathrm{Var}(M_{n_1} - M_{n_2}) \\ &= \sigma^2 \frac{1-\mu^{-n_1}}{\mu(\mu-1)} + \sigma^2 \frac{1-\mu^{-n_2}}{\mu(\mu-1)} - 2\frac{\sigma^2}{\mu^{n_1+n_2}}\mu^{n_2-1}\frac{1-\mu^{-n_1}}{1-\mu} \\ &= \sigma^2 \frac{\mu^{-n_1} - \mu^{-n_2}}{\mu(\mu-1)} \to 0 \end{aligned} \tag{4.3}$$

as $n_1, n_2 \to \infty$. Therefore, $\{M_n\}$ converges to a limit in the mean square (m.s.). This limit is of course M. Since $\sigma^2 < \infty$, Eq. (1.8) gives $E(M_n^2) < \infty$ for each n. The results concerning $E(M)$ and $\mathrm{Var}(M)$ follow from properties of the m.s. convergence (see Sec. 4.5 of Chap. 4), since

$$E(M_n) = 1, \quad \mathrm{Var}(M_n) = \sigma^2 \frac{1-\mu^{-n}}{\mu(\mu-1)} \to \frac{\alpha^2}{\mu(\mu-1)}. \tag{4.4}$$

To obtain the Laplace transform $\phi(\theta)$, we denote $\phi_n(\theta) = E(e^{-\theta M_n}) = F_n(e^{-\theta/\mu^n})$, where $F_n(s)$ is the p.g.f. of X_n. From (1.4) we have

$F_{n+1}(s) = K \circ F_n(s)$. Putting $s = e^{-\theta/\mu^n}$ in this we obtain $\phi_{n+1}(\mu\theta) = K \circ \phi_n(\theta)$, or

$$\phi_{n+1}(\theta) = K \circ \phi_n\left(\frac{\theta}{\mu}\right) \quad (n \geq 0). \tag{4.5}$$

Since $M_n \to M$, $\phi_n(\theta) \to E(e^{\theta M}) = \phi(\theta)$. Since K is continuous, (4.5) gives (4.2) in the limit as $n \to \infty$. Since $E(M) = 1$ we must have

$$\frac{1 - \phi(\theta)}{\theta} \to 1 \quad \text{as } \theta \to 0+\,. \tag{4.6}$$

Thus $\phi(\theta)$ obtained as the limit of $\phi_n(\theta)$ in (4.5) is a solution of (4.2). To prove the uniqueness of this solution, let $\psi(\theta)$ be a second solution which is the Laplace transform of a distribution on $[0, \infty)$ satisfying the condition

$$\frac{1 - \psi(\theta)}{\theta} \to 1 \quad \text{as } \theta \to 0+\,. \tag{4.7}$$

Since $K'(x) \leq K'(1) = \mu$, we have

$$|\phi(\theta) - \psi(\theta)| = \left|K \circ \phi\left(\frac{\theta}{\mu}\right) - K \circ \psi\left(\frac{\theta}{\mu}\right)\right| \leq \mu\left|\phi\left(\frac{\theta}{\mu}\right) - \psi\left(\frac{\theta}{\mu}\right)\right|.$$

Using this repeatedly we obtain

$$|\phi(\theta) - \psi(\theta)| \leq \mu^n\left|\phi\left(\frac{\theta}{\mu^n}\right) - \psi\left(\frac{\theta}{\mu^n}\right)\right| \to 0$$

as $n \to \infty$, on account of (4.6) and (4.7). Thus $\phi(\theta) = \psi(\theta)$, as was required to be proved. □

Example 4.1 Suppose that the offspring distribution is the modified geometric (2.1) of Example (2.1), with mean $\mu = (1 - a)/p$. We assume that $\mu > 1$, so that $1 - a > p$ or $q > a$. From (2.4) it follows that the extinction probability is given by $\zeta = a/q$. The mean and variance of the limit random variable M is given by

$$E(M) = 1, \quad \text{Var}(M) = \frac{1 + \zeta}{1 - \zeta} > 0. \tag{4.8}$$

It is difficult to solve the Eq. (4.2) for the Laplace transform $\phi(\theta) = E(e^{-\theta M})$, but we make a guess and claim the distribution function of M to be

$$G(x) = 1 - (1 - \zeta)e^{-(1-\zeta)x} \quad (x \geq 0). \tag{4.9}$$

This has density $(1-\zeta)^2 e^{-(1-\zeta)x}$ in $(0,\infty)$ and a probability ζ at the origin. Its Laplace transform is given by

$$\phi(\theta) = \zeta + \int_0^\infty e^{-\theta x}(1-\zeta)^2 e^{-(1-\zeta)^x}\,dx = \frac{\theta\zeta + 1 - \zeta}{\theta + 1 - \zeta} \quad (\theta > 0). \qquad (4.10)$$

To verify that this expression satisfies (4.2) we substitute it in (4.2) and find that its right side becomes

$$\frac{a + (p-a)\phi\left(\frac{p\theta}{(1-a)}\right)}{1 - q\phi\left(\frac{p\theta}{(1-a)}\right)} = \frac{[a + (p-a)\zeta]\theta + (1-a)(1-\zeta)}{(1-q\zeta)\theta + (1-a)(1-\zeta)}. \qquad (4.11)$$

From $K(\zeta) = \zeta$ we obtain

$$a + (p-a)\zeta = (1-q\zeta)\zeta = (1-a)\zeta$$

and so the second expression in (4.11) reduces to $\phi(\theta)$. Thus $\phi(\theta)$ given by (4.10) is a solution of (4.2) and Theorem 4.1 guarantees that it is indeed the Laplace transform $E(e^{\theta M})$.

8.5 Estimation

We shall now consider the problem of estimating the offspring distribution $\{k_j\}$ and its mean μ. Suppose we have observed the initial generation and N generations that descended from it. For each generation we classify the data according to the number of offsprings of individuals of that generation. Thus, let x_{nj} be the number of individuals of the nth generation who have j offsprings ($j = 0, 1, 2, \ldots,\ n = 0, 1, \ldots, N$). Clearly,

$$\sum_{j=0}^{\infty} x_{nj} = x_n, \quad \sum_{j=0}^{\infty} j x_{nj} = x_{n+1} \quad (n = 0, 1, \ldots, N), \qquad (5.1)$$

where x_n is the total observed number of individuals in the nth generation. Also,

$$w_j = \sum_{n=0}^{N} x_{nj} \quad (j = 0, 1, 2, \ldots), \qquad (5.2)$$

where w_j is the total number of observed individuals with j offsprings. Let

$$z_i = x_0 + x_1 + \cdots + x_i \quad (0 \le i \le N+1). \qquad (5.3)$$

Table 8.2.

Generation n	No. of individuals x_{nj} with j offsprings (j) 0	1	2	·	·	·	·	·	Total
0	x_{00}	x_{01}	x_{02}	·	·	·	·	·	x_0
1	x_{10}	x_{11}	x_{12}	·	·	·	·	·	x_1
·	·	·	·	·	·	·	·	·	·
·	·	·	·	·	·	·	·	·	·
n	x_{n0}	x_{n1}	x_{n2}	·	·	·	·	·	x_n
·	·	·	·	·	·	·	·	·	·
·	·	·	·	·	·	·	·	·	·
N	x_{N0}	x_{N1}	x_{N2}	·	·	·	·	·	x_N
Total	w_0	w_1	w_2	·	·	·	·	·	z_N

Then the total progeny observed is z_N, but our data also contain information concerning z_{N+1}. The data may be exhibited as in Table 8.2. We have the following.

Theorem 5.1 *The maximum likelihood estimators of the offspring distribution* $\{k_j\}$ *and its mean* μ, *based on the data of Table* 8.2, *are respectively given by*

$$\hat{k}_j = \frac{w_j}{z_N} \quad (j = 0, 1, 2, \ldots), \quad \hat{\mu} = \frac{z_{N+1} - z_0}{z_N}. \tag{5.4}$$

Proof. The offspring produced in each generation may be viewed as the result of multinomial trials with probability k_0 for no offsprings, probability k_1 for exactly one offspring, etc. Therefore, the conditional probability of obtaining $x_{n0}, x_{n1}, \ldots$ such individuals, given the total size x_n of the nth generation equals

$$\frac{x_n!}{x_{n0}!x_{n1}!\cdots} k_0^{x_{n0}} k_1^{x_{n1}} \cdots$$

The likelihood function of the observed data is therefore given by

$$\begin{aligned} L &= \prod_{n=0}^{N} \frac{x_n!}{x_{n0}!x_{n1}!\cdots} k_0^{x_{n0}} k_1^{x_{n1}} \cdots \\ &= c_1 k_0^{w_0} k_1^{w_1} \cdots \end{aligned} \tag{5.5}$$

where c_1 is a function of the observations, but not depending on $k_0, k_1, \ldots$. We have

$$\log L = c_2 + \sum_{j=0}^{\infty} w_j \log k_j \quad (c_2 = \log c_1)$$

which can be written as

$$\log L = c_2 + z_N \sum_{j=0}^{\infty} a_j \log k_j \tag{5.6}$$

with

$$a_j = \frac{w_j}{z_N} \ (j \geq 0), \quad a_j \geq 0, \quad \sum_0^{\infty} a_j = 1. \tag{5.7}$$

We need to maximize $\log L$ in (5.6) with respect to the probability distribution $\{k_j\}$ that is, subject to the constraints $k_j \geq 0$, $k_0 + k_1 + k_2 + \cdots = 1$. In order to do this we define a random variable X such that

$$P\left\{X = \frac{k_j}{a_j}\right\} = a_j \quad (j \geq 0). \tag{5.8}$$

For this we have

$$E(X) = \sum_0^{\infty} a_j \frac{k_j}{a_j} = 1 \tag{5.9}$$

$$E(\log X) = \sum_0^{\infty} a_j \log \frac{k_j}{a_j} = \sum_0^{\infty} a_j \log k_j - \sum_0^{\infty} a_j \log a_j. \tag{5.10}$$

Since $\log x$ is a convex function, Jensen's inequality gives

$$E(\log X) \leq \log E(X) = \log 1 = 0$$

or, in view of (5.10),

$$\sum_0^{\infty} a_j \log k_j \leq \sum_0^{\infty} a_j \ \log a_j. \tag{5.11}$$

This shows that the expression $\sum_0^{\infty} a_j \log k_j$ is a maximum with respect to $\{k_j\}$ when $k_j = a_j = w_j/z_N (j \geq 0)$. The maximum likelihood estimators of k_j are thus given by (5.4) as desired. The maximum likelihood estimator of μ is given by

$$\hat{\mu} = \sum_0^\infty j\hat{k}_j = \sum_0^\infty j\frac{w_j}{z_N}.$$

Using (5.1)–(5.3) we find that

$$\begin{aligned}\hat{\mu} &= \frac{1}{z_N}\sum_0^\infty j\sum_0^N x_{nj} = \frac{1}{z_N}\sum_0^N x_{n+1} \\ &= \frac{1}{z_N}(z_{N+1} - z_0).\end{aligned}$$

□

Example 5.1 In the binary splitting model of Example 1.1, of the particles of the nth generation, x_{n2} split into two and the remaining x_{n0} particles remain in their current state. We have

$$x_{n0} + x_{n2} = x_n \quad \text{and} \quad 2x_{n2} = x_{n+1}$$

and

$$w_0 = z_N - \frac{1}{2}(z_{N+1} - z_0), \quad w_2 = \frac{1}{2}(z_{N+1} - z_0).$$

Therefore, we need only the observation $(x_0, x_1, \ldots, x_N)$. The maximum likelihood estimator of p is given by

$$\hat{p} = \frac{z_{N+1} - z_0}{2z_N}.$$

Example 5.2 This example is not without its amusement value, but is relevant, because the theory of branching processes had its origin in the problem of extinction of noble families. Table 8.3 is the genealogical table of the Bernoulli family, whose members have made fundamental contributions to probability theory and other branches of science. (Bernoulli trials are named after Jacob I, and Nicolaus I and Daniel tried to resolve the St. Petersburg paradox.)

The data of Table 8.3 can be arranged in the from of Table 8.2.

Table 8.4 gives the estimates of k_j as

$$\hat{k}_0 = \frac{7}{11}, \quad \hat{k}_1 = \hat{k}_2 = \hat{k}_3 = \hat{k}_4 = \frac{1}{11}$$

Table 8.3.

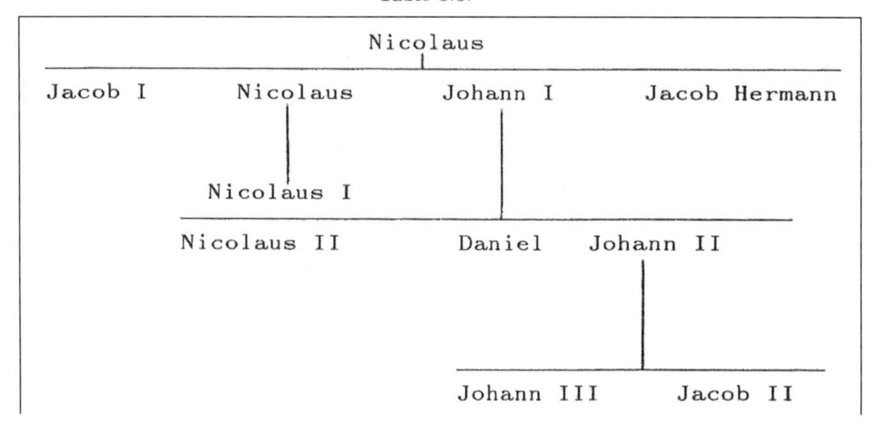

Table 8.4.

Generation	No. of individuals x_{nj}, j = 0	1	2	3	4	Total
0	0	0	0	0	1	1
1	2	1	0	1	0	4
2	3	0	1	0	0	4
3	2	0	0	0	0	2
Total	7	1	1	1	1	11

and the estimate of the mean μ as

$$\hat{\mu} = \frac{11-1}{11} = \frac{10}{11} = 0.91 < 1.$$

As this estimate $\hat{\mu}$ indicates, extinction is a sure event, as was evidently the case, since the Bernoulli dynasty came to an abrupt end with the drowning death in 1789 of Jacob II.

8.6 Problems for Solution

1. If $\{X_n\}$ is a branching process and k a positive integer, show that $\{X_{nk}, n=0,1,2,\ldots\}$ is also a branching process whose offspring distribution has generating function $F_k(s)$.

2. Let $\phi_n(s_1, s_2, \ldots, s_n)$ denote the joint p.g.f. of $X_1, X_2, \ldots, X_n$, given $X_0 = 1$. Show that

$$\phi_{n+1}(s_1, s_2, \ldots, s_{n+1}) = \phi_n(s_1, s_2, \ldots, s_{n-1}, s_n K(s_{n+1})).$$

3. Let $\{Y_n, n \geq 1\}$ be a sequence of independent and identically distributed random variables with a finite mean μ and finite variance σ^2. Also, let N be a non-negative integer-valued random variable which is independent of $\{Y_n\}$ and has finite mean and variance. Finally, let

$$X = Y_1 + Y_2 + \cdots + Y_N.$$

Show that

$$E(X) = \mu E(N)$$
$$\mathrm{Var}(X) = \sigma^2 E(N) + \mu^2 \ \mathrm{Var}(N).$$

4. Use the results of problem 3 to prove Theorem 1.2.
5. For the total progeny Z_n up to and including the nth generation show that

$$E(Z_n) = \frac{1-\mu^{n+1}}{1-\mu} \quad \text{if } \mu \neq 1$$
$$= n+1 \quad \text{if } \mu = 1$$

and

$$\mathrm{Var}(Z_n) = \frac{\sigma^2}{(1-\mu)^3}[1 - \mu^{2n+1} - (2n+1)\mu^n(1-\mu)] \quad \text{if } \mu \neq 1$$
$$= \frac{\sigma^2}{6}(2n+1)(n+1)n \quad \text{if } \mu = 1.$$

6. For the branching process with binary splitting (Example 1.1) prove:
(i) The joint distribution of $X_1, X_2, \ldots, X_n$, given $X_0 = x_0$ is given by

$$P\{X_1 = x_1, X_2 = x_2, \ldots, X_n = x_n | X_0 = x_0\}$$
$$= p^{(z_n - z_0)/2} q^{z_{n-1}} - \frac{z_n - z_0}{2} \prod_{r=1}^{n} \binom{x_{r-1}}{x_{r/2}},$$

where $z_i = x_0 + x_1 + \cdots + x_i \ (0 \leq i \leq n)$.
(ii) Without using Therorem 5.1 show that the maximum likelihood estimator of p based on the observations $(x_0, x_1, \ldots, x_n)$ of

$X_0, X_1, \ldots, X_n$ is given by

$$\hat{p} = \frac{z_n - z_0}{2z_{n-1}},$$

and find the asymptotic variance of this estimate.

7. If each individual produces at most one successor, so that $k_0 = q, k_1 = p$, where $0 < p < 1$ and $q = 1 - p$, prove:

   (i) $P\{X_n = 0 | X_0 = 1\} = 1 - p^n \ (n \geq 1)$.

   (ii) The extinction time T has the distribution

$$P\{T = n | X_0 = i\} = \sum_{j=1}^{n} \binom{i}{j} (-1)^{j-1} (1 - p^j)(p^j)^{n-1} \quad (n \geq 1).$$

   (iii) The distribution of the progeny Z_n up to and including the nth generation is given by

$$\begin{aligned} P\{Z_n = j | X_0 = 1\} &= qp^{j-1} \quad && (1 \leq j \leq n) \\ &= p^n && (j = n + 1). \end{aligned}$$

8. Suppose that the offspring distribution is geometric, namely, $k_j = pq^j \ (j = 0, 1, 2, \ldots)$ and that there were $i \ (\geq 1)$ ancestors. Prove:
   (i) The probability of extinction is given by

$$\zeta = \min\left\{1, \left(\frac{p}{q}\right)^i\right\}.$$

   (ii) The distribution of the total progeny is given by $\{g_n\}$, where

$$g_n = \frac{i}{n} \binom{2n - i - 1}{n - i} p^n q^{n-i} \quad (n \geq 1).$$

9. For a branching process with one ancestor and the Poisson offspring distribution $k_j = e^{-\lambda} \lambda^j / j! \ (j = 0, 1, 2, \ldots)$, show that the distribution of the total progeny is given by

$$g_n = e^{-n\lambda} \frac{(n\lambda)^{n-1}}{n!} \sim (2\pi)^{-1/2} e^{-n(\lambda - 1)} n^{-3/2} \lambda^{n-1} \quad (n \to \infty).$$

Also, show that if $\lambda > 1$, the probability of ultimate extinction is given by $\zeta \sim 1 - 2(\lambda - 1)\lambda^{-2}$.

10. If the offspring distribution has the generating function $K(s) = (\alpha + \beta s)/(\gamma + \delta s)$ with $B\gamma - \alpha\delta \neq 0$, show that the p.g.f. of X_n is given by

$$F_n(s) = \frac{s_0(1-\mu^n) - (1-\mu^n s_0)s}{s_0 - \mu^n - (1-\mu^n)s} \quad \text{if } \mu \neq 1$$
$$= \frac{n\delta - (\gamma + \delta + n\delta)s}{n\delta - \delta - \gamma - n\delta s} \quad \text{if } \mu = 1,$$

where $\mu = K'(1)$ and $s_0 = (-\alpha)\delta$. Also show that the probability of ultimate extinction is 1 if $\mu \leq 1$ and is s_0 if $\mu > 1$. (The branching processes of Example 2.1 and problem 8 are special cases of the present one.)

11. For the branching process of problem 10 show that the distribution of extinction time T is given by the following:
(i) Let $\mu \neq 1$. Then

$$P\{T = n | X_0 = 1\} = E(1-Q)Q^{n-1} \quad (n \geq 1),$$

where Q is a random variable with the distribution

$$P\{Q = \mu^j\} = \left(1 - \frac{1}{s_0}\right)\left(\frac{1}{s_0}\right)^{j-1} \quad \text{if } \mu < 1,$$
$$P\{Q = \mu^{-j}\} = (1 - s_0)s_0^j \quad \text{if } \mu > 1$$

for $j \geq 1$.
(ii) Let $\mu = 1$. Then

$$P\{T = n | X_0 = 1\} = \frac{(-\delta)(\gamma + \delta)}{[(n-1)\delta - \gamma][(n-2)\delta - \gamma]} \quad (n \geq 1).$$

12. For the branching process of problem 10, prove the following:
(i) If $\mu > 1$, then $\lim_{n\to\infty} P\left\{\frac{X_n}{\mu^n} \leq x\right\} = 1 - (1 - s_0)e^{(1-s_0)x}$.
(ii) If $\mu = 1$, then $\lim_{n\to\infty} P\left\{\frac{X_n}{n} \leq x | X_n > 0\right\} = 1 - e^{(1+\frac{\gamma}{\delta})x}$.
(iii) If $\mu < 1$, then $\lim_{n\to\infty} P\{X_n = j | X_n > 0\}$

$$= \left(1 - \frac{1}{s_0}\right)\left(\frac{1}{s_0}\right)^{j-1} \quad (j \geq 1).$$

13. *Branching processes with immigration.* Suppose there also exists some immigration into the population during each generation, described

by the p.g.f. $h(s)$. Show that the p.g.f. $G_n(s)$ of X_n satisfies the relation

$$G_{n+1}(s) = h(s)G_n(K(s)) \quad (n \geq 0).$$

Investigate the limit distribution of X_n as $n \to \infty$.

14. In the branching process of problem 7, there is an immigration during each generation, described by the p.g.f. $h(s) = e^{-\lambda+\lambda s}$. Show that as $n \to \infty$, x_n has the limit distribution given by

$$e^{-\lambda/q}(\lambda/q)^j/j! \quad (j = 0, 1, 2, \ldots).$$

15. *Extension of the lemma on branching processes.* If the p.g.f. $K(s)$ exists for all $s \geq 0$, and $K(s)/s \to \infty$, then Lemma 2.1 can be extended as follows: Apart from the obvious root $s = 1$, the equation $K(s) = s$ has a second root ζ such that $0 < \zeta < 1$ if $\mu > 1$, and $1 < \zeta < \infty$ if $\mu < 1$. These two roots coincide, that is $\zeta = 1$, if and only if $\mu = 1$.
16. In a model proposed for the spread of an infectious disease in a closed homogeneously mixing population, the following assumptions are made. During a time-interval of length h (i) adequate contact between any two specific individuals occurs with probability $\lambda h + o(h)$ and (ii) a specific infective is removed from circulation (by isolation, death, or recovery) with probability $\mu h + o(h)$. In order to study the total number of susceptibles eventually infected, the following branching process is formulated. Identify the susceptibles infected by a specific infective as his descendants (offspring). The susceptibles infected by the infectives of any generation constitute the next generation. Initially there are a (> 0) infectives. Let X_t denote the number of infectives who belong to the tth generation $(t \geq 0)$. The total number of susceptibles eventually infected is identical with the total progeny (excluding ancestors). Prove the following:

    (a) The distribution of the number of susceptibles eventually infected is given by

    $$P_w(\lambda, \mu) = \frac{a(2w + a - 1)!}{w!(a + w)!} \cdot \frac{\lambda^w \mu^{a+w}}{(\lambda + \mu)^{a+2w}} \quad (w = 0, 1, 2, \ldots).$$

    (b) The number of susceptibles eventually infected is finite with probability ζ^a where $\zeta = \min(1, \rho)$, ρ being the removal rate μ/λ.

Further Reading

Asmussen, S and Hering, H (1983). *Branching Processes.* Boston: Birkhauser.

Athreya, KB and Ney, PE (1972). *Branching Processes.* New York: Springer-Verlag.

Gutlorp, P (1991). *Statistical Inference for Branching Processes.* New York: John Wiley.

Harris, TE (1963). *Branching Processes.* New York: Springer-Verlag.

Heyde, CC, Aeydo, CC and Senata, E (1977), IJ Bienaymé: *Statistical Theory Anticipated.* New York: Springer-Verlag.

Jagers, P (1975). *Branching Processes with Biological Applications.* New York: John Wiley.

Sankaranarayanan, G (1989). *Branching Processes and Its Estimation Theory.* New York: John Wiley.

CHAPTER 9
REGENERATIVE PHENOMENA

9.1 Introduction

Let the index set T be either $[0, \infty)$ or $\{0, 1, 2, \ldots\}$, and (Ω, F, P) a probability space

Definition 1.1 A regenerative phenomenon $Z = \{Z(t), t \in T\}$ on a probability space (Ω, F, P) is a stochastic process taking values 0 or 1 and such that for $0 = t_0 < t_1 < \cdots < t_r$ $(r \geq 1)$ we have

$$\begin{aligned} P\{Z(t_1) &= Z(t_2) = \cdots = Z(t_r) = 1\} \\ &= P\{Z(t_1) = 1\} P\{Z(t_2 - t_1) = 1\} \cdots P\{Z(t_r - t_{r-1}) = 1\}. \end{aligned} \tag{1.1}$$

This relation determines all finite-dimensional distributions (Theorem 1.1). The term *regenerative* is justified by Theorem 1.2.

Theorem 1.1 *The property* (1.1) *determines all finite-dimensional distributions of* Z.

Proof. For $\alpha_r \in \{0, 1\}$ $(1 \leq r \leq n)$ consider the event

$$A = \{Z(t_1) = \alpha_1, \quad Z(t_2) = \alpha_2, \ldots, Z(t_n) = \alpha_n\}.$$

Its indicator function can be written as

$$\begin{aligned} 1_A &= \prod_{r=1}^{n} (-1)^{\alpha_r + 1} [Z(t_r) + \alpha_r - 1] \\ &= \sum c(k_1, k_2, \ldots, k_r) Z(t_{k_1}) Z(t_{k_2}) \cdots Z(t_{k_r}), \end{aligned}$$

where the sum is taken over $0 < k_1 < k_2 < \cdots < k_r < n$ $(r \geq 1)$ and $c(k_1, k_2, \ldots, k_r)$ is a constant. Therefore,

$$P(A) = E(1_A) = \sum c(k_1, k_2, \ldots, k_r) P\{Z(t_{k_1}) = Z(t_{k_2}) = \cdots = Z(t_{k_r}) = 1\}$$

and this last probability is determined by (1.1). □

Theorem 1.2 *For $\tau > 0$ let F_τ^- be the smallest σ-field with respect to which $Z(t)$ is measurable for all $t < \tau$, and F_τ^+ the smallest σ-field with respect to which $Z(t)$ is measurable for all $t > \tau$. Let $A \in F_\tau^-$ and $B \in F_\tau^+$. Then we have*

$$P\{A \cap [Z(\tau) = 1] \cap B\} = P\{A \cap [Z(\tau) = 1]\} P\{B'\}, \qquad (1.2)$$

where B' is obtained from B by the shift $u \to u - \tau$.

Proof. Let $0 = t_0 < t_1 < \cdots < t_m < \tau < u_1 < u_2 < \cdots < u_n$ and $\alpha_r, \beta_s \in \{0, 1\}$. Then we shall prove that

$$\begin{aligned} P\{Z(t_r) &= \alpha_r (1 \leq r \leq m), Z(\tau) = 1, Z(u_s) = \beta_s (1 \leq s \leq n)\} \\ &= P\{Z(t_r) = \alpha_r (1 \leq r \leq m), Z(\tau) = 1\} \\ &\quad \cdot P\{Z(u_s - \tau) = \beta_s (1 \leq s \leq n)\}. \end{aligned} \qquad (1.3)$$

The proof is then completed by using the Kolmogorov extension theorem. To prove (1.3), we note that the indicator of the event C on its left side can be expressed as

$$\begin{aligned} 1_C &= \prod_{r=1}^{m} (-1)^{\alpha_r + 1} [Z(t_r) + \alpha_r - 1] Z(\tau) \prod_{s=1}^{n} (-1)^{\beta_s + 1} [Z(u_s) + \beta_s - 1] \\ &= \sum c(k_1, k_2, \ldots, k_r) Z(t_{k_1}) Z(t_{k_2}) \ldots Z(t_{k_r}) Z(\tau) \\ &\quad \times \sum d(\ell_1, \ell_2, \ldots, \ell_s) Z(u_{\ell_1}) Z(u_{\ell_2}) \ldots Z(u_{\ell_s}) \\ &= \sum \sum c(k_1, k_2, \ldots, k_r) d(\ell_1, \ell_2, \ldots, \ell_s) \\ &\quad \times Z(t_{k_1}) Z(t_{k_2}) \ldots Z(t_{k_r}) Z(\tau) Z(u_{\ell_1}) \ldots Z(u_{\ell_s}). \end{aligned}$$

Therefore,

$$\begin{aligned}P(A) = E(1_A) &= \sum\sum c(k_1, k_2, \ldots, k_r)d(\ell_1, \ell_2, \ldots, \ell_s)\\ &\quad\times P\{Z(t_{k_1} = 1)\}P\{Z(t_{k_2} - t_{k_1}) = 1\}\cdots P\{Z(\tau - t_{k_r}) = 1\}\\ &\quad\times P\{Z(u_{\ell_1} - \tau) = 1\}P\{Z(u_{\ell_2} - u_{\ell_1}) = 1\}\cdots P\{Z(u_{\ell_s} - u_{\ell_{s-1}}) = 1\}\\ &= E\prod_{r=1}^{m}(-1)^{\alpha_r+1}[Z(t_r) + \alpha_r - 1]Z(\tau)\\ &\quad\cdot E\prod_{s=1}^{n}(-1)^{\beta_s+1}[Z(u_s - \tau) + \beta_s - 1]\\ &= P\{Z(t_r) = \alpha_r(1 \le r \le m),\ z(\tau) = 1\}\\ &\quad\cdot P\{Z(u_s - \tau) = \beta_s(1 \le s \le m)\}\end{aligned}$$

as required. □

According as $T = \{0, 1, 2, \ldots\}$ or $T = [0, \infty)$ we shall denote the phenomenon as

$$Z = \{Z_n, n \ge 0\} \quad \text{or} \quad Z = \{Z(t), t \ge 0\}. \tag{1.4}$$

The discrete time phenomenon is sometimes called a *recurrent phenomenon.* In this case, let

$$u_n = P\{Z_n = 1\}. \tag{1.5}$$

We shall take $Z_0 = 1$ a.s., so that $u_0 = 1$. The class of sequences $u = \{u_n, n \ge 0\}$ that arises from a recurrent phenomenon will be denoted by R

In the continuous time case, let

$$p(t) = P\{Z(t) = 1\}; \tag{1.6}$$

this is called the p-function of the regenerative phenomenon Z.

Definition 1.2 The continuous time regenerative phenomenon (or its p-function) is standard if

$$p(t) \to 1 \quad \text{as} \quad \mathrm{t} \to 0+\,. \tag{1.7}$$

For a standard phenomenon we shall define $Z(0) = 1$ a.s., so that $p(0) = 1$. The class of standard p-functions will be denoted by P.

In Secs. 9.2–9.7 we discuss recurrent phenomena and in the remaining sections continuous time regenerative phenomena.

9.2 Discrete Time Regenerative Phenomena

Definition 2.1 Let $Z = \{Z_n, n \geq 0\}$ be a recurrent phenomenon. The set

$$\zeta = \{n \geq 0; Z_n = 1\} \tag{2.1}$$

is called the recurrence set associated with Z.

Definition 2.2 Let $T_0 = 0$ and for $k \geq 1$,

$$T_k = \min\{n > T_{k-1} \colon Z_n = 1\}. \tag{2.2}$$

T_k are called the successive recurrence times of Z. The distribution $\{f_n, n \geq 1\}$ of T_1 is called the (first) recurrence time distribution.

In view of Theorem 1.2, the random variables $T_k - T_{k-1}$ $(k \geq 1)$ are mutually independent and have a common distribution $\{f_n\}$. Thus $\{T_k, k \geq 0\}$ is a renewal process with lifetime distribution $\{f_n\}$. We can write

$$\zeta = \{T_0, T_1, T_2, \ldots\} \tag{2.3}$$

and

$$u_n = P\{n \in \zeta\} = \sum_{k=1}^{\infty} P\{T_k = n\} \quad (n \geq 1). \tag{2.4}$$

We have the following result.

Theorem 2.1 *Let $\{f_n\}$ be the recurrence time distribution of a phenomenon Z, and $u_n = P\{Z_n = 1\}$. Then we have the recurrence relation*

$$u_n = \sum_{m=1}^{n} f_m u_{n-m} \ (n \geq 1), \quad \text{with } u_0 = 1. \tag{2.5}$$

For a given $\{f_n\}$, (2.5) yields a unique solution $\{u_n\}$ with $0 < u_n < 1$ $(n \geq 1)$.

Proof. The result (2.5) follows from Eq. (2.29) of Chap. 6, or could be derived directly from (2.4). To show uniqueness, let

$$F(s) = \sum_{1}^{\infty} f_n s^n, \quad U(s) = \sum_{0}^{\infty} u_n s^n \ (0 < s < 1). \tag{2.6}$$

Then, (2.5) gives

$$U(s) = \frac{1}{1 - F(s)} \tag{2.7}$$

and thus $F(s)$ determines $U(s)$ and hence $\{u_n\}$ uniquely. □

We have thus seen that a recurrent phenomenon gives rise to a renewal process with lifetime distribution concentrated on the set $\{1, 2, \ldots\}$. The following theorem shows that this is the only way that a recurrent phenomenon can occur.

Theorem 2.2 (i) *Let $\{T_k, k \geq 0\}$ be a renewal process induced by a discrete distribution $\{f_n, n \geq 1\}, R = \{n \geq 0; T_k = n$ for some $k \geq 0\}$ its range, and $Z'_n = 1_{\{n \in R\}}$ $(n \geq 0)$. Then $Z' = \{Z'_n,\ n \geq 0\}$ is a recurrent phenomenon.*

(ii) *Conversely, any recurrent phenomenon Z is equivalent to a phenomenon Z' generated in the above manner in the sense that Z and Z' have the same $\{u_n\}$ sequence.*

Proof. (i) Let $v_0 = 1$ and for $n \geq 1$,

$$v_n = P\{n \in R\} = \sum_{k=1}^{\infty} P\{T_k = n\}. \tag{2.8}$$

For $0 = n_0 < n_1 < \cdots < n_r$ $(r \geq 1)$ we have

$$\begin{aligned} &P\{Z'_{n_1} = Z'_{n_2} = \cdots = Z'_{n_r} = 1\} \\ &\quad = P\left\{\bigcup_{k_1}\{T_{k_1} = n_1\}, \bigcup_{k_2}\{T_{k_2} = n_2\}, \ldots, \bigcup_{k_r}\{T_{k_r} = n_r\}\right\} \\ &\quad = \sum_{0<k_1<k_2<\cdots<k_r} P\{T_{k_1} = n_1, T_{k_2} = n_2, \ldots, T_{k_r} = n_r\}. \end{aligned}$$

Now

$$\begin{aligned} P\{T_{k_\ell} &= n_\ell \big| T_{k_1} = n_1, T_{k_2} = n_2, \ldots, T_{k_{\ell-1}} = n_{\ell-1}\} \\ &= P\{T_{k_\ell} - T_{k_{\ell-1}} = n_\ell - n_{\ell-1} \big| T_{k_{\ell-1}} = n_{\ell-1}\} \\ &= P\{T_{k_\ell} - T_{k_{\ell-1}} = n_\ell - n_{\ell-1}\} = P\{T_{k_\ell - k_{\ell-1}} = n_\ell - n_{\ell-1}\}. \end{aligned}$$

Therefore,

$$\begin{aligned}
&P\{Z'_{n_1} = Z'_{n_2} = \cdots = Z'_{n_r} = 1\} \\
&= \sum_{0<k_1<k_2<\cdots<k_r} P\{T_{k_1} = n_1\}P\{T_{k_2-k_1} = n_2 - n_1\}\cdots \\
&\quad \times P\{T_{k_r-k_{r-1}} = n_r - n_{r-1}\} \\
&= \sum_{k_1>0} P\{T_{k_1} = n_1\} \sum_{k_2>k_1} P\{T_{k_2-k_1} = n_2 - n_1\}\cdots \\
&\quad \times \sum_{k_r>k_{r-1}} P\{T_{k_r-k_{r-1}} = n_r - n_{r-1}\} = v_{n_1}v_{n_2-n_1}\cdots v_{n_r-n_{r-1}}.
\end{aligned}$$

This shows that Z' is a recurrent phenomenon. From the convolution formula

$$P\{T_{k+1} = n\} = \sum_{m=1}^{n-1} P\{T_1 = m\}P\{T_{k+1} - T_1 = n - m\} \quad (k \geq 1)$$

we obtain the relation

$$v_n = \sum_{m=1}^{n} f_m v_{n-m} \quad (n \geq 1). \tag{2.9}$$

(ii) Conversely, let Z be a recurrent phenomenon with the associated sequence $\{u_n, n \geq 0\}$. Let $\{f_n\}$ be the recurrence time distribution, so that by Theorem 2.1, the recurrence relation (2.5) holds. Let Z' be the recurrent phenomenon constructed as in (i) from the recurrence times $\{T_k\}$. Then $\{v_n\}$ satisfies (2.9). Because of the uniqueness of the solution of (2.9) we find that $v_n = u_n$, as required.

□

Corollary 2.1 *A sequence* $u = \{u_n, n \geq 0\} \in R$ *iff it is a renewal sequence; that is, iff there exists a discrete distribution* $\{f_n, n \geq 1\}$ *such that*

$$u_0 = 1, \quad u_n = \sum_{m=1}^{n} f_m u_{n-m} \ (n \geq 1), \tag{2.10}$$

with $f_n \geq 0, \sum_1^\infty f_n \leq 1$.

Proof. (i) Let $u \in R$; that is, there exists a recurrent phenomenon Z such that $u_n = P\{Z_n = 1\}$ $(n \geq 0)$. Theorem 2.1 shows that there exists a distribution $\{f_n, n \geq 1\}$ such that (2.10) holds.

(ii) Conversely, let u satisfy (2.10). Let $\{T_k\}$ be a renewal process induced by $\{f_n\}$, and $Z' = \{Z'_n, n \geq 0\}$ be the recurrent phenomenon constructed as in Theorem 2.2(i). For this we have $P\{Z'_n = 1\} = u_n$ $(n \geq 0)$, so that $u \in R$.

□

Theorem 2.2 establishes the connection between the theory of recurrent phenomena and renewal theory with lifetime distribution concentrated on $\{1, 2, \ldots\}$. However, in dealing with applications it is useful to have a slightly broader conceptual framework, as provided by the following definitions and results.

Definition 2.3 If $T_1 < \infty$ a.s. we say that Z is a persistent phenomenon, and otherwise it is transient.

Definition 2.4 Suppose $T_1 < \infty$ a.s., and let $E(T_1) = \mu \leq \infty$. If $\mu < \infty$ we say that Z is non-null, and if $\mu = \infty$ it is null.

Definition 2.5 Let $d = g$ c.d. $\{n\colon u_n > 0\}$. If $d = 1$ we say that Z is aperiodic, and if $d > 1, Z$ is periodic with period d.

Theorem 2.3 *A recurrent phenomenon Z with the associated sequence $\{u_n\}$ is persistent or transient according as*

$$\sum_0^\infty u_n = \infty \quad \text{or} \quad \sum_0^\infty u_n < \infty. \tag{2.11}$$

Proof. Let $P\{T_1 < \infty\} = p$. From (2.4) we find that

$$\sum_0^\infty u_n = 1 + \sum_1^\infty P\{T_k < \infty\} = \sum_0^\infty p^k$$

and the last sum is finite iff $p < 1$. □

Theorem 2.4 *Let Z be a recurrent phenomenon and ζ its recurrent set. If Z is persistent, then ζ is non-terminating a.s., while if Z is transient, then ζ is terminating a.s.*

Proof. Let $P\{T_1 < \infty\} = p$. We can write

$$\zeta = \{T_0, T_1, T_2, \ldots, T_N\}, \tag{2.12}$$

where the random variable N is such that $T_N < \infty$, $T_{N+1} = \infty$. We have $P\{N > k\} = p^{k+1}$, which equals 1 if $p = 1$, and otherwise $\to 0$ as $k \to \infty$. This means that $N = \infty$ a.s. if $p = 1$, and $N < \infty$ a.s. if $p < 1$. □

We also define the random variable

$$N(n) = \{k\colon k \in \zeta \cap (0, n]\}. \tag{2.13}$$

This is the number of times Z 'occurs' upto the epoch n (not counting the initial occurrence). From Theorem 2.4 we see that $N(n) \to N \leq \infty$ where $N = \infty$ a.s. if Z is persistent and $N < \infty$ a.s. otherwise.

Theorem 2.5 *Let Z be a recurrent phenomenon with the associated sequence $\{u_n\}$. Then as $n \to \infty$*

$$\begin{aligned} u_n &\to 0 \quad \textit{if } Z \textit{ is transient or persistent null} \\ u_n &\to \frac{1}{\mu} \quad \textit{if } Z \textit{ is persistent aperiodic non-null} \\ u_{nd} &\to \frac{d}{\mu} \quad \textit{if } Z \textit{ is persistent non-null and has period } d. \end{aligned} \tag{2.14}$$

Proof. If Z is transient, then $\Sigma u_n < \infty$ by Theorem 2.3, and so $u_n \to 0$ as $n \to \infty$. The other results follow from discrete renewal theory. □

We have so far assumed that $Z_0 = 1$. Suppose that $P\{Z_0 = 1\} < 1$. Such a phenomenon is said to be *delayed.* For this we have

$$P\{Z_n = 1\} = u_n^{(0)} \quad (n \geq 1) \tag{2.15}$$

and for $0 < n_1 < n_2 < \cdots < n_r$ $(r \geq 2)$ we have

$$P\{Z_{n_1} = Z_{n_2} = \cdots = Z_{n_r} = 1\} = u_{n_1}^{(0)} u_{n_2 - n_1} \cdots u_{n_r - n_{r-1}}, \tag{2.16}$$

where u_n are as before, and in particular, satisfy the recurrence relation (2.5). Denoting by T_k $(k \geq 1)$ the successive recurrence times of Z we find that the random variables $T_k - T_{k-1}$ $(k \geq 1)$ are mutually independent; T_1 has a distribution $\{k_n, n \geq 1\}$; and that $T_k - T_{k-1}$ $(k \geq 2)$ have a common distribution $\{f_n, n \geq 1\}$. Further we have

$$u_n^{(0)} = \sum_{m=1}^{n} k_m u_{n-m} \quad (n \geq 1). \tag{2.17}$$

Proceeding as in the proof of Theorem 2.2 we can establish the connection between delayed recurrent phenomena and renewal processes induced by the distributions $\{(k_n), (f_n),\ n \geq 1\}$. The classification of Z is based on the properties of the distribution $\{f_n\}$. Thus Z is persistent or transient depending upon whether the distribution $\{f_n\}$ is proper or defective. A persistent phenomenon is non-null or null according as $\mu = \Sigma_1^\infty n f_n < \infty$ or $= \infty$. If the distribution $\{f_n\}$ is arithmetic with span d, we shall assume that $\{k_n\}$ is also arithmetic with the same span and say that Z is *periodic with span d*. Results analogous to Theorems 2.3 and 2.4 hold for delayed phenomena. The limit behavior of $u_n^{(0)}$ is given by the following theorem which follows from (2.17) and the known behavior of u_n as $n \to \infty$.

Theorem 2.6 *Let Z be a delayed recurrent phenomenon with the associated sequence $\{u_n^{(0)}\}$. Then as $n \to \infty$,*

$$\begin{aligned}
&u_n^{(0)} \to 0 && \text{if } Z \text{ is transient or persistent null} \\
&u_n^{(0)} \to \frac{1}{\mu}\left(\sum_1^\infty k_n\right) && \text{if } Z \text{ is persistent aperiodic non-null} \\
&u_{nd}^{(0)} \to \frac{d}{\mu} && \text{if } Z \text{ is persistent non-null and has period } d.
\end{aligned}$$

9.3 Subordination of Renewal Counting Processes

Let $S = \{S_n, n \geq 0\}$ be a renewal process induced by the distribution F and $S' = \{S'_n, n \geq 0\}$ an independent Poisson renewal process with parameter $\lambda (0 < \lambda < \infty)$. Denote by $N = \{N(t), t \geq 0\}$ the renewal counting process associated with S, and

$$C_n = N(S'_n) = \max\{k \colon S_k \leq S'_n\} \quad (n \geq 1). \tag{3.1}$$

We are interested in the process $C = \{C_n, n \geq 1\}$. This process was investigated by J.F.C. Kingman, who called C the Poisson count of S. We prefer to view C as a process *subordinate* to the renewal counting process N, the *subordinator* being the Poisson process S'.

If S is a Poisson renewal process we have the following elementary result.

Theorem 3.1 *The subordinate process of a Poisson renewal process with parameter μ $(0 < \mu < \infty)$ is a renewal process induced by the distribution $\{qp^n, n \geq 0\}$, where*

$$p = \frac{\mu}{\lambda + \mu}, \quad q = \frac{\lambda}{\lambda + \mu}. \tag{3.2}$$

Proof. By Theorem 5.1 of Chap. 6, the counting process of S has stationary independent increments. Because of the independence of S and S', the random variables $N(S_1'), N(S_2') - N(S_1'), \dots$ are also mutually independent and have a common distribution. We have

$$P\{N(S_1') = n\} = \int_0^\infty e^{-\mu t} \frac{(\mu t)^n}{n!} \lambda e^{-\lambda t} dt = qp^n \quad (n \geq 0). \tag{3.3}$$

Since $C_n = \sum_{k=1}^n \left[N(S_k') - N(S_{k-1}')\right]$ a.s., it follows that $\{C_n, n \geq 1\}$ is a renewal process. □

In the general case when S is an arbitrary renewal process, we shall prove that C is a *compound renewal process* of the type described by (3.7) below.

In order to do this we observe that the range R_0^+ of the process S, excluding the origin can be represented as $R_0^+ = \cup_1^\infty R_n$, where

$$R_n = R_0^+ \cap \left[S_{n-1}', S_n'\right] \quad (n \geq 1). \tag{3.4}$$

For $n \geq 1$ define

$$\begin{aligned} Z_n &= 0 \quad \text{if } R_n \text{ is empty} \\ &= 1 \quad \text{if } R_n \text{ is non-empty.} \end{aligned} \tag{3.5}$$

It turns out that $Z = \{Z_n, n \geq 1\}$ is a delayed recurrent phenomenon (Theorem 3.3). Let $\{T_k, k \geq 1\}$ be the successive recurrence times of Z, and let

$$Y_k = \#\{m > 0 \colon S_m \in R_{T_k}\} \quad (k \geq 1). \tag{3.6}$$

Then $\{(T_k - T_{k-1}, Y_k), k \geq 1\}$ with $T_0 = 0$ is a sequence of mutually independent random vectors with a common distribution for $k > 2$ (Theorem 3.4). Given these facts the following result is obvious.

Theorem 3.2 *The subordinate process C is a compound renewal process, namely,*

$$C_n = \sum_1^{N(n)} Y_k \quad (n \geq 1), \tag{3.7}$$

where $N(n) = \max\{k \colon T_k \leq n\}$ and an empty sum is interpreted as zero.

It remains to prove the results indicated above. Let us denote by U the renewal function of S, so that $U(t) = \sum_1^\infty F_n(t)$, where F_n is the distribution of S_n. Also, let

$$a_n = \int_0^\infty e^{-\lambda t} \frac{(\lambda t)^n}{n!} F\{dt\} \quad (n \geq 0). \tag{3.8}$$

Theorem 3.3 *The sequence Z is a delayed recurrent phenomenon, with*

$$u_n^{(0)} = (1 - a_0) \int_0^\infty e^{-\lambda t} \frac{(\lambda t)^{n-1}}{(n-1)!} U\{dt\} \quad (n \geq 1) \tag{3.9}$$

$$u = (1 - a_0) \int_0^\infty e^{-\lambda t} \frac{(\lambda t)^n}{n!} U\{dt\} \quad (n > 1). \tag{3.10}$$

If F is defective, then Z is transient. If F is proper with mean μ, then Z is persistent, non-null or null according as $\mu < \infty$ or $\mu = \infty$.

Proof. If R_{n_t} is non-empty we shall denote by S_{ℓ_t} the last S_k in it. For $0 = n_0 < n_1 < \cdots < n_r$ $(r \geq 1)$ we have

$$\begin{aligned} &P\{Z_{n_1} = Z_{n_2} = \cdots = Z_{n_r} = 1\} \\ &\quad = \sum P\{S'_{n_t - 1} \leq S_{\ell_t} < S'_{n_t} \leq S_{\ell_t + 1} \ (1 \leq t \leq r)\}, \end{aligned}$$

the sum being taken over $\ell_1, \ell_2, \ldots, \ell_r$ such that $0 < \ell_1 < \ell_2 < \cdots < \ell_r$. This last expression equals

$$\begin{aligned} &\sum P\{N'(S_{\ell_t}) = n_t - 1,\ N'(S_{\ell_t + 1}) \geq n_t \ (1 \leq t \leq r)\} \\ &\quad = \sum P\{N'(S_{\ell_1}) = n_1 - 1\} \cdot \prod_{t=1}^{r-1} P\{N'(S_{\ell_t + 1}) - N'(S_{\ell_t}) \geq 1, \\ &\qquad N'(S_{\ell_{t+1}}) - N'(S_{\ell_t}) = n_{t+1} - n_t\} \cdot P\{N'(S_{\ell_r + 1}) - N'(S_{\ell_r}) \geq 1\}, \end{aligned}$$

where N' is the renewal counting process of S'. Easy calculations reduce this expression to

$$u_{n_1}^{(0)} u_{n_2 - n_1} \cdots u_{n_r - n_{r-1}} \tag{3.11}$$

with $u_n^{(0)}$ and u_n defined by (3.9) and (3.10), respectively. Therefore, Z is a recurrent phenomenon. We have

$$\sum_1^\infty u_n^{(0)} = (1 - a_0) U(\infty), \tag{3.12}$$

where $U(\infty) < \infty$ or $= \infty$ according as F is defective or proper. Also, using Theorem 2.6 and Theorem A4.4 we obtain

$$\begin{aligned}\lim_{n\to\infty} u_n^{(0)} &= \lim_{z\to 1-}(1-z)\sum_{n=1}^{\infty} u_n^{(0)} z^n \\ &= \frac{1-a_0}{\lambda}\lim_{\theta\to 0+}\theta\int_0^{\infty} e^{-\theta t}U\{dt\} = \frac{1-a_0}{\lambda\mu} \end{aligned} \tag{3.13}$$

by the Elementary Renewal Theorem. Here, μ is the mean of F and the limit is non-zero or zero according as $\mu < \infty$ or $\mu = \infty$. □

Theorem 3.4 *Let $\{T_k, k \geq 0\}$ be the recurrence times of Z, and Y_k be defined by (3.6). Then $\{(T_k, Y_1 + Y_2 + \cdots + Y_k), k \geq 0\}$ is a renewal process induced by the distributions*

$$P\{T_1 = n, Y_1 = r\} = a_{n-1}(1-a_0)a_0^{r-1} \tag{3.14}$$

$$P\{T_k - T_{k-1} = n, Y_k = r\} = a_n a_0^{r-1} \quad (k \geq 2) \tag{3.15}$$

for $n \geq 1, r \geq 1$.

Proof. We consider the case $k=3$, the general case being similar. For $k = 3$ we have

$$\begin{aligned}
&P\{T_1 = m, Y_1 = r, T_2 = m+n, Y_2 = s, T_3 = m+n+p, Y_3 = t\} \\
&\quad = P\{S'_{m-I} \leq S_1 \leq S_r < S'_m, S'_{m+n-1} \leq S_{r+1} \leq S_{r+s} < S'_{m+n}, S'_{m+n+p-1} \\
&\qquad \leq S_{r+s+l} \leq S_{r+s+t} < S'_{m+n+p} \leq S_{r+s+t+1}\} \\
&\quad = P\{N'(S_1) = m-1, N'(S_r) = m-1, N'(S_{r+1}) = m+n-1, N'(S_{r+s}) \\
&\qquad\quad = m+n-1, N'(S_{r+s+1}) = m+n+p-1, N'(S_{r+s+t}) \\
&\qquad\quad = m+n+p-1, N'(S_{r+s+t+1}) \geq m+n+p\} \\
&\quad = P\{N'(S_1) = m-1, N'(S_r) - N'(S_1) = 0, N'(S_{r+1}) - N'(S_r) = n, \\
&\qquad\quad N'(S_{r+s}) - N'(S_{r+1}) = 0, N'(S_{r+s+1}) - N'(S_{r+s}) = p, \\
&\qquad\quad N'(S_{r+s+t}) - N'(S_{r+s+1}) = 0, N'(S_{r+s+t+1}) - N'(S_{r+s+t}) \geq 1\} \\
&\quad = a_{m-1}a_0^{r-1}a_n a_0^{s-1}a_p a_0^{t-1}(1-a_0)
\end{aligned} \tag{3.16}$$

for $m \geq 1, r \geq 1, n \geq 1, s \geq 1, p \geq 1, t \geq 1$. From this we find that the marginal distributions of $(T_1, Y_1), (T_2 - T_1, Y_2), (T_3 - T_2, Y_3)$ are as given by (3.14) and (3.15). We can therefore write (3.16) as

$$P\{T_1 = m, Y_1 = r, T_2 - T_1 = n, Y_2 = s, T_3 - T_2 = p, Y_3 = t\}$$
$$= P\{T_1 = m, Y_1 = r\} \cdot P\{T_2 - T_1 = n, Y_2 = s\} \cdot P\{T_3 - T_2 = p, Y_3 = t\}\cdot$$

This shows that (T_1, Y_1), $(T_2 - T_1, Y_2)$, $(T_3 - T_2, Y_3)$ are mutually independent. □

9.4 The Simple Random Walk in D Dimensions

Let $X_k = \{X_k^{(1)}, X_k^{(2)}, \ldots, X_k^{(d)}\}$ $(k \geq 1)$ be a sequence of mutually independent random vectors with

$$P\{X_k = e_i\} = \frac{1}{2d}, \quad P\{X_k = -e_i\} = \frac{1}{2d} \ (1 \leq i \leq d), \tag{4.1}$$

where $e_1 = (1, 0, \ldots, 0), e_2 = (0, 1, 0, \ldots, 0), \ldots$ and $e_d = (0, 0, \ldots, 0, 1)$. Thus X_k takes the $2d$ possible values $\pm e_i$ with equal probabilities. Let $S_0 \equiv 0, S_n = X_1 + X_2 + \cdots + X_n$ $(n \geq 1)$; we call $\{S_n, n \geq 0\}$ a random walk. For $\ell = (\ell_1, \ell_2, \ldots, \ell_d)$ we have

$$P\{S_n = \ell\} = \sum \frac{n!}{k_1!(k_1 + \ell_1)! \cdots k_d!(k_d + \ell_d)!} \left(\frac{1}{2d}\right)^n, \tag{4.2}$$

the sum being taken over $(k_1, k_2, \ldots, k_d)$ such that

$$k_i \geq 0 \, (1 \leq i \leq d), 2\,(k_1 + k_2 + \cdots + k_d) + (\ell_1 + \ell_2 + \cdots + \ell_d) = n.$$

By a return to the origin at epoch n we mean the event $\{S_n = 0\}$. For its probability u_n we have $u_{2n+1} = 0$, while

$$\begin{aligned} u_{2n} &= \sum_{\Sigma k_i = n} \frac{(2n)!}{(k_1! k_2! \cdots k_d!)^2} \left(\frac{1}{2d}\right)^{2n} \\ &= \binom{2n}{n} \left(\frac{1}{2}\right)^{2n} \sum_{\Sigma k_i = n} p(k_1, k_2, \ldots, k_d)^2, \end{aligned} \tag{4.3}$$

where $p(k_1, k_2, \ldots, k_d)$ is the multinomial probability

$$p(k_1, k_2, \ldots, k_d) = \frac{n!}{k_1! k_2! \cdots k_d!} \left(\frac{1}{d}\right)^n. \tag{4.4}$$

We have the following result.

Theorem 4.1 *With probability* 1 *the simple random walk in* d *dimensions returns to the origin infinitely often if* $d \leq 2$, *but only finitely often if* $d \geq 3$.

Proof. Let $Z_n = 1_{\{S_n=0\}}(n \geq 0)$, so that $P\{Z_n = 1\} = u_n$. For $0 = n_0 < n_1 < \cdots < n_r$ $(r \geq 1)$ we have

$$\begin{aligned} &P\{Z_{n_1} = Z_{n_2} = \cdots = Z_{n_r} = 1\} \\ &\quad = P\{S_{n_1} = S_{n_2} = \cdots = S_{n_r} = 0\} \\ &\quad = P\{S_{n_1} = 0, S_{n_2} - S_{n_1} = 0, \ldots, S_{n_r} - S_{n_r - 1} = 0\} \\ &\quad = P\{S_{n_1} = 0\} P\{S_{n_2 - n_1} = 0\} \ldots P\{S_{n_r - n_{r-1}} = 0\} \end{aligned}$$

because the increments $S_{n_1}, S_{n_2} - S_{n_1}, \ldots, S_{n_r} - S_{n_{r-1}}$ are mutually independent and $S_{n_t} - S_{n_{t-1}}$ has the same distribution as $S_{n_t - n_{t-1}}(1 \leq t \leq r)$. Thus

$$P\{Z_{n_1} = Z_{n_2} = \cdots = Z_{n_r} = 1\} = u_{n_1} u_{n_2 - n_1} \cdots u_{n_r - n_{r-1}}, \tag{4.5}$$

which shows that $Z = \{Z_n, n \geq 0\}$ is a recurrent phenomenon. Clearly, Z has period 2. We proceed to investigate its properties.

(i) For $d = 1$, (4.4) gives

$$u_{2n} = \binom{2n}{n} \left(\frac{1}{2}\right)^{2n} \sim \frac{1}{\sqrt{n\pi}} \tag{4.6}$$

so that $\Sigma u_{2_n} = \infty$. By Theorem 2.3, Z is persistent and the desired result follows from Theorem 2.4.

(ii) For $d = 2$, (4.4) gives

$$u_{2n} = \binom{2n}{n} \left(\frac{1}{2}\right)^{4n} \sum_0^n \binom{n}{k}^2 = \left[\binom{2n}{n} \left(\frac{1}{2}\right)^{2n}\right]^2 \sim \frac{1}{n\pi}. \tag{4.7}$$

Here again $\Sigma u_{2n} = \infty$, and the conclusion is as before.

(iii) Let $d \geq 3$. We have

$$\sum p(k_1, k_2, \ldots, k_d)^2 \leq M \sum p(k_1, k_2, \ldots, k_d) = M,$$

where

$$\begin{aligned} M &= \max_{k_i} p(k_1, k_2, \ldots, k_d) \sim p(k_1', k_2', \ldots, k_d') \quad \text{with } k_1' = k_2' \\ &= \cdots = k_d' = \left[nd^{-1}\right] \\ &= \frac{n!}{\left(\left[\frac{n}{d}\right]!\right)^d} \cdot \left(\frac{1}{d}\right)^n. \end{aligned}$$

Therefore,

$$u_{2n} \le M\binom{2n}{n}\left(\frac{1}{2}\right)^{2n} \sim \sqrt{2}\left(\frac{d}{2n\pi}\right)^{d/2}.$$

Since $\sum_1^\infty n^{-d/2} < \infty$ for $d > 2$, Z is transient by Theorem 2.3 and the desired result follows by Theorem 2.4. □

9.5 The Bernoulli Random Walk

Let $\{X_k, k \ge 1\}$ be a sequence of mutually independent random variables with

$$P\{X_k = +1\} = p \quad \text{and} \quad P\{X_k = -1\} = q, \tag{5.1}$$

where $0 < p < 1$ and $q = 1-p$. Let $S_0 \equiv 0$, $S_n = X_1+X_2+\cdots+X_n$ $(n \ge 1)$; we call $\{S_n; n \ge 0\}$ the Bernoulli random walk. Let $Z_n = 1_{\{S_n=0\}}$ $(n \ge 0)$. Then as in Sec. 9.4 we find that $Z = \{Z_n, n \ge 0\}$ is a recurrent phenomenon with

$$u_n = P\{S_n = 0\} = \binom{n}{\frac{n}{2}}(pq)^{\frac{n}{2}} \quad (n \ge 0). \tag{5.2}$$

Here Z has period 2. We have the following.

Theorem 5.1 *The phenomenon Z is transient if $p \ne q$ and persistent null otherwise.*

Proof. We have

$$u_{2n} = \binom{2n}{n}(pq)^n \sim \frac{(4pq)^n}{\sqrt{n\pi}}. \tag{5.3}$$

If $p \ne q$, then $4pq < 1$ and $\sum u_{2n}$ converges faster than the geometric series $\Sigma(4pq)^n$. If $p = q$, then $u_{2n} \sim (n\pi)^{-1/2}$, so that $\Sigma u_{2n} = \infty$, but $u_{2n} \to 0$ as $n \to \infty$. The desired results therefore follow from Theorems 2.3 and 2.5. □

A consequence of Theorem 5.1 is that with probability 1 the Bernoulli random walk returns to the origin only finitely often if $p \ne q$ and infinitely often if $p = q$. The latter part of this statement is contained in Theorem 4.1 (the case $d=1$). For the symmetric random walk $(p=q)$ we have the following further result.

Theorem 5.2 *Let $N(2n)$ be the number of returns to the origin in the symmetric Bernoulli random walk. Then*

$$EN(2n) \sim 2\left(\frac{n}{\pi}\right)^{1/2} \quad (n \to \infty) \tag{5.4}$$

$$\lim_{n\to\infty} P\left\{\frac{N(2n)}{\sqrt{2n}} < x\right\} = N_+(x). \tag{5.5}$$

Proof. Let $\{T_k, k \geq 0\}$ be the recurrence times of the phenomenon Z. These are the successive return times of the random walk to the origin, and we have

$$T_0 \equiv 0, \quad T_k = \min\{n > T_{k-1} \colon S_n = 0\} \ (k \geq 1). \tag{5.6}$$

Theorem 5.1 implies that T_1 is a proper random variable with $E(T_1) = \infty$. From Theorem 2.1 we find that

$$F(z) = E(z^{T_1}) = 1 - \left(\sum_0^\infty u_n z^n\right)^{-1} = 1 - \sqrt{1 - z^2}, \tag{5.7}$$

which confirms these statements concerning T_1. We have

$$\sqrt{1-z} \cdot \frac{1 - F(z)}{1 - z} = \sqrt{1 + z} \to \sqrt{2} \quad \text{as } z \to 1-,$$

so that

$$\frac{1 - F(z)}{1 - z} \sim \sqrt{2}(1 - z)^{-1/2} \quad (z \to 1-). \tag{5.8}$$

By Theorem A4.3(ii) we obtain

$$P\{T_1 > n\} \sim \sqrt{\frac{2}{\pi}} n^{-1/2} \quad (n \to \infty). \tag{5.9}$$

Thus the condition 6.1 of Chap. 6 holds for $\alpha = 1/2$ and $L(n) = \sqrt{2}$. Lemma 6.2 of Chap. 6 therefore yields (5.4) and Theorem 6.1 of Chap. 6 yields the result

$$\lim_{n\to\infty} P\left\{\sqrt{2}\frac{N(2n)}{\sqrt{2n}} \geq cx^{-1/2}\right\} = G_{1/2}(x).$$

Here we choose $c = \sqrt{2}$, so that $G_{1/2}$ has the Laplace transform $e^{-\sqrt{2\theta}}$. With this standardization $G_{1/2}(x) = 2\left[1 - N\left(\frac{1}{\sqrt{x}}\right)\right]$. Therefore,

$$\begin{aligned}\lim_{n\to\infty} P\left\{\frac{N(2n)}{\sqrt{2n}} < x\right\} &= 1 - G_{1/2}\left(\frac{1}{x^2}\right) = 1 - 2[1 - N(x)] \\ &= 2N(x) - 1 = N_+(x),\end{aligned}$$

which is the result (5.5). □

Let us now return to the general case and denote

$$M_n = \max(0, S_1, S_2, \ldots, S_n), \quad m_n = \min(0, S_1, S_2, \ldots, S_n) \quad (n \geq 0). \tag{5.10}$$

For arbitrary random walks on the real line, these functionals are investigated in Chap. 6 but the results for the Bernoulli case follow in an elementary manner from renewal theory. To see the connection, we consider the hitting times

$$\tau_k = \min\{n\colon S_n = k\}. \tag{5.11}$$

It turns out that the sequence $\{\tau_k, k \geq 0\}$ is a renewal process and

$$M_n = \max\{k\colon \tau_k \leq n\} \quad (n \geq 0), \tag{5.12}$$

so that $\{M_n, n \geq 0\}$ is the renewal counting process associated with $\{\tau_k\}$. A similar statement holds for m_n. We have the following.

Theorem 5.3 *Let M_n be the maximum functional of the Bernoulli random walk.*

(i) *If $p < q$, then as $n \to \infty$, $M_n \to M < \infty$ a.s., where the random variable M has the geometric distribution*

$$P\{M = k\} = \left(1 - \frac{p}{q}\right)\left(\frac{p}{q}\right)^k \quad (k \geq 0). \tag{5.13}$$

(ii) *If $p \geq q$, then as $n \to \infty$*

$$\frac{M_n}{n} \to p - q \text{ a.s.} \quad \text{and} \quad E\left(\frac{M_n}{n}\right) \to (p - q) \tag{5.14}$$

(iii) *If $p > q$, then*

$$E(M_n) = n(p-q) + \frac{q}{p-q} + o(1), \quad \mathrm{Var}(M_n) = 4npq + o(n), \tag{5.15}$$

and

$$\lim_{n\to\infty} P\left\{\frac{M_n - n(p-q)}{2\sqrt{npq}} < x\right\} = N(x). \tag{5.16}$$

(iv) *If $p = q$, then*

$$\lim_{n\to\infty} P\left\{\frac{M_n}{\sqrt{n}} < x\right\} = N_+(x). \tag{5.17}$$

Proof. We shall prove (below) that $\{\tau_k\}$ is a renewal process for which

$$P\{\tau_1 < \infty\} = \min\left(1, \frac{p}{q}\right) \tag{5.18}$$

$$E(\tau_1) = \frac{1}{p-q} \ (p \geq q), \quad \mathrm{Var}(\tau_1) = \frac{4pq}{(p-q)^3} \quad (p > q). \tag{5.19}$$

Moreover if $p = q$, then

$$P\{\tau_1 > n\} \sim \sqrt{\frac{2}{\pi}} n^{-1/2} \quad (n \to \infty). \tag{5.20}$$

We have $M_{n+1} = \max(M_n, S_{n+1}) \geq M_n$, so that the sequence $\{M_n\}$ is non-decreasing. Therefore, $M_n \to M \leq \infty$ a.s. If $p < q$, then renewal process is terminating (Example (6.1.1.)) and so $M < \infty$ a.s. and has the distribution (5.13). If $p \geq q$, the results (5.14) follow from Theorems 6.2.3 and 6.2.4. If $p > q$, the results (5.15) and (5.16) can be proved similarly. The proof of (5.17) is similar to that of (5.5) in view of (5.20).

It remains to prove the results (5.18)–(5.20) concerning the process of hitting times $\{\tau_k, k \geq 0\}$. Since X_k are independently and indentically distributed, we have

$$E[z^{\tau_k} | S_m = j] = E[z^{m+\tau_{k-j}}] \quad (j < k, \ m \leq k). \tag{5.21}$$

The (strong) Markov property of $\{S_n\}$ gives

$$E[z^{\tau_k - \tau_{k-1}} | \tau_1, \tau_2, \ldots, \tau_{k-1}] = E[z^{\tau_1}] \quad (k \geq 1), \tag{5.22}$$

so that the increments $\tau_k - \tau_{k-1}$ $(k \geq 1)$ are mutually independent and have a common distribution. Thus $E[z^{\tau_k}] = \xi^k$ $(k > 0)$, where $\xi = \xi(z)$ has to be determined. By considering the possible values of S_1 (+1 or −1) we find that

$$\begin{aligned} E[z^{\tau_k}] &= pE[z^{\tau_k}|S_1 = 1] + qE[z^{\tau_k}|S_1 = -1] \\ &= pE[z^{1+\tau_{k-1}}] + qE[z^{1+\tau_{k+1}}] \end{aligned}$$

using (5.21). This yields the equation $\xi^k = pz\,\xi^{k-1} + qz\,\xi^{k+1}$, or

$$qz\,\xi^2 - \xi + pz = 0. \tag{5.23}$$

The unique continuous root of (5.23) with $\xi(0+) = 0$ is given by

$$\xi = \frac{1 - \sqrt{1 - 4pqz^2}}{2qz}. \tag{5.24}$$

From this generating function of τ_1, the properties (5.18) and (5.19) follow in the usual manner. Finally, when $p = q$,

$$\xi = \frac{1 - \sqrt{1 - z^2}}{z}$$

so that

$$\sqrt{1-z} \cdot \frac{1-\xi}{1-z} = \frac{\sqrt{1+z} - \sqrt{1-z}}{z} \to \sqrt{2} \quad \text{as } z \to 1-$$

or

$$\frac{1 - E(z^{\tau_1})}{1-z} \sim \sqrt{2}(1-z)^{-1/2} \quad (z \to 1-). \tag{5.25}$$

Theorem A4.3(ii) now yields (5.20). □

Combining Theorem 5.3 with similar results for m_n we obtain the following.

Theorem 5.4 *The symmetric Bernoulli random walk is oscillating in the sense that as* $n \to \infty$,

$$M_n \to \infty, \quad m_n \to -\infty \quad \text{a.s.} \tag{5.26}$$

In the non-symmetric case, either

$$M_n \to \infty, \quad m_n \to m > -\infty \quad \text{a.s. } \textit{if } p > q \tag{5.27}$$

(the random walk drifts to $+\infty$*) or*

$$M_n \to M < \infty, \quad m_n \to -\infty \quad \text{a.s. } \textit{if } p < q \tag{5.28}$$

(the random walk drifts to $-\infty$*).*

9.6 Ladder Sets of Random Walks on the Real Line

Let $\{X_k,\ k \geq 1\}$ be a sequence of independent random variables with a common distribution $F, S_0 = 0, S_n = X_1 + X_2 + \cdots + X_n$ $(n \geq 1)$. We define $\{S_n, n \geq 0\}$ to be the random walk induced by F. If F is concentrated on $[0, \infty)$ or on $(-\infty, 0], \{S_n\}$ reduces to renewal process investigated in Chap. 6 or its negative counterpart. We shall exclude this from our consideration. For $n \geq 0$ let

$$M_n = \max(0, S_1, S_2, \ldots, S_n), \quad m_n = \min(0, S_1, S_2, \ldots, S_n) \tag{6.1}$$

and consider the sets

$$\zeta_+ = \{n \geq 0\colon M_n - S_n = 0\}, \quad \zeta_- = \{n \geq 0\colon S_n - m_n = 0\}. \tag{6.2}$$

Also denote

$$Z_n = 1_{\{n \in \zeta_+\}}, \quad \bar{Z}_n = 1_{\{n \in \zeta_-\}}\ (n \geq 0) \tag{6.3}$$

and

$$Z_+ = \{Z_n, n \geq 0\}, \quad Z_- = \{\bar{Z}_n, n \geq 0\}. \tag{6.4}$$

We first prove that Z_+ and Z_- are recurrent phenomena.

Theorem 6.1 *The sequences Z_+ and Z_- are recurrent phenomena. The associated recurrence sets are given respectively by*

$$\zeta_+ = \{T_0, T_1, T_2, \ldots\}, \quad \zeta_- = \{\bar{T}_0, \bar{T}_1, \bar{T}_2, \ldots\}, \tag{6.5}$$

where the random variables $\{T_k\}$ and $\{\bar{T}_k\}$ are defined as follows: $T_0 = \bar{T}_0 = 0$ a.s., and for $k \geq 1$

$$T_k = \min\{n > T_{k-1}\colon S_n \geq S_{T_{k-1}}\} \tag{6.6}$$

$$\bar{T}_k = \min\{n > \bar{T}_{k-1}\colon S_n \leq S_{\bar{T}_{k-1}}\} \tag{6.7}$$

Proof. It suffices to prove the statements concerning ζ_+. Let $W_n = M_n - S_n$ $(n \geq 0)$. We have $W_0 = 0$ a.s. and for $n \geq 0$

$$W_{n+1} = \max(M_n, S_{n+1}) - S_{n+1} = \max(0, W_n - X_{n+1}).$$

This shows that $\{W_n, n \geq 0\}$ is a time-homogeneous Markov chain on the state space $[0, \infty)$. For the state 0 the Markov property yields the relation

$$\begin{aligned} &P\{W_{n_1} = W_{n_2} = \cdots = W_{n_r} = 0\} \\ &\quad = P\{W_{n_1} = 0\}P\{W_{n_2 - n_1} = 0\} \cdots P\{W_{n_r - n_{r-1}} = 0\} \end{aligned}$$

for $0 = n_0 < n_1 < \cdots < n_r$ $(r \geq 1)$. This can be written as

$$\begin{aligned} &P\{Z_{n_1} = Z_{n_2} = \cdots = Z_{n_r} = 1\} \\ &\quad = P\{Z_{n_1} = 1\}P\{Z_{n_2 - n_1} = 1\} \cdots P\{Z_{n_r - n_{r-1}} = 1\} \end{aligned}$$

which shows that Z_+ is a recurrent phenomenon. For its first recurrence time T_1 we have

$$T_1 = \min\{n > 0 \colon n \in \zeta_+\} = \min\{n > 0 \colon M_n - S_n = 0\}.$$

Now it can be easily verified that

$$\begin{aligned} &\{M_1 > S_1, M_2 > S_2, \ldots, M_{n-1} > S_{n-1}, M_n = S_n\} \\ &\quad = \{S_1 < 0, S_2 < 0, \ldots, S_{n-1} < 0, S_n \geq 0\}. \end{aligned}$$

We can therefore write

$$T_1 = \min\{n > 0 \colon S_n \geq 0\}. \tag{6.8}$$

Similarly, for $k \geq 2$

$$\begin{aligned} T_k &= \min\{n > T_{k-1} \colon n \in \zeta_+\} \\ &= \min\{n > T_{k-1} \colon S_n \geq S_{T_{k-1}}\}. \end{aligned}$$

□

We shall call ζ_+ and ζ_- the *ascending* and *descending ladder sets*, respectively of the random walk $\{S_n\}$. We note that since the T_k are stopping times of the random walk, $\{(T_k, S_{T_k}), k \geq 0\}$ is a renewal process. In view of (6.5) we have the following result, which expresses M_n as a compound renewal process of the type described by (3.7).

Lemma 6.1 *For the maximum M_n of a random walk $\{S_n\}$ we have*

$$M_n = \sum_{k=1}^{N(n)} \left(S_{T_k} - S_{T_{k-1}}\right), \tag{6.9}$$

where

$$N(n) = \#\{k\colon T_k \in \zeta_+ \cap (0,n]\} = \max\{k\colon T_k \le n\}, \qquad (6.10)$$

and an empty sum is interpreted as zero.

Theorem 6.2 *As $n \to \infty, M_n \to M$ a.s. where $M = \infty$ a.s. or $M < \infty$ a.s. according as Z_+ is persistent or transient. Similar statements hold for m_n.*

Proof. Since $M_{n+1} = \max(M_n, S_{n+1}) \ge M_n$ $(n \ge 0)$, the sequence $\{M_n\}$ is monotone non-decreasing and $M_n \to M \le \infty$ a.s. as $n \to \infty$. Since $S_{T_1} \ge \max(0, X_1), S_{T_1} = 0$ a.s. would imply $X_1 \le 0$ a.s., which is not the case. Therefore, $S_{T_1} > 0$ with positive probability. If Z_+ is persistent, then $N(n) \to \infty$ a.s. as $n \to \infty$ by Theorem 2.4, and $M_n \to \infty$ a.s. by Lemma 6.1. If Z_+ is transient, then $N(n) \to N < \infty$ a.s., and $M_n \to \sum_1^N (S_{T_k} - S_{T_{k-1}}) < \infty$ a.s. □

Theorem 6.3 *(i) It is impossible for the phenomena Z_+, Z_- to be both transient.*

(ii) *If only one is persistent, then it is non-null.*

(iii) *If both are persistent, then both are null.*

Proof. (i) Suppose Z_+ and Z_- are both transient. Then by Theorem 6.2, $M_n \to M < \infty$ a.s. and $m_n \to m > -\infty$ a.s. Since $\{M_n\}$ and $\{m_n\}$ are monotone we find that

$$m \le S_n \le M \quad (n \ge 0) \text{ a.s.}, \qquad (6.11)$$

which is impossible by a result in Sequential Analysis (see Section 7 of Chap. 7).

(ii) Suppose that Z_+ is persistent and Z_- transient. Then by Theorem 6.2, $M_n \to \infty$ and $m_n \to m > -\infty$ a.s. By Lemma 6.2 (below) we find that

$$\begin{aligned} P\{n \in \zeta_+\} &= P\{M_n - S_n = 0\} = P\{m_n = 0\} \to P\{m = 0\} \\ &= P\{S_1 \ge 0, S_2 \ge 0, \dots\} \\ &\ge P\{S_1 > 0, S_2 > 0, \dots\} = P\{\bar{T}_1 = \infty\} > 0. \end{aligned}$$

By Theorem 2.5, Z_+ is non-null.

(iii) Suppose that Z_+ and Z_- are both persistent. Then $M_n \to \infty$ and $m_n \to -\infty$ a.s. Therefore,

$$P\{n \in \zeta_+\} \to P\{m = 0\} = 0$$
$$P\{n \in \zeta_-\} \to P\{M = 0\} = 0.$$

Therefore, Z_+ and Z_- are both null, again by Theorem 2.5. □

Lemma 6.2 *We have*

$$(M_n, M_n - S_n) \stackrel{d}{=} (S_n - m_n, -m_n). \tag{6.12}$$

Proof. The permutation

$$(X_1, X_2, \ldots, X_n) \to (X_n, X_{n-1}, \ldots, X_1) \tag{6.13}$$

leaves the distributions of M_n, m_n, and S_n invariant. This results in the permutation of partial sums

$$(S_0, S_1, \ldots, S_n) \to (S_0', S_1', \ldots, S_n'), \tag{6.14}$$

where $S_m' = S_n - S_{n-m}$ $(0 \leq m \leq n)$. Denoting by M_n' and m_n' the maximum and minimum functionals of $\{S_n'\}$ we find that

$$M_n \stackrel{d}{=} M_n' = \max_{0 \leq m \leq n} (S_n - S_{n-m}) = S_n - \min_{0<m<n} (S_m) = S_n - m_n$$

and

$$M_n - S_n \stackrel{d}{=} M_n' - S_n = S_n - m_n - S_n = -m_n.$$ □

We shall use the properties of the ladder sets derived here to establish the fluctuation theory of random walks in Chap. 6. We also need a third set, which is actually a subset of ζ_+. To define this, we observe that the maximum M_n may be achieved at any of the epochs $m = 0, 1, \ldots, n$. Let us denote by π_n the first epoch at which the maximum is achieved. Thus

$$\pi_n = \min\{m \leq n \colon S_m = M_n\}. \tag{6.15}$$

We now define the set

$$\hat{\zeta}_+ = \{n \geq 0 \colon M_n - S_n = 0, \pi_n = n\}. \tag{6.16}$$

As before, let $\hat{Z}_n = 1_{\{n \in \hat{\zeta}_+\}}$ and $\hat{Z}_+ = \{\hat{Z}_n, n \geq 0\}$. We then have the following.

Theorem 6.4 *The sequence $\hat{Z}_+$ is a recurrent phenomenon. The associated recurrence set can be expressed as*

$$\hat{\zeta}_+ = \{\hat{T}_0, \hat{T}_1, \hat{T}_2, \ldots\}, \tag{6.17}$$

where $\hat{T}_0 = 0$ a.s., and for $k \geq 1$,

$$\hat{T}_k = \min\{n \colon S_n > S_{\hat{T}_{k-1}}\}. \tag{6.18}$$

Proof. The fact that $\hat{Z}_+$ is a recurrent phenomenon follows as in the proof of Theorem 6.1. Since $\{\pi_n = n\} \subset \{M_n - S_n = 0\}$ we have

$$\hat{T}_1 = \min\{n > 0 \colon n \in \hat{\zeta}_+\} = \min\{n > 0 \colon \pi_n = n\}. \tag{6.19}$$

Since

$$\begin{aligned} &\{\pi_1 < 1, \pi_2 < 2, \ldots, \pi_{n-1} < n-1, \pi_n = n\} \\ &\quad = \{S_1 \leq 0, S_2 \leq 0, \ldots, S_{n-1} \leq 0,\ S_n > 0\} \end{aligned}$$

we can write (6.19) as

$$\hat{T}_1 = \min\{n \colon S_n > 0\}. \tag{6.20}$$

Similarly, $\hat{T}_k$ $(k \geq 2)$ are as in (6.18). □

Comparison of (6.6) and (6.18) shows that the $\hat{T}_k$ are essentially the same as T_k except that the inequality in (6.18) is stronger. We shall call $\hat{\zeta}_+$ the *strong ascending ladder set* and ζ_+ the *weak ascending ladder set*. We may also define the strong descending ladder set in an analogous manner. However, consideration of $\hat{\zeta}_+$ and ζ_- together will yield almost all relevant information concerning the fluctuation theory of the random walk developed in Chap. 6. However, for future reference we remark that the results of this section can be reformulated in terms of $\{(\hat{T}_k, S_{\hat{T}_k}), k \geq 0\}$ with obvious modifications. In particular, the following result will be used in the sequel.

Theorem 6.5 (i) *The maximum M_n of a random walk $\{S_n\}$ can be represented as*

$$M_n = \sum_{k=1}^{\hat{N}(n)} \left(S_{\hat{T}_k} - S_{\hat{T}_{k-1}}\right), \tag{6.21}$$

where $\bar{N}(n) = \max\{k \colon \hat{T}_k \leq n\}$.

(ii) *If $\hat{Z}_+$ is transient, then $M_n \to M < \infty$ a.s., where*

$$M = \sum_{k=1}^{\hat{N}} \left(S_{\hat{T}_k} - S_{\hat{T}_{k-1}} \right), \tag{6.22}$$

with $\hat{N} = \#\{k > 0 \colon T_k \in \hat{\zeta}_+\}, P\{\hat{N} = n\} = p(1-p)^n$ $(n \geq 0)$ and $p = P\{M = 0\} = P\{\hat{T}_1 = \infty\} > 0$. In (6.22) the conditional distribution of $S_{\hat{T}_k} - S_{\hat{T}_{k-1}}$ given $\hat{N} \geq k$ is independent of k.

Similar results hold for m_n and lim $m_n = m$, when m is a.s. finite.

Proof. The representations (6.21) and (6.22) follow as in Lemma 6.1 and the concluding part of the proof of Theorem 6.2. Here $\hat{N} = \hat{N}(\infty)$ and its distribution is geometric as we found in Example 6.1.1. □

Remark 6.1 The above theorem implies that if $\hat{Z}_+$ is transient, then

$$M \stackrel{d}{=} \sum_{k=1}^{\hat{N}} U_k, \tag{6.23}$$

where $\{U_k, k \geq 1\}$ are mutually independent random variables with a common distribution and moreover, independent of $\hat{N}$, while $\hat{N}$ is geometrically distributed as above. Also, the distribution of U_1 equals the conditional distribution of $S_{\hat{T}_1}$ given $\{\hat{T}_1 < \infty\}$.

Remark 6.2 A comparison of the sums in (6.9) and (6.21) shows that since $P\{S_{T_k} - S_{T_{k-1}} > 0\} \geq P\{X_1 > 0\} > 0$, the former sum consists of a geometric number of zero terms followed by a positive term, followed by a geometric number of zero terms, etc. The terms in (6.21) are just the positive terms in (6.9). It also follows that $\hat{Z}_+$ and Z_+ are either both persistent or else both transient.

9.7 Further Examples of Recurrent Phenomena

Example 7.1 Let $\{S_n, n \geq 0\}$ be a discrete renewal process induced by the distribution $\{k_j, j \geq 0\}$, with the generating function

$$K(z) = \sum_{0}^{\infty} k_j z^j \quad (0 < z < 1). \tag{7.1}$$

If $k_0 > 0$ we know from the theory of branching processes that the equation $\xi \equiv zK(\xi)$ has a unique continuous root $\xi \equiv \xi(z)$ in (0,1) such that $\xi(0+) = 0$, and moreover, that as $z \to 1-, \xi(z) \to \zeta$, where $\zeta < 1$ iff $K'(1) = \rho > 1$. We are interested in the sequence $Z = \{Z_n, n \geq 0\}$, where $Z_n = 1_{\{S_n=n\}}$. We ignore the trivial case where $k_1 = 1$.

Theorem 7.1 *The sequence Z is a recurrent phenomenon.*

(i) *If $k_0 = 0, k_1 < 1$, then Z is transient and its recurrence set is $\{0, 1, 2, \ldots, N\}$, where the random variable N has the distribution*

$$P\{N = n\} = (1 - k_1)k_1^n \quad (n \geq 0). \tag{7.2}$$

(ii) *If $k_0 > 0$, then the recurrence time distribution $\{f_n\}$ has the generating function*

$$F(z) = \sum_1^\infty f_n z^n = zK'(\xi) \quad (0 < z < 1). \tag{7.3}$$

The phenomenon is persistent null if $\rho = 1$ and transient otherwise.

Proof. Let $u_n = P\{Z_n = 1\}$. Proceeding as in Sec. 9.4 we find that

$$P\{Z_{n_1} = Z_{n_2} = \cdots = Z_{n_r} = 1\} = u_{n_1} u_{n_2-n_1} \cdots u_{n_r-n_r-1} \tag{7.4}$$

for $0 = n_0 < n_1 < \cdots < n_r$ $(r \geq 1)$. This shows that Z is a recurrent phenomenon.

(i) If $k_0 = 0$, then $S_n \geq n$ $(n \geq 1)$. If $S_1 = 1$, then Z occurs at $n = 1$, while if $S_1 > 1$, then $S_n > n$ $(n \geq 2)$, so Z does not occur. This means that the recurrence time distribution is given by

$$f_1 = k_1, \quad f_n = 0 \ (n > 1). \tag{7.5}$$

Since this is a defective distribution, Z is transient. The recurrence set is given by $\{0, 1, 2, \ldots, N\}$, where N is such that $T_N < \infty$, but $T_{N+1} = \infty$. In view of (7.5) this gives (7.2).

(ii) Let $k_0 > 0$. The recurrence time distribution $\{f_n\}$ can be evaluated by direct but cumbersome probability arguments, but we prefer to use the following analytical technique. The generating function $U(z) = \sum_0^\infty u_n z^n$ is the term free of ω in the expansion of

$$\chi(\omega) = \sum_0^\infty z^n E\left(\omega^{S_n - n}\right) = \sum_0^\infty z^n \left[\frac{K(\omega)}{\omega}\right]^n. \tag{7.6}$$

Now let α be a real number $> \xi$. In the annulus $\alpha < |\omega| < 1$ we have

$$\left|\frac{K(\omega)}{\omega}\right| < \frac{K(\alpha)}{\alpha} < \frac{K(\xi)}{\xi} = \frac{1}{z},$$

so that $z|K(\omega)/\omega| < 1$ and

$$\chi(\omega) = \frac{\omega}{\omega - zK(\omega)}. \tag{7.7}$$

Therefore,

$$U(z) = \frac{1}{2\pi i}\int_{|\omega|=\alpha} \frac{\chi(\omega)}{\omega} d\omega = \frac{1}{2\pi i}\int_{|\omega|=\alpha} \frac{d\omega}{\omega - zK(\omega)}. \tag{7.8}$$

Since $z|K(\omega)| < |\omega|$ for $|\omega| = \alpha$, by Rouche's theorem, $\omega = \xi$ is the only pole of the integrand in (7.8) within the circle $|\omega| \leq \alpha$ and the integral has therefore the value

$$2\pi i \lim_{\omega\to\xi} \frac{\omega - \xi}{\omega - zK(\omega)} = \frac{2\pi i}{1 - zK'(\xi)}. \tag{7.9}$$

Therefore, $U(z) = [1 - zK'(\xi)]^{-1}$ and

$$F(z) = 1 - U(z)^{-1} = zK'(\xi)$$

which is the result (7.3). We have $F(1) = K'(1) = \rho$ if $\rho \leq 1$ and

$$\begin{aligned} F(1) &= K'(1) = \rho \quad \text{if } \rho \leq 1 \\ &= K'(\zeta) < 1 \quad \text{if } \rho > 1. \end{aligned}$$

This shows that Z is persistent iff $\rho = 1$, in which case

$$F'(1) = \rho + k''(1)\xi'(1) = \infty,$$

so that Z is null. □

Example 7.2 (The (s, S) inventory model). Let $\{X_k, k \geq 1\}$ be a sequence of independent non-negative random variables with a common distribution G, and $0 < s < S < \infty$. We define the sequence of random variables $\{I_n, n \geq 0\}$ as follows: $I_0 < s$ a.s. and for $n \geq 0$,

$$\begin{aligned} I_{n+1} &= I_n - X_{n+1} \quad \text{for } s \leq I_n \leq S \\ &= S - X_{n+1} \quad \text{for } I_n < s. \end{aligned} \tag{7.10}$$

These equations describe a model in which at epoch $n > 0$, I_n represents the inventory level of a certain material stocked at a warehouse, X_k is the demand for this material during the time interval $(k-1, k]$, the warehouse meets this demand exactly, and whenever the inventory level falls below a level s, places an order for the material to bring up the level to S. This is the so-called (s, S) inventory model.

Let E_n be the event: an order is placed at epoch n, and $Z_n = 1_{E_n}$ $(n \geq 0)$. We have the following.

Theorem 7.2 *The sequence $Z = \{Z_n, n \geq 0\}$ is a recurrent phenomenon, which is persistent non-null. Its recurrence time distribution is given by*

$$f_n = G_{n-1}(\Delta) - G_n(\Delta) \quad (n \geq 1), \tag{7.11}$$

where G_n is the n-fold convolution of G with itself $(n \geq 1), G_0(x) = 0$ for $x < 0$ and $= 1$ for $x \geq 0$, and $\Delta = S - s > 0$.

Proof. For $0 = n_0 < n_1 < \cdots < n_r$ $(r \geq 1)$ we have

$$\begin{aligned} P\{Z_{n_1} &= Z_{n_2} = \cdots = Z_{n_r} = 1\} \\ &= \prod_{t=1}^{r} P\{I_{n_t} < s | I_{n_1} < s, I_{n_2} < s, \ldots, I_{n_{t-1}} < s\}. \end{aligned} \tag{7.12}$$

Now, given $I_{n_1} < s, I_{n_2} < s, \ldots, I_{n_{t-1}} < s$, Eqs. (7.10) show that I_{n_t} depends only on $X_{n_{t-1}+m}$ $(1 \leq m \leq n_t - n_{t-1})$, and these random variables have the same distribution as $X_1, X_2, \ldots, X_{n_t - n_{t-1}}$. Therefore,

$$\begin{aligned} P\{I_{n_t} &< s | I_{n_1} < s, I_{n_2} < s, \ldots, I_{n_{t-1}} < s\} \\ &= P\{I_{n_t - n_{t-1}} < s\} = P\{Z_{n_t - n_{t-1}} = 1\}. \end{aligned} \tag{7.13}$$

Using (7.13) in (7.12) we find that

$$P\{Z_{n_1} = Z_{n_2} = \cdots = Z_{n_r} = 1\} = \prod_{t=1}^{r} P\{Z_{n_t - n_{t-1}} = 1\} \tag{7.14}$$

which shows that Z is a recurrent phenomenon. Its recurrence time distribution is given by

$$\begin{aligned} f_n &= P\{I_1 \geq s, I_2 \geq s, \ldots, I_{n-1} \geq s, I_n < s\} \\ &= P\{S - X_1 \geq s, S - X_1 - X_2 \geq s, \ldots, S - X_1 - X_2 - \cdots - X_{n-1} \\ &\qquad \geq s, S - X_1 - X_2 - \cdots - X_n < s\} \\ &= P\{S_{n-1} \leq \Delta, S_n > \Delta\} \end{aligned}$$

where $S_0 = 0$, $S_n = X_1 + X_2 + \cdots + X_n$ $(n \geq 1)$. Since G_n is the distribution of S_n, we thus obtain (7.11). Now $U(\Delta) = \Sigma_1^\infty G_n(\Delta)$ is the renewal function (evaluated at Δ) of the process $\{S_n\}$ and we know from Theorem 2.1 of Chap. 6, that $\sum_1^\infty G_n(\Delta) < \infty$. Therefore,

$$\sum_1^N f_n = 1 - G_N(\Delta) \to 1 \quad \text{as } N \to \infty \tag{7.15}$$

and

$$\mu = \sum_1^\infty n f_n = 1 + U(\Delta) < \infty. \tag{7.16}$$

Thus Z is persistent non-null. □

9.8 Regenerative Phenomena in Continuous Time

Lemma 8.1 *For $s > 0, t > 0$ we have*

$$p(s)p(t) \leq p(s+t) \leq 1 + p(s)p(t) - \max\{p(s), p(t)\}. \tag{8.1}$$

Proof. For $\alpha, \beta \in \{0, 1\}$ we have

$$\begin{aligned}
P\{Z(s) = \alpha, Z(s+t) = \beta\} &= E(-1)^{\alpha+1}[Z(s) + \alpha - 1](-1)^{\beta+1}[Z(s+t) + \beta - 1] \\
&= E(-1)^{\alpha+\beta}[Z(s)Z(s+t) + (\beta - 1)Z(s) + (\alpha - 1)Z(s+t) \\
&\quad + (\alpha - 1)(\beta - 1)] \\
&= (-1)^{\alpha+\beta}[p(s)p(t) + (\beta - 1)p(s) + (\alpha - 1)p(s+t) + (\alpha - 1)(\beta - 1)].
\end{aligned}$$

Therefore,

$$(-1)^{\alpha+\beta}[p(s)p(t) + (\beta - 1)p(s) + (\alpha - 1)p(s+t) + (\alpha - 1)(\beta - 1)] \geq 0.$$

When $\alpha = 0, \beta = 1$, this yields $p(s)p(t) - p(s+t) \leq 0$, as desired and when $\alpha = \beta = 0$ we obtain $1 - p(s+t) + p(s)p(t) \geq p(s)$. Similarly $1 - p(s+t) + p(s)p(t) \geq p(t)$, and so

$$1 - p(s+t) + p(s)p(t) \geq \max\{p(s), p(t)\},$$

as desired. □

Theorem 8.1 *Let $p \in P$. Then we have the following.*

(i) *$p(t) > 0$ for $t > 0$,*
(ii) *$p(t)$ is uniformly continuous in $(0 < t < \infty)$, and*
(iii) *the limit*

$$q = \lim_{t \to 0+} \frac{1 - p(t)}{t} \tag{8.2}$$

exists, where $0 \le q \le \infty$, and $q = 0$ iff $p(t) \equiv 1$. If $q < \infty$, then $p(t) \ge e^{-qt}$ for $t > 0$.

Proof. (i) From Lemma 8.1 we have $p(s+t) \ge p(s)p(t)$. This gives

$$p(t) \ge p\left(\frac{t}{n}\right)^n. \tag{8.3}$$

Since $p(t) \to 1$ as $t \to 0+$, the right side of (8.3) is strictly positive for sufficiently large n. This gives $p(t) > 0$.

(ii) From Lemma 8.1 we obtain

$$-p(t)[1 - p(s)] \le p(t+s) - p(t) \le [1 - p(s)][1 - p(t)]$$

so that

$$|p(t+s) - p(t)| \le 1 - p(s). \tag{8.4}$$

This shows that $p(t)$ is uniformly continuous in $(0 < t < \infty)$.

(iii) Let $\phi(t) = -\log p(t)$. From $p(s+t) \ge p(s)p(t)$ we find that

$$\phi(s+t) \le \phi(s) + \phi(t). \tag{8.5}$$

This ϕ is sub-additive in $t > 0$. By Lemma 8.2 (below)

$$\lim_{t \to 0+} \frac{\phi(t)}{t} = \sup_{t>0} \frac{\phi(t)}{t} = q \le \infty. \tag{8.6}$$

Since $p(t) \to 1$ as $t \to 0+$, we have $\phi(t) \to 0$ as $t \to 0+$ and therefore

$$\lim_{t \to 0+} \frac{1 - p(t)}{t} = \lim_{t \to 0+} \frac{1 - e^{-\phi(t)}}{t} \lim_{t \to 0+} \frac{\phi(t)}{t} = q,$$

as was to be proved. The other statements in (iii) are obvious. □

Lemma 8.2 *Let ϕ be a finite sub-additive function on $(0, \infty)$, such that $\phi(t) \to 0$ as $t \to 0+$, and*

$$\sup_{t>0} \frac{\phi(t)}{t} = q \leq \infty. \tag{8.7}$$

Then

$$\lim_{h \to 0+} \frac{\phi(h)}{h} = q. \tag{8.8}$$

Proof. If $q < \infty$ we can choose an a such that $\phi(a) > (q - \varepsilon)a$. Let $h < a$ and define the positive integer n and $\delta \in [0, h)$ by the relation $a = nh + \delta$. Then

$$q - \varepsilon < \frac{\phi(a)}{a} \leq \frac{\phi(nh)}{a} + \frac{\phi(\delta)}{a} \leq \frac{nh}{a} . \frac{\phi(h)}{h} + \frac{\phi(\delta)}{a}. \tag{8.9}$$

In this let $h \to 0$. Then $nh/a \to 1$ and $\phi(\delta) \to 0$, whence

$$q - \varepsilon \leq \liminf \frac{\phi(h)}{h} \leq \limsup \frac{\phi(h)}{h} \leq q. \tag{8.10}$$

Since ε is arbitrary, the desired result follows. If $q = \infty$ we can use the same argument, replacing $q - \varepsilon$ by M, where M is any given large number, and arrive at the same result. □

Definition 8.1 A standard regenerative phenomenon is called stable or instantaneous according as $q < \infty$ or $q = \infty$.

Example 8.1 Let $N = \{N(\tau), \tau \geq 0\}$ be a simple Poisson process with the parameter $\lambda\, (0 < \lambda < \infty)$ and $Y(\tau) = N(\tau) - \tau (\tau \geq 0)$. Then $Y = \{Y(\tau), \tau \geq 0\}$ is a Lévy process on the state space $(-\infty, \infty)$. Let $Z(\tau) = 1_{\{Y(\tau)=0\}}$ $(\tau \geq 0)$. For $0 = \tau_0 < \tau_1 < \cdots < \tau_r$ $(r \geq 1)$ we have

$$\begin{aligned}
&P\{Z(\tau_1) = Z(\tau_2) = \cdots = Z(\tau_r) = 1\} \\
&\quad = P\{Y(\tau_1) = Y(\tau_2) = \cdots = Y(\tau_r) = 0\} \\
&\quad = P\{Y(\tau_1) = 0, Y(\tau_2) - Y(\tau_1) = 0, \ldots, Y(\tau_r) - Y(\tau_{r-1}) = 0\} \\
&\quad = p(\tau_1)p(\tau_2 - \tau_1) \ldots p(\tau_r - \tau_{r-1}),
\end{aligned}$$

where

$$\begin{aligned}
p(t) = P\{Y(t) = 0\} &= e^{-\lambda t} \frac{(\lambda t)^t}{t!} && \text{for } t = 0, 1, 2, \ldots \\
&= 0 && \text{otherwise.}
\end{aligned} \tag{8.11}$$

Therefore, the process $Z = \{Z(\tau), \tau \geq 0\}$ is a regenerative phenomenon. Since $p(\tau) \to 0$ as $\tau \to 0+$, Z is *not standard.*

Let S_n be the epoch of the nth jump $(n \geq 0)$ in the Y process and $\hat{Z} = \{Z(S_n), n \geq 0\}$. The process $\hat{Z}$ is a recurrent phenomenon, which is said to be subordinate to the original phenomenon. Since $Y(S_n) = n - S_n$, the probability $u_n = P\{Z(S_n) = 1\}$ is given by

$$u_0 = 1, \quad u_n = 0 \ (n \geq 1). \tag{8.12}$$

Thus $\hat{Z}$ is a trivial phenomenon. This should be compared with the phenomenon described in Example 7.1, where S_n has a discrete distribution.

Example 8.2 Let N be the Poisson process of Example 8.1, $Y(\tau) = N(\tau) + \tau$ $(\tau \geq 0)$, and R the range of Y, namely,

$$R = \{t \geq 0\colon Y(\tau) = t \quad \text{for some } \tau \geq 0\}. \tag{8.13}$$

Let $Z(t) = 1_{\{t \in R\}}$ and $Z = \{Z(t), t \geq 0\}$. If $Y(\tau) = t$, then $N(\tau) = t - \tau$ and so we must have $\tau = t, t-1, \ldots, t - [t]$. Therefore,

$$p(t) = P\{Z(t) = 1\} = \sum_{n=0}^{[t]} e^{-\lambda(t-n)} \frac{\lambda^n (t-n)^n}{n!}. \tag{8.14}$$

For $0 = t_0 < t_1 < \cdots < t_r$ $(r \geq 1)$ we have

$$\begin{aligned}
&P\{Z(t_1) = Z(t_2) = \cdots = Z(t_r) = 1\} \\
&\quad = P\left\{ \bigcup_{\tau_1 \geq 0} \{Y(\tau_1) = t_1\}, \bigcup_{\tau_2 \geq 0} \{Y(\tau_2) = t_2\}, \ldots, \bigcup_{\tau_r \geq 0} \{Y(\tau_r) = t_r\} \right\} \\
&\quad = \sum_{0 < \tau_1 < \tau_2 < \cdots < \tau_r} P\{Y(\tau_1) = t_1, Y(\tau_2) = t_2, \ldots, Y(\tau_r) = t_r\} \\
&\quad = \sum P\{Y(\tau_1) = t_1, Y(\tau_2) - Y(\tau_1) = t_2 - t_1, \ldots, Y(\tau_r) \\
&\qquad\qquad - Y(\tau_{r-1}) = t_r - t_{r-1}\} \\
&\quad = \sum_{\tau_1 > 0} P\{Y(\tau_1) = t_1\} \sum_{\tau_2 > \tau_1} P\{Y(\tau_2 - \tau_1) = t_2 - t_1\} \ldots \\
&\qquad \times \sum_{\tau_r > \tau_{r-1}} P\{Y(\tau_r - \tau_{r-1}) = t_r - t_{r-1}\} \\
&\quad = p(t_1) p(t_2 - t_1) \cdots p(t_r - t_{r-1}).
\end{aligned}$$

This shows that Z is a regenerative phenomenon. It is *standard* and *stable*, since $p(h) \to 1$ as $h \to 0+$ and $[1 - p(h)]h^{-1} \to \lambda$ as $h \to 0+$. Now

$$p(1) = \sum_0^1 e^{-\lambda(1-n)} \frac{\lambda^n (1-n)^n}{n!} = e^{-\lambda}$$

$$p(1+h) = \sum_0^1 e^{-\lambda(1+h-n)} \frac{\lambda^n (1+h-n)^n}{n!} = e^{-\lambda-\lambda h} + e^{-\lambda h}\lambda h$$

$$p(1-h) = \left\{ e^{-\lambda(1-h-n)} \frac{\lambda^n (1-h-n)^n}{n!} \right\}_{n=0} = e^{-\lambda+\lambda h}.$$

Therefore, as $h \to 0+$

$$\frac{p(1+h) - p(1)}{h} = e^{-\lambda h}\lambda + e^{-\lambda} \frac{e^{-\lambda h} - 1}{h} \quad \to \quad \lambda - \lambda e^{-\lambda}$$

and

$$\frac{p(1-h) - p(1)}{-h} = e^{-\lambda} \frac{e^{\lambda h} - 1}{-h} \quad \to \quad -\lambda e^{-\lambda}.$$

Thus, $D_+ p(1) = \lambda - \lambda e^{-\lambda}$, while $D_- p(1) = -\lambda e^{-\lambda}$.

Theorem 8.2 *Let $p \in P$. Then there exists a unique positive measure μ on $(0, \infty]$ with*

$$\int_0^{\infty+} (1 - e^{-x})\mu\{dx\} < \infty \tag{8.15}$$

such that for $\theta > 0$,

$$p^*(\theta) = \int_0^\infty e^{-\theta t} p(t) dt = \left[\theta + \int_0^{\infty+} (1 - e^{-\theta x})\mu\{dx\} \right]^{-1}. \tag{8.16}$$

Proof. For $h > 0$ consider $Z_h = \{Z(nh), n \geq 0\}$, which is the discrete skeleton of Z at scale h. Clearly, $\{p(nh)\} \in R$, and by Theorem 2.1, there exists a sequence $\{f_n(h)\}$ such that $f_n(h) \geq 0$, $\sum_1^\infty f_n(h) \leq 1$, and

$$\sum_0^\infty p(nh)s^n = \frac{1}{1 - \sum_1^\infty f_n(h)s^n} = \frac{1}{\sum_1^{\infty+} f_n(h)(1 - s^n)},$$

where $f_\infty(h) = 1 - \sum_1^\infty f_n(h) \geq 0$. For $\theta > 0$ choose $s = e^{-\theta h}$. Then

$$\sum_0^\infty p(nh)e^{-\theta nh} = \left[\sum_0^{\infty+} f_n(h)(1 - e^{-\theta nh}) \right]^{-1}.$$

Now, by the dominated convergence theorem we have

$$p^*(\theta) = \lim_{h\to 0+} \int_0^\infty p\left(\left[\frac{t}{h}\right] h\right) e^{-\theta t} dt.$$

Here the integral can be written as

$$\begin{aligned}
&\sum_{n=0}^{\infty} \int_{nh}^{nh+h} p\left(\left[\frac{t}{h}\right] h\right) e^{-\theta t} dt \\
&\qquad = \theta^{-1} \sum_{n=0}^{\infty} p(nh) \left[e^{-\theta nh} - e^{-\theta(n+1)h}\right] \\
&\qquad = \frac{1-e^{-\theta h}}{\theta} \sum_{0}^{\infty} p(nh) e^{-\theta nh} \\
&\qquad = \frac{1-e^{-\theta h}}{\theta h} \left[\int_{0-}^{\infty+} k(x,\theta)\lambda_h\{dx\}\right]^{-1},
\end{aligned}$$

where

$$k(x,\theta) = \frac{1-e^{-\theta x}}{1-e^{-x}} \quad (0 \le x \le \infty),$$

the function being defined by continuity at $x = 0$ and $x = \infty$, and

$$\lambda_h\{nh\} = h^{-1} f_n(h)(1-e^{-nh}) \quad (1 \le n \le \infty).$$

We have thus proved that

$$p^*(\theta)^{-1} = \lim_{h\to 0+} \left[\int_{0-}^{\infty+} k(x,\theta)\lambda_h\{dx\}\right]. \tag{8.17}$$

Clearly, λ_h is a finite positive measure, and

$$\infty > p^*(1)^{-1} = \lim_{h\to 0+} \lambda_h\{[0,\infty]\} \tag{8.18}$$

so that the total measure of λ_h remains bounded as $h \to 0+$. An important property of the function k is that if for two positive measures λ, λ' on $[0,\infty]$

$$\int_{0-}^{\infty+} k(x,\theta)\lambda\{dx\} = \int_{0-}^{\infty+} k(x,\theta)\lambda'\{dx\} \quad \text{for all } \theta > 0, \tag{8.19}$$

then $\lambda \equiv \lambda'$. This is seen as follows. From (8.19) we find that

$$\int_{0-}^{\infty+} [k(x,\theta+1) - k(x,\theta)]\lambda\{dx\} = \int_{0-}^{\infty+} [k(x,\theta+1) - k(x,\theta)]\lambda'\{dx\}.$$

This reduces to

$$\int_{0-}^{\infty} e^{-\theta x}\lambda\{dx\} = \int_{0-}^{\infty} e^{-\theta x}\lambda'\{dx\}$$

so that $\lambda \equiv \lambda'$ on $[0, \infty)$. A second implication of (8.19) is that

$$\int_{0-}^{\infty+} k(x,1)\lambda\{dx\} = \int_{0-}^{\infty+} k(x,1)\lambda'\{dx\},$$

which gives $\lambda\{[0,\infty]\} = \lambda'\{[0,\infty]\}$. Hence $\lambda \equiv \lambda'$ on $[0,\infty]$, as was to be proved. From this fact, (8.17) and (8.18), $\lambda = \lim_{h\to 0+} \lambda_h$ exists by Lemma 8.3 (below) and

$$p^*(\theta)^{-1} = \left[\int_{0-}^{\infty+} k(x,\theta)\lambda\{dx\}\right]. \tag{8.20}$$

It remains to express (8.20) in the desired form. Now

$$1 = \lim_{t\to 0+} p(t) = \lim_{\theta\to\infty} \theta p^*(\theta) = \lim_{\theta\to\infty}\left[\int_{0-}^{\infty+} \frac{1-e^{-\theta x}}{\theta(1-e^{-x})}\lambda\{dx\}\right]^{-1} = \lambda\{0\}^{-1}$$

by the dominated convergence theorem. Therefore (8.20) can be written as

$$\begin{aligned} p^*(\theta) &= \left[\theta + \int_{0+}^{\infty+} \frac{1-e^{-\theta x}}{1-e^{-x}}\lambda\{dx\}\right]^{-1} \\ &= \left[\theta + \int_{0}^{\infty+} (1-e^{-\theta x})\mu\{dx\}\right]^{-1}, \end{aligned}$$

where $\mu\{dx\} = (1-e^{-x})^{-1}\lambda\{dx\}$, for $x > 0$. We have thus established (8.16), and (8.15) is satisfied in view of (8.18). To prove the uniqueness of μ, suppose that there is a second measure μ' with $p^*(\theta)$ expressed in terms of μ' as in (8.16). Then

$$\int_{0}^{\infty+} (1-e^{-\theta x})\mu\{dx\} = \int_{0}^{\infty+} (1-e^{-\theta x})\mu'\{dx\}. \tag{8.21}$$

Define λ' by setting $\lambda'\{dx\} = (1-e^{-x})\mu'\{dx\}$ for $x > 0, \lambda'\{0\} = 1$. We can then write (8.21) in the form (8.19), and this means that $\lambda \equiv \lambda'$. □

Lemma 8.3 *Let I be a fixed interval of the form $[a,b]$, where $-\infty \le a < b \le \infty$, M the class of finite positive Borel measures on I, and C a class of* *continuous real-valued functions f such that whenever two measures $\lambda, \lambda' \in M$ have*

$$\int_I f(x)\lambda\{dx\} = \int_I f(x)\lambda'\{dx\} \tag{8.22}$$

for all $f \in C$, then $\lambda \equiv \lambda'$. Let $\{\lambda_n\}$ be a sequence of measures in M for which

$$\sup_n \lambda_n\{I\} < \infty. \tag{8.23}$$

If

$$\lim_{n\to\infty} \int_I f(x)\lambda_n\{dx\} \tag{8.24}$$

exists for all $f \in C$, then

$$\lambda = \lim_{n\to\infty} \lambda_n \tag{8.25}$$

exists and hence

$$\lim_{n\to\infty} \int_I f(x)\lambda_n\{dx\} = \int_I f(x)\lambda\{dx\}. \tag{8.26}$$

For proof, see Kingman (1972).

Corollary 8.1 *We have*

$$\mu\{(0,\infty]\} = q \leq \infty \tag{8.27}$$

where q was defined in Theorem 8.1(iii).

Proof. For $\theta > 0$ we have

$$\begin{aligned}\int_0^\infty \frac{1-p(x/\theta)}{x/\theta} xe^{-x}dx &= \int_0^\infty \frac{1-p(t)}{t}\theta te^{-\theta t}\theta dt \\ &= \theta^2\left[\frac{1}{\theta} - p^*(\theta)\right] \\ &= \theta p^*(\theta)\int_0^{\infty+}(1-e^{-\theta x})\mu\{dx\}.\end{aligned}$$

Now

$$\lim_{\theta\to\infty} \theta p^*(\theta) = \lim_{t\to 0+} p(t) = 1$$

since $p \in P$, and

$$\lim_{\theta \to \infty} \int_0^{\infty+} (1 - e^{-\theta x})\mu\{dx\} = \mu\{(0, \infty]\}$$

by the monotone convergence theorem. Therefore,

$$\mu\{(0, \infty]\} = \lim_{\theta \to \infty} \int_0^{\infty} \frac{1 - p(x/\theta)}{x/\theta} x e^{-x} dx. \tag{8.28}$$

(i) Let $q < \infty$. We have

$$\frac{1 - p(x/\theta)}{x/\theta} \leq q,$$

so by the dominated convergence theorem

$$\lim_{\theta \to \infty} \int_0^{\infty} \frac{1 - p(x/\theta)}{x/\theta} x e^{-x} dx \to \int_0^{\infty} q x e^{-x} dx = q. \tag{8.29}$$

(ii) Let $q = \infty$. By Fatou's lemma

$$\liminf_{\theta \to \infty} \int_0^{\infty} \frac{1 - p(x/\theta)}{x/\theta} x e^{-x} dx \geq \int_0^{\infty} \infty \cdot x e^{-x} dx = \infty$$

so that

$$\lim_{\theta \to \infty} \int_0^{\infty} \frac{1 - p(x/\theta)}{x/\theta} x e^{-x} dx = \infty. \tag{8.30}$$

We have thus shown that the limit in (8.28) is $q \leq \infty$. □

Definition 8.2 The measure μ obtained in Theorem 8.2 is called the canonical measure of the phenomenon Z.

Corollary 8.1 shows that Z is stable iff its canonical measure is totally finite.

Definition 8.3 A standard regenerative phenomenon Z is called

(i) transient if $\mu\{\infty\} > 0$;
(ii) persistent null if $\mu\{\infty\} = 0$ and $\int_0^{\infty} x\mu\{dx\} = \infty$;
(iii) persistent non-null if $\mu\{\infty\} = 0$ and $\int_0^{\infty} x\mu\{dx\} < \infty$;

Theorem 8.3 *Let Z be a standard regenerative phenomenon with the p-function p. Then, Z is transient iff*

$$\int_0^{\infty} p(t)dt < \infty. \tag{8.31}$$

Proof. By the monotone convergence theorem we have

$$\begin{aligned}\infty \geq \int_0^\infty p(t)dt &= \lim_{\theta\to 0+} \int_0^\infty e^{-\theta t} p(t)dt \\ &= \lim_{\theta\to 0+} \left[\theta + \int_0^{\infty+} (1-e^{-\theta x})\mu\{dx\}\right]^{-1} \\ &= [\mu\{\infty\}]^{-1} < \infty \quad \text{if } Z \text{ is transient} \\ &= \infty \quad \text{if } Z \text{ is persistent.}\end{aligned}$$

□

Theorem 8.4 *Let $p \in P$. Then*

$$\begin{aligned}\lim_{t\to\infty} p(t) &= 0 && \textit{if } Z \textit{ is transient or persistent null} \\ &= \left[1 + \int_0^\infty x\mu\{dx\}\right]^{-1} && \textit{if } Z \textit{ is transient non-null.}\end{aligned}$$

Proof. As in the proof of Theorem 8.2, the sequence $\{p(nh)\} \in R$ and as $n \to \infty$, $p(nh)$ converges to a limit by Theorem 2.5. Therefore, given $\varepsilon > 0$, there exists an N such that

$$|p(nh) - p(n'h)| < \frac{1}{3}\varepsilon \quad \text{for } n, n' > N. \tag{8.32}$$

Since p is uniformly continuous by Theorem 8.1(ii), we may choose h so that the oscillation of p in any interval of length h is less than $\frac{1}{3}\varepsilon$. It follows that

$$|p(t) - p(t')| < \varepsilon \quad \text{for } t, t' > Nh, \tag{8.33}$$

proving the existence of the limit of $p(t)$ as $t \to \infty$. The actual value of the limit is given by

$$\begin{aligned}\lim_{t\to\infty} p(t) &= \lim_{\theta\to 0+} \theta p^*(\theta) \\ &= \lim_{\theta\to 0+} \left[1 + \int_0^{\infty+} \frac{1-e^{-\theta x}}{\theta}\mu\{dx\}\right]^{-1} \\ &= \left[1 + \int_0^{\infty+} x\mu\{dx\}\right]^{-1}\end{aligned} \tag{8.34}$$

by the monotone convergence theorem. This leads to the desired results. □

Theorems 8.3 and 8.4 are the analogues of Theorems 2.3 and 2.5 for discrete phenomena. It should be noted in view of Theorem 8.1(i), that continuous time phenomena cannot be periodic.

To investigate further properties of the p-function we introduce the function

$$m(t) = \mu\{(t, \infty]\} \quad (t > 0). \tag{8.35}$$

The following properties of m will be used in the sequel:

(i) $0 \leq m(t) < \infty$ for $t > 0$.

(ii) m is non-increasing: for $0 < a < b < \infty$ we have

$$m(a) - m(b) = \mu\{(a, b]\} \geq 0.$$

(iii) m is right-continuous: as $h \to 0+$, we have

$$m(t) - m(t + h) = \mu\{(t, t + h]\} \to 0$$

while

$$m(t - h) - m(t) = \mu\{(t - h, t]\} \to \mu\{t\}.$$

(iv) $\int_0^1 m(t)dt < \infty$, since

$$\int_0^1 m(t)dt = \int_0^\infty \min(1, x)\mu\{dx\} \leq \int_0^\infty 2(1 - e^{-x})\mu\{dx\} < \infty.$$

(v) Let $m_n(t)$ be the n-fold convolution of $m(t)$ with itself: that is, $m_1(t) = m(t)$, and for $n \geq 2$

$$m_n(t) = t^{n-1} \int_{U_n} m(ty_1)m(ty_2) \cdots m(ty_n)\lambda_{n-1}\{dy\}, \tag{8.36}$$

where U_n is the simplex

$$U_n = \left\{y = (y_1, y_2, \ldots, y_n)\colon y_j \geq 0, \sum_1^n y_j = 1\right\}$$

and λ_{n-1} is the normalized $(n-1)$-dimensional Lebesgue measure on U_n. For $n \geq 2$ the integrand is continuous almost everywhere modulo λ_{n-1} since m has at most a countable number of discontinuities. Hence m_n is continuous for $n \geq 2$.

Lemma 8.4 (a) *The series* $\sum_{n=1}^\infty \int_0^t m_n(s)ds$ *converges for every* $t \in (0, \infty)$. (b) *The series* $\sum_{n=1}^\infty (-1)^{n-1} m_n(t)$ *is absolutely convergent, uniformly in any compact sub-interval of* $(0, \infty)$.

Proof. For $0 < t < \infty$, denote

$$A = \int_0^t m(s)ds,$$

which is finite by property (iv) above, Define

$$\begin{aligned} f(s) &= \frac{m(s)}{A} \quad \text{for } 0 < s < t \\ &= 0 \qquad\quad \text{for } s \geq t. \end{aligned}$$

Then, f is a probability density. For the density f_n of the sum $S_n = X_1 + X_2 + \cdots + X_n$ $(n \geq 1)$ we have $f_n(s) = A^{-n} m_n(s)$ for $0 < s < t$, so

$$\int_0^t m_n(s)ds = A^n P\{S_n \leq t\}.$$

Now for $\theta > 0$ we have

$$E\big(e^{-\theta S_n}\big) \geq e^{-\theta t} P\{S_n \leq t\} = e^{-\theta t} A^{-n} \int_0^t m_n(s)ds,$$

and therefore

$$\begin{aligned} \int_0^t m_n(s)ds \leq e^{\theta t} A^n E(e^{-\theta S}n) &= e^{\theta t}[E(Ae^{-\theta X_1})]^n \\ &= e^{\theta t} m^*(\theta)^n, \end{aligned}$$

where

$$m_t^*(\theta) = \int_0^t e^{-\theta s} m(s)ds.$$

Choosing θ so large that $m_t^*(\theta) < \delta < 1$, we obtain

$$\int_0^t m_n(s)ds < e^{\theta t}\delta^n,$$

which gives

$$\int_0^t m_n(s)ds = 0(\delta^n). \tag{8.37}$$

This leads to the result (a), and the proof of (b) is similar. See Kingman (1972). □

Theorem 8.5 (The Integral Equation of Regenerative Phenomena) *Let Z be a standard regenerative phenomenon with the p-function p and canonical measure μ. Let the function m be defined by (8.35). Then p is the unique solution of the integral equation*

$$p(t) + \int_0^t p(t-s)m(s)ds = 1 \tag{8.38}$$

which is bounded over all finite intervals. We can express this solution as

$$p(t) = 1 - \int_0^t \sum_{n=1}^{\infty} (-1)^{n-1} m_n(s)ds. \tag{8.39}$$

Proof. For $\theta > 0$ let

$$m^*(\theta) = \int_0^\infty e^{-\theta t} m(t)dt. \tag{8.40}$$

We have

$$\begin{aligned}\theta m^*(\theta) &= \theta \int_0^\infty e^{-\theta t} \int_t^{\infty+} \mu\{dx\}dt = \int_0^{\infty+} \mu\{dx\} \int_0^x \theta e^{-\theta t}dt \\ &= \int_0^{\infty+} (1 - e^{-\theta x})\mu\{dx\} = p^*(\theta)^{-1} - \theta,\end{aligned}$$

which can be written as

$$p^*(\theta) + p^*(\theta)m^*(\theta) = \frac{1}{\theta}. \tag{8.41}$$

Inverting this transform we find that p satisfies the integral equation (8.38). Proceeding as in the proof of Theorem 2.6 of Chap. 6 and using Lemma 8.4(a) we arrive at the unique solution (8.39), which is bounded over finite intervals. □

Theorem 8.6 *Let $p \in P$. Then*

(i) *p is absolutely continuous in $t \geq 0$, and*
(ii) *the left and right derivatives of $p(t)$ exist and are finite for all $t > 0$, and are given respectively by*

$$D_-p(t) = D_+p(t) - \mu\{t\}, \quad D_+p(t) = \sum_1^\infty (-1)^n m_n(t). \tag{8.42}$$

Thus, p is differentiable almost everywhere. It is differentiable in $t > 0$ iff the measure μ is continuous. In this case $p'(t)$ is continuous in $t > 0$ and is given by

$$p'(t) = \sum_{1}^{\infty} (-1)^n m_n(t). \tag{8.43}$$

Proof. We can write (8.39) as

$$p(t) = 1 - \int_0^t m(s)ds - \int_0^t c(s)ds, \tag{8.44}$$

where

$$c(t) = \sum_{n=2}^{\infty} (-1)^{n-1} m_n(t). \tag{8.45}$$

Since m_n is continuous for $n \geq 2$, it follows from Lemma 8.4(b) that c is continuous. From (8.44) we find that p has finite left and right derivatives

$$D_- p(t) = -m(t-) - c(t), \quad D_+ p(t) = -m(t) - c(t).$$

Since the discontinuities of m are at the atoms of μ the desired results (ii) follow. The other statements of the theorem are obvious. □

Example 8.3 (Continuation of Example 8.2) We found that

$$p(t) = \sum_{n=0}^{[t]} e^{-\lambda(t-n)} \frac{\lambda^n}{n!} (t-n)^n$$

(Eq. (8.14)). The Laplace transform of this p-function is given by

$$\begin{aligned}
p^*(\theta) &= \sum_{n=0}^{\infty} \int_n^{n+1} e^{-\theta t} \sum_{m=0}^{n} e^{-\lambda(t-m)} \frac{\lambda^m}{m!} (t-m)^m dt \\
&= \sum_{m=0}^{\infty} \sum_{n=m}^{\infty} \int_n^{n+1} e^{-\theta t - \lambda(t-m)} \frac{\lambda^m}{m!} (t-m)^m dt \\
&= \sum_{m=0}^{\infty} \int_m^{\infty} e^{-\theta t - \lambda(t-m)} \frac{\lambda^m}{m!} (t-m)^m dt \\
&= \sum_{m=0}^{\infty} \frac{1}{\theta + \lambda} \left(\frac{\lambda e^{-\theta}}{\theta + \lambda} \right)^m = \frac{1}{\theta + \lambda - \lambda e^{-\theta}} \quad (\theta > 0).
\end{aligned} \tag{8.46}$$

Comparing this with the result of Theorem 8.2 we see that the canonical measure in this case has a single atom at 1 with weight λ. The phenomenon is thus stable (which we already know). It is obviously persistent non-null and

$$\lim_{t\to\infty} p(t) = \frac{1}{1+\lambda}. \tag{8.47}$$

We have already found that $p'(t)$ is discontinuous at $t = 1$. This is in agreement with Theorem 8.6(ii).

9.9 Stable Regenerative Phenomena

Stable regenerative phenomena are especially easy to deal with. In fact such phenomena arise from a generalization of Example 8.2. We have the following important result.

Theorem 9.1 (i) *Let $Y = \{Y(\tau), \tau \geq 0\}$ be a compound Poisson process on $[0, \infty)$ with unit positive draft, and R its range:*

$$R = \{t \geq 0 \colon Y(\tau) = t \quad \text{for some } \tau \geq 0\}. \tag{9.1}$$

Let $Z(t) = 1_{t\in R}$ $(t \geq 0)$ and $Z = \{Z(t)\}$. Then Z is a stable regenrative phenomenon.

(ii) *Conversely, any stable regenerative phenomenon is equivalent to a phenomenon constructed as in (i) in the sense that they have the same p-function.*

Proof. (i) We can express $Y(\tau)$ as

$$Y(\tau) = X_1 + X_2 + \cdots + X_{N(\tau)} + \tau, \tag{9.2}$$

where $\{X_k, k \geq 1\}$ is a sequence of mutually independent random variables with a common distribution F concentrated on $(0, \infty]$, and $N = \{N(\tau), \tau \geq 0\}$ is a simple Poisson process with parameter λ $(0 < \lambda < \infty)$, independent of $\{X_k\}$. Let $T_0 = 0$ and for $n \geq 1$, let T_n be the epoch of the nth jump in the N process. Denote

$$S_0 = 0, \quad S_n = T_n + X_1 + X_2 + \cdots + X_n \ (n \geq 1). \tag{9.3}$$

We can then write

$$R = \bigcup_{n=0}^{\infty} [S_n, S_n + \tau_{n+1}), \tag{9.4}$$

where $\tau_n = T_n - T_{n-1} (n \geq 1)$. Proceeding as in Example 8.2 we can show that Z is a regenerative phenomenon. Its p-function will be expressed in terms of the renewal measure

$$U\{I\} = \sum_{n=0}^{\infty} P\{S_n \in I\}. \tag{9.5}$$

Since the random variables T_n have an absolutely continuous distribution, so does S_n, and consequently U has a density u in $(0, \infty)$. Now

$$\begin{aligned}
p(t) &= P\left\{ t \in \bigcup_{n=0}^{\infty} [S_n, S_n + \tau_{n+1}) \right\} \\
&= \sum_{n=0}^{\infty} P\{S_n \leq t < S_n + \tau_{n+1}\} \\
&= \sum_{n=0}^{\infty} \int_{0-}^{t} P\{S_n \in ds\} P\{S_n + \tau_{n+1} > t | S_n = s\} \\
&= \int_{0-}^{t} U\{ds\} e^{-\lambda(t-s)} \\
&= e^{-\lambda t} + \int_0^t e^{-\lambda(t-s)} u(s) ds.
\end{aligned} \tag{9.6}$$

This shows that $p(t) \to 1$ as $t \to 0+$, so that the phenomenon is *standard*. Also,

$$\frac{1-p(t)}{t} = \frac{1-e^{-\lambda t}}{t} - \frac{1}{t} \int_0^t e^{-\lambda(t-s)} u(s) ds \quad \to \quad \lambda \quad \text{as } t \to 0+,$$

which shows that Z is *stable*. The transformation of the p-function is given by

$$\begin{aligned}
p^*(\theta) &= \frac{1}{\theta+\lambda} \int_{0-}^{\infty} e^{-\theta t} U\{dt\} \\
&= \frac{1}{\theta+\lambda} \cdot \frac{1}{1 - \frac{\lambda}{\theta+\lambda} F^*(\theta)} = \frac{1}{\theta + \lambda - \lambda F^*(\theta)},
\end{aligned} \tag{9.7}$$

where $F^*(\theta)$ is the Laplace transform of the distribution F.

(ii) To prove the converse we suppose that Z is a stable regenerative phenomenon. For its p-function p we have by Theorem 8.2

$$p^*(\theta) = \int_0^\infty e^{-\theta t} p(t) dt = \frac{1}{\theta + \int_0^{\infty+}(1 - e^{-\theta x})\mu\{dx\}} \tag{9.8}$$

with $\mu\{(0, \infty]\}$ finite. Let us denote $\mu\{(0, \infty]\} = \lambda$ $(0 < \lambda < \infty)$ and define a probability distribution F by setting

$$\lambda F\{I\} = \mu\{I \cap (0, \infty]\}. \tag{9.9}$$

Proceeding as in (i) we can obtain a standard a regenerative phenomenon Z', whose p-function p' satisfies the relation

$$\int_0^\infty e^{-\theta t} p'(t) dt = \frac{1}{\theta + \lambda - \lambda F^*(\theta)}.$$

Using (9.9) we find that

$$\int_0^\infty e^{-\theta t} p'(t) dt = \frac{1}{\theta + \int_0^{\infty+}(1 - e^{-\theta x})\mu\{dx\}}. \tag{9.10}$$

Thus,

$$\int_0^\infty e^{-\theta t} p(t) dt = \int_0^\infty e^{-\theta t} p'(t) dt.$$

Since p-functions are continuous, this gives $p \equiv p'$, so that Z and Z' are equivalent. □

In view of Theorem 9.1 we shall speak of a stable regenerative phenomenon induced by (λ, F), where $0 < \lambda < \infty$, and F is a distribution concentrated on $(0, \infty]$.

Theorem 9.2 *A stable regenerative phenomenon Z induced by (λ, F) is persistent iff F is proper. In this case Z is non-null or null according as $\nu < \infty$ or $\nu = \infty$, where*

$$\nu = \int_0^\infty x F\{dx\}. \tag{9.11}$$

In the non-null case

$$p(t) \rightarrow \frac{1}{1 + \lambda\nu} \quad \text{as } t \rightarrow \infty. \tag{9.12}$$

Proof. We recall that $\lambda F\{I\} = \mu\{I \cap (0,\infty]\}$. In particular, $F\{\infty\} = \lambda^{-1}\mu\{\infty\}$, so that Z is persistent iff $F\{\infty\} = 0$. The other results follow from Definition 8.3 and Theorem 8.4. (The result (9.12) also follows from (9.6) by the Key Renewal Theorem.) □

9.10 Problems for Solution

1. The recurrence time distribution is given by $f_n = pq^{n-1}$ $(n \geq 1)$, where $0 < p < 1$ and $q = 1 - p$. Show that $N(n)$ has the binomial distribution $\binom{n}{k}p^k q^{n-k}$ $(0 \leq k \leq n)$.
2. In Sec. 9.3 suppose that $\{S_n\}$ is a Poisson renewal process with parameter μ $(0 < \mu < \infty)$, and let

$$p = \frac{\mu}{\lambda + \mu} \quad \text{and} \quad q = \frac{\lambda}{\lambda + \mu}.$$

Show that, for $k \geq 1$,

$$P\{T_k - T_{k-1} = n, Y_k = r\} = q^n p^r \quad (n \geq 1,\ r \geq 1).$$

Use this to show that

$$P\{C_n = k\} = q^n \binom{-n}{k} (-p)^k \quad (k \geq 0).$$

3. For $d \geq 2$, let $X_k = \{X_k^{(1)}, X_k^{(2)}, \ldots, X_k^{(d)}\}$ $(k \geq 1)$ be a sequence of independent random vectors for which

$$P\{X_k = e_i\} = \frac{1}{d} \quad (i = 1, 2, \ldots, d),$$

where $e_1 = (1,0,0,\ldots,0)$, $e_2 = (0,1,0,\ldots,0), \ldots,$ and $e_d = (0,0,\ldots,0,1)$. For $n \geq 1, i \geq 1$, let $S_n^i = X_1^{(i)} + X_2^{(i)} + \cdots + X_n^{(i)}$ and Z_n be the indicator of the event

$$E_n = \{S_n^{(1)} = S_n^{(2)} = \cdots = S_n^{(d)}\}.$$

Prove that $Z = \{Z_n, n \geq 1\}$ is a periodic recurrent phenomenon which is persistent null if $d \leq 3$ and transient otherwise. (This is the general problem of returns to equilibrium. The case $d = 2$ corresponds to cointossing and the case $d = 6$ to tossing of a die.)
4. Let d (≥ 2) unbiased coins be tossed repeatedly, and Z be the phenomenon that the accumulated number of heads is the same for all d coins. Show that Z is a recurrent phenomenon which is persistent

null if $d \leq 3$ but transient otherwise. (This is the problem of ties in multiple coin games.)

5. Let d unbiased coins be tossed repeatedly, and Z be the phenomenon that for each of the d coins the accumulated number of heads and tails is equal. Show that Z is a recurrent phenomenon, which is persistent null if $d \leq 2$ and transient if $d \geq 3$. For $d = 3$ show that the probability of recurrence is approximately $1/3$.
6. In Sec. 9.6, let $\zeta_0 = \{n \geq 0 \colon M_n = S_n = 0\}$ and $Z_n^0 = 1_{\{n \in \zeta_0\}}$. Show that

   (i) $Z_0 = \{Z_n^0, n \geq 0\}$ is a recurrent phenomenon; and
   (ii) except when $X_k = 0$ a.s., Z_0 is transient.

7. *The* (s, S) *inventory model.* Show that the distribution of I_n can be expressed in the form

$$P\{I_n \leq x\} = \sum_{m=1}^{n} U_{n-m} H_m(x) \quad (x \leq s)$$

   and find $H_m(x)$. Also, find the limit distribution of I_n as $n \to \infty$.
8. In Example 7.1, by choosing the Poisson distribution $e^{-\lambda}\lambda^j/j!$ $(j = 0, 1, \dots)$ for the X_k show that if $\lambda \neq 1$,

$$\sum_{n=0}^{\infty} e^{-n\lambda} \frac{(n\lambda)^n}{n!} = (1 - \lambda\zeta)^{-1},$$

   where ζ is the least positive root of the equation $e^{-\lambda + \lambda\zeta} = \zeta$.
9. (Continuation). Show more generally that if $\lambda a \neq 1$,

$$\sum_{n=0}^{\infty} e^{-\lambda(x+na)} \frac{\lambda^n}{n!}(x + na)^n = e^{-\lambda x(1-\zeta)}(1 - \lambda a\zeta)^{-1},$$

   where ζ is the least positive root of the equation $e^{-\lambda a(1-\zeta)} = \zeta$. (The case $\lambda a > 1$ corresponds to the result of J.L.W.V. Jensen.)
10. Let $\{T_n, n \geq 0\}$ be a renewal process induced by the density $\lambda e^{-\lambda t}$ $(0 < \lambda < \infty)$, and $\{Y_n, n \geq 1\}$ an independent sequence of independent positive random variables with a common distribution function F. Let

$$\begin{aligned} Z(t) &= 0 \quad \text{if} \quad t \in \bigcup_{n=1}^{\infty} [T_n,\ T_n + Y_n) \\ &= 1 \quad \text{otherwise.} \end{aligned}$$

Show that $Z(t), t > 0$ is a regenerative phenomenon with the p-function

$$p(t) = e^{-\lambda \int_0^t [1-F(x)]dx}.$$

11. Let $Z' = \{Z'_n,\ n \geq 0\}$ be a recurrent phenomenon with the recurrence time distribution $\{f_n, n \geq 1\}$ and $\{T_n, n \geq 0\}$ a renewal process induced by the density $\lambda e^{-\lambda t}(0 < \lambda < \infty)$, independently of Z'. Define $Z(t)$ as follows:

$$Z(t) = Z'_n \quad \text{for } T_n \leq t < T_{n+1} \quad (n \geq 0).$$

Show that the process $Z = \{Z(t),\ t \geq 0\}$ is a stable regenerative phenomenon and that its canonical measure μ is absolutely continuous in $(0, \infty)$ with density

$$\sum_{n=2}^{\infty} f_n e^{-\lambda t} \lambda^n \frac{t^{n-2}}{(n-2)!}$$

and has an atom at ∞ with weight $\mu[\infty] = \lambda(1 - \sum_1^\infty f_n) \geq 0$.

12. Let $Z = \{Z(t),\ t \geq 0\}$ be a standard regenerative phenomenon and $\{T_n, n \geq 0\}$ a renewal process induced by the density $\lambda e^{-\lambda t}$ $(0 < \lambda < \infty)$, independently of Z.

    (i) Show that $\{Z(T_n), n \geq 0\}$ is a recurrent phenomenon whose recurrence time distribution $\{f_n\}$ is given by

    $$\sum_1^{\infty} f_n z^n = \frac{\lambda r(\lambda) z}{1 - r(\lambda) \sum_1^\infty \mu_n z^n},$$

    where

    $$\mu_n = \int_0^\infty e^{-\lambda t} \frac{(\lambda t)^n}{n!} \mu(dt) \quad (n \geq 1),$$

    μ being the canonical measure of Z and r the Laplace transform of its p-function.

    (ii) Show that $\{Z(T_n)\}$ is persistent if and only if Z is so.

Further Reading

Feller, W (1968). *An introduction to Probability Theory and Its Applications*: Volume 1, 3rd Edition. New York: John Wiley.

Kingman, JFC (1972). *Regenerative Phenomena*. New York: John Wiley.

CHAPTER 10

Markov Chains

10.1 Introduction

In this chapter we consider time-homogeneous Markov chains $\{X(t), t \in T\}$ where the index set T may be $[0, \infty)$ or $\{0, 1, 2, \ldots\}$. For these we have

$$P\{X(t) = j \mid X(t_1) = i_1,\ X(t_2) = i_2, \ldots, X(t_r) = i_r\}$$
$$= P\{X(t) = j \mid X(t_r) = i_r\}, \quad (0 \leq t_1 < t_2 < \cdots < t_r < t)\ (r \geq 1), \tag{1.1}$$

where the second probability is,

$$= P\{X(t - t_r) = j \mid X(o_r) = i_r\} \tag{1.2}$$

because of time-homogeneity. When the index set is $T = \{0, 1, 2, \ldots\}$ we shall use the notation $\{X_n,\ n \geq 0\}$ for the Markov chain (the discrete time case).

Finite Markov chains are those on the state space $\{0, 1, 2, \ldots, N\}$ where $0 < N < \infty$. These can be treated by techniques based on matrix algebra. The general case of the state space as also discussed in discrete as well as continuous time.

10.2 Discrete Time Markov Chains

The transition probabilities of the Markov chain $\{X_n, n \geq 0\}$ are defined by

$$P_{ij}^{(n)} = P\{X_{m+n} = j \mid X_m = i\} \quad (m \geq 0, n \geq 0) \tag{2.1}$$

and $P_{ij}^{(0)} = \delta_{ij}$, where δ_{ij} is Kronecker's delta. For $n = 1$ we simplify the notation to P_{ij}. We have for $i \geq 0$

$$P_{ij}^{(n)} \geq 0, \quad \sum_{j=0}^{\infty} P_{ij}^{(n)} = 1 \quad (n \geq 0). \tag{2.2}$$

The Chapman–Kolmogorov equations are given by

$$P_{ij}^{(m+n)} = \sum_{k=0}^{\infty} P_{ik}^{(m)} P_{kj}^{(n)} \quad (m \geq 0,\ n \geq 0). \tag{2.3}$$

In particular

$$P_{ij}^{(n+1)} = \sum_{k=0}^{\infty} P_{ik} P_{kj}^{(n)} \quad (n \geq 0). \tag{2.4}$$

For $n = 1, 2, \ldots$ this last relation yields $P_{ij}^{(2)}, P_{ij}^{(3)}, \ldots$ successively.

We denote $P = (P_{ij})$ as the transition probability matrix of the Markov Chain. Relations (2.4) shows that $P^n = (P_{jk}^{(n)})$ $(n \geq 0)$. In order to complete this description we need to specify the initial distribution $\{a_i^{(0)}\}$, where

$$a_i^{(0)} = P\{X_0 = i\} \quad (i \geq 0). \tag{2.5}$$

The unconditional (absolute) distribution of X_n is then given by

$$a_j^{(n)} = P\{X_n = j\} = \sum_{i=0}^{\infty} a_i^{(0)} P_{ij}^{(n)} \quad (n \geq 1) \tag{2.6}$$

while for $0 \leq n_1 < n_2 < \cdots < n_r \quad (r \geq 1)$ the joint distribution of $(X_{n_1}, X_{n_2}, \ldots, X_{n_r})$ by

$$\begin{aligned} P\{X_{n_1} &= j_1, X_{n_2} = j_2, \ldots, X_{n_r} = j_r\} \\ &= a_{j_1}^{(n_1)} P_{j_1 j_2}^{(n_2 - n_1)} \cdots P_{j_{r-1} j_r}^{(n_r - n_{r-1})}. \end{aligned} \tag{2.7}$$

It is easily verified that the finite dimensional distributions (2.7) satisfy the symmetry and compatibility conditions.

Property (2.1) can also be characterized by stating that the transition probabilities are stationary. However, the process $\{X_n\}$ itself is not stationary in general. For if in (2.7) we replace $(n_1, n_2, \ldots, n_r)$ by $(n_1 + m, n_2 + m, \ldots, n_r + m)$, then its right side remains unchanged except for the first factor, which becomes $a_{i_1}^{(n_1+m)}$. Therefore a necessary and sufficient condition for $\{X_n\}$ to be stationary is that the absolute distribution of X_n should be independent of n. In view of (2.6) this condition reduces to an appropriate choice of the initial distribution. We are thus led to the following definition and theorem.

Definition 2.1 The distribution $\{u_j\}$ is stationary for the Markov chain $\{X_n\}$ if

$$u_j = \sum_i u_i P_{ij}. \tag{2.8}$$

Theorem 2.1 *A time-homogeneous Markov chain is stationary if and only if its initial distribution is stationary.*

Proof. In view of the above remarks it remains to prove that the distribution of X_n does not depend on n if we choose $\{a_i^{(0)}\}$ to be a stationary distribution. We have

$$a_j^{(n+1)} = \sum_i a_i^{(n)} P_{ij} \quad (n \geq 0). \tag{2.9}$$

Putting $n = 0, 1$ in this we find successively that

$$a_j^{(1)} = \sum_i a_i^{(0)} P_{ij} = a_j^{(0)},$$
$$a_j^{(2)} = \sum_i a_i^{(1)} P_{ij} = \sum_i a_i^{(0)} P_{ij} = a_j^{(0)}.$$

It follows by induction that $a_j^{(n)} = a_j^{(0)}$ for $n \geq 1$, as required. □

10.3 Examples of Finite Markov Chains

10.3.1 *Markov Trials*

We consider a series of trials, each of which results in the occurrence or non-occurance of an event E (success or failure). The trials are dependent so that the outcome of any trial depends on the outcomes of other trials. A simple case of dependence is the one in which the outcome of any trial depends on the previous trial (Markov dependence). Here the probability of success in the $(n+1)$th trial is equal to α if it is known that the nth trial resulted in success, and equal to β if the nth trial resulted in failure, these two probabilities being independent of the outcomes of trials preceding the nth. In addition, the probabilities of success and failure at the initial trial are given to be α_0 and $1 - \alpha_0$, respectively.

We let $X_n = 0$ if the nth trial is a success and $X_n = 1$ if it is a failure. Then our assumptions imply that the random variables $\{X_n, n \geq 0\}$ form a Markov chain on the state space $\{0, 1\}$. Its transition probability matrix

P is given by

$$P = \begin{bmatrix} \alpha & 1-\alpha \\ \beta & 1-\beta \end{bmatrix}. \tag{3.1}$$

We might be interested in the outcome of an indefinitely long series of trials. Clearly, the probabilities of the outcomes of such a series are given by $\lim_{n\to\infty} P_{ij}^{(n)}$, and these limits are the elements of the matrix $\lim_{n\to\infty} P^n$. The case where $|\alpha - \beta| = 1$ is trivial. Here we must have either (i) $\alpha = 1$, $\beta = 0$, or (ii) $\alpha = 0,\ \beta = 1$.

(i) If $\alpha = 1,\ \beta = 0$, then a success is followed by a success and a failure by a failure. The transition probability matrix $P = I = (\delta_{ij})$, and therefore $P^n = I$ for $n \geq 1$ and also $\lim_{n\to\infty} = P^n = I$.

(ii) If $\alpha = 0, \beta = 1$, then a success is followed by a failure, and vice versa. Here

$$P = \begin{bmatrix} 0 & 1 \\ 1 & 0 \end{bmatrix}.$$

We have $P^2 = I$, and it follows by induction that $P^{2n} = I$ and $P^{2n+1} = P$ for $n \geq 0$. Clearly $\lim_{n\to\infty} P^n$ does not exist in this case. In view of the above discussion it remains to consider the case where $|\alpha - \beta| < 1$. The required results are given by the following theorem.

Theorem 3.1 *Let* $|\alpha - \beta| < 1$.

(a) *The conditional probabilities of the outcomes of the nth trial are given by*

$$P_{00}^{(n)} = \frac{\beta}{1-\alpha+\beta} + \frac{1-\alpha}{1-\alpha+\beta}(\alpha-\beta)^n, \quad P_{01}^{(n)} = 1 - P_{00}^{(n)}, \tag{3.2}$$

$$P_{10}^{(n)} = \frac{\beta}{1-\alpha+\beta} - \frac{\beta}{1-\alpha+\beta}(\alpha-\beta)^n, \quad P_{11}^{(n)} = 1 - P_{10}^{(n)}. \tag{3.3}$$

(b) *The absolute probabilities of the outcomes of the nth trial are given by*

$$a_0^{(n)} = \frac{\beta}{1-\alpha+\beta} + \left(\alpha_0 - \frac{\beta}{1-\alpha+\beta}\right)(\alpha-\beta)^n \tag{3.4}$$

for a success and $a_1^{(n)} = 1 - a_0^{(n)}$ *for a failure.*

(c) *The limit probabilities are given by*

$$u_0 = \frac{\beta}{1-\alpha+\beta'} \quad u_1 = \frac{1-\alpha}{1-\alpha+\beta} \tag{3.5}$$

for a success and a failure, respectively, independently of the outcomes of the initial trial.

Proof. Let us write matrix (3.1) as $P = I + Q$, where

$$Q = \begin{bmatrix} -(1-\alpha) & 1-\alpha \\ \beta & -\beta \end{bmatrix}.$$

The relation $Q^2 = -(1-\alpha+\beta)Q$ can be verified easily, and leads to $Q^r = (-1)^{r-1}(1-\alpha+\beta)^{r-1}Q$ $(r \geq 2)$. Further, we have the binomial expansion

$$P^n = (I+Q)^n = I + \sum_{r=1}^{n} \binom{n}{r} Q^r \quad (n \geq 1),$$

which can be proved by induction. Thus

$$\begin{aligned} P^n &= I + Q\sum_{r=1}^{n}(-1)^{r-1}\binom{n}{r}(1-\alpha+\beta)^{r-1} = I - \frac{(\alpha-\beta)^n - 1}{1-\alpha+\beta}Q \\ &= \left(I + \frac{1}{1-\alpha+\beta}Q\right) - \frac{(\alpha-\beta)^n}{1-\alpha+\beta}Q \\ &= \frac{1}{1-\alpha+\beta}\left\Vert \begin{matrix} \beta & 1-\alpha \\ \beta & 1-\alpha \end{matrix} \right\Vert - \frac{(\alpha-\beta)^n}{1-\alpha+\beta}\left\Vert \begin{matrix} -(1-\alpha) & 1-\alpha \\ \beta & -\beta \end{matrix} \right\Vert. \end{aligned}$$

This leads to probabilities (3.2) and (3.3). The absolute probability of a success at the nth trial is given by

$$\alpha_0 P_{00}^{(n)} + (1-\alpha_0)P_{10}^{(n)}$$

which simplifies to (3.4). The limit probabilities (3.5) follow directly from (3.2) and (3.3). □

As special cases of Markov trials we have the following:

(i) Each of a series of urns contains a white balls and b black balls. One ball is transferred from the first urn to the second, then one ball from the second to the third, and so on. We need to find the probability that the ball drawn from the nth urn will be white.

Let us denote the drawing of a white ball as a success, and a black ball as a failure. It is clear that the trials are Markov dependent, and matrix P is given by

$$P = \begin{bmatrix} \frac{a+1}{a+b+1} & \frac{b}{a+b+1} \\ \frac{a}{a+b+1} & \frac{b+1}{a+b+1} \end{bmatrix}.$$

Here $\alpha = (a+1)/a+b+1$, and $\beta = a/a+b+1$. We therefore obtain

$$P^n = \begin{bmatrix} p & q \\ p & q \end{bmatrix} - (a+b+1)^{-n} \begin{bmatrix} -q & q \\ p & -p \end{bmatrix},$$

where $p = a/a+b$ and $q = b/a+b$. The probability of drawing a white ball from the nth urn is therefore

$$p + q(a+b+1)^{-n} \quad \text{or} \quad p - p(a+b+1)^{-n},$$

depending on whether the ball drawn from the first urn was white or black. The unconditional probability of drawing a white ball from the nth urn is

$$p[p + q(a+b+1)^{-n}] + q[p - p(a+b+1)^{-n}] = p,$$

which is the same as from the first urn.

(ii) Two urns contain, respectively, a white and b black balls, and c white and d black balls. A series of drawings is made according to the following rules. Each time only one ball is drawn and immediately replaced in the same urn it came from. If the ball drawn is white, the next drawing is made from the first urn, and if it is black, it is made from the second. We need to find the probability that the nth ball drawn will be white.

Here again, we denote a white ball as a success and a black ball as failure. Matrix P in this case is given by

$$P = \begin{bmatrix} \frac{a}{a+b} & \frac{b}{a+b} \\ \frac{c}{c+d} & \frac{d}{c+d} \end{bmatrix}.$$

Here $\alpha = a/a+b$, $\beta = c/c+d$, and hence

$$P^n = \frac{1}{ac+bd+2bc} \begin{bmatrix} ac+bc & bc+bd \\ ac+bc & bc+bd \end{bmatrix} - \frac{(ad-bc)^n}{(ac+bd+2bc)(a+b)^n(c+d)^n} \begin{bmatrix} -bc-bd & bc+bd \\ ac+bc & -ac-bc \end{bmatrix}.$$

10.3.2 *The Bernoulli–Laplace Diffusion Model*

We have a total of $2a$ balls, of which a are white and a black. These are distributed equally in two urns, and a series of trials are made. At each trial one ball is selected at random from each urn, and these two balls are interchanged. Let X_n denote the number of white balls in the first urn after the nth interchange ($n \geq 1$). Initially, this urn contains a white balls so that $X_0 = a$.

It is clear that for $n \geq 0$

$$\begin{aligned} X_{n+1} &= X_n \qquad \text{with probability } \frac{2}{a^2} X_n(a - X_n) \\ &= X_n - 1 \quad \text{with probability } \frac{1}{a^2} X_n^2 \\ &= X_n + 1 \quad \text{with probability } \frac{1}{a^2}(a - X_n)^2. \end{aligned} \tag{3.6}$$

This shows that given $X_0, X_1, \ldots, X_n$, X_{n+1} depends only on X_n, so the sequence of random variables $\{X_n, n \geq 0\}$ is a Markov chain on the state space $\{0, 1, 2, \ldots, a\}$. Its transition probabilities are given by

$$P_{ii} = 2\frac{i}{a}\left(1 - \frac{i}{a}\right), P_{i,i-1} = \left(\frac{i}{a}\right)^2, P_{ii+1} = \left(1 - \frac{i}{a}\right)^2. \tag{3.7}$$

To obtain $P_{ij}^{(n)}$ we can use the results of matrix algebra. We illustrate this in the simple case $a = 2$, for which the transition probability matrix is given by

$$P = \begin{bmatrix} 0 & 1 & 0 \\ \frac{1}{4} & \frac{1}{2} & \frac{1}{4} \\ 0 & 1 & 0 \end{bmatrix}. \tag{3.8}$$

The eigenvalues λ of P are the roots of the equation $|\lambda I - P| = 0$, which reduces to

$$\lambda(2\lambda^2 - \lambda - 1) = 0. \tag{3.9}$$

This gives $\lambda = 0, 1, -1/2$. The eigenvectors x_1 and y_1 corresponding to $\lambda = 1$, respectively, satisfy the equations

$$(\lambda I - P)x_1' = 0, \quad y_1(\lambda I - P) = 0.$$

We easily find that $x_1 = (1, 1, 1)$ and $y_1 = (1, 4, 1)$. Finally, the corresponding idempotent matrix Z_1 is given by $Z_1 = c_1 x_1' y_1$, where c_1 is the normalizing constant $(y_1 x_1')^{-1} = 6^{-1}$. Thus

$$Z_1 = \begin{bmatrix} \frac{1}{6} & \frac{2}{3} & \frac{1}{6} \\ \frac{1}{6} & \frac{2}{3} & \frac{1}{6} \\ \frac{1}{6} & \frac{2}{3} & \frac{1}{6} \end{bmatrix}.$$

For the eigenvalue $\lambda = -1/2$ the corresponding quantities are $x_2 = (2, -1, 2)$, $y_2 = (1, -2, 1), c_2 = 6^{-1}$, and

$$Z_2 = \begin{bmatrix} \frac{1}{3} & -\frac{2}{3} & \frac{1}{3} \\ -\frac{1}{6} & \frac{1}{3} & -\frac{1}{6} \\ \frac{1}{3} & -\frac{2}{3} & \frac{1}{3} \end{bmatrix}.$$

These calculations lead to the result

$$P^n = Z_1 + \left(-\frac{1}{2}\right)^n Z_2 \tag{3.10}$$

or equivalently

$$\begin{aligned} P_{i0}^{(n)} = P_{i2}^{(n)} &= \frac{1}{6} + \frac{1}{3}\left(-\frac{1}{2}\right)^n, \quad P_{i1}^{(n)} = \frac{2}{3} - \frac{2}{3}\left(-\frac{1}{2}\right)^n \quad (1 = 0, 2), \\ P_{10}^{(n)} = P_{12}^{(n)} &= \frac{1}{6} - \frac{1}{6}\left(-\frac{1}{2}\right)^n, \quad P_{11}^{(n)} = \frac{2}{3} + \frac{1}{3}\left(-\frac{1}{2}\right)^n. \end{aligned} \tag{3.11}$$

It follows that as $n \to \infty$

$$P_{i0}^{(n)} \to \frac{1}{6}, \quad P_{i1}^{(n)} \to \frac{2}{3}, \quad P_{i2}^{(n)} \to \frac{1}{6} \tag{3.12}$$

independently of i. (The general case $a \geq 2$ is discussed in Sec. 10.4.)

10.4 The Limit Distribution of Finite Markov Chains

We now investigate the conditions under which the limit probabilities $\lim_{n\to\infty} P_{ij}^{(n)}$ exist in a Markov chain with a finite number of states $(0, 1, 2, \ldots, N)$. We have the following.

Theorem 4.1 *If for some value of $r > 0$*

$$P_{ij}^{(r)} > 0 \quad (i, j = 0, 1, \ldots, N), \tag{4.1}$$

then the limit probabilities

$$\lim_{n\to\infty} P_{ij}^{(n)} = u_j > 0 \quad (j = 0, 1, \ldots, N) \tag{4.2}$$

exist independently of the initial state i.

Proof. Let $M_j^{(n)}, m_j^{(n)}$ denote, respectively, the maximum and the minimum element in the jth column of the matrix P^n. Thus

$$M_j^{(n)} = \max_{0<i<N} P_{ij}^{(n)}, \quad m_j^{(n)} = \min_{0<i<N} P_{ij}^{(n)}. \tag{4.3}$$

We have

$$P_{ij}^{(n+1)} = \sum_{k=0}^{N} P_{ik} P_{kj}^{(n)} \leq \sum_{k=0}^{N} P_{ik} M_j^{(n)}$$

so that

$$M_j^{(n+1)} \leq M_j^{(n)} \quad (n \geq 1). \tag{4.4}$$

Similarly

$$m_j^{(n+1)} \geq m_j^{(n)} \quad (n \geq 1). \tag{4.5}$$

Thus $\{M_j^{(n)}, n \geq 1\}$ is a monotone nonincreasing sequence, and $\{m_j^{(n)}, n \geq 1\}$ is a monotone nondecreasing sequence. Both are bounded, since $0 \leq m_j^{(n)} \leq M_j^{(n)} \leq 1$. Further, we have

$$M_j^{(n+r)} - m_j^{(n+r)} = \max_{i,\ell} \sum_{k=0}^{N} \{P_{ik}^{(r)} - P_{\ell k}^{(r)}\} P_{kj}^{(n)}. \tag{4.6}$$

Let us put $\beta_{i\ell k}^{(r)} = |P_{ik}^{(r)} - P_{\ell k}^{(r)}|$. Let K_1 denote the set of values of k such that $P_{ik}^{(r)} > P_{\ell k}^{(r)}$, and K_2 the complementary set. Then

$$\begin{aligned} \sum_{k\in K_1} \beta_{i\ell k}^{(r)} - \sum_{k\in K_2} \beta_{i\ell k}^{(r)} &= \sum_{k\in K_1} \{P_{ik}^{(r)} - P_{\ell k}^{(r)}\} + \sum_{k\in K_2} \{P_{ik}^{(r)} - P_{\ell k}^{(r)}\} \\ &= \sum_{k=1}^{N} P_{ik}^{(r)} - \sum_{k=1}^{N} P_{\ell k}^{(r)} = 1 - 1 = 0. \end{aligned} \tag{4.7}$$

Further, by our assumption there exists a $\delta > 0$ such that $P_{ij}^{(r)} \geq \delta > 0$, so

$$0 < \sum_{k\in K_1} \beta_{i\ell k}^{(r)} = 1 - \sum_{k\in K_2} P_{ik}^{(r)} - \sum_{k\in K_1} P_{\ell k}^{(r)} \leq 1 - N\delta. \tag{4.8}$$

From (4.7) and (4.8) it follows that

$$\begin{aligned}\sum_{k=0}^{N}\{P_{ik}^{(r)}-P_{\ell k}^{(r)}\}P_{kj}^{(n)} &= \sum_{k\in K_1}\beta_{i\ell k}^{(r)}P_{kj}^{(n)}-\sum_{k\in K_2}\beta_{i\ell k}^{(r)}P_{kj}^{(n)} \\ &\le \sum_{k\in K_1}\beta_{i\ell k}^{(r)}M_j^{(n)}-\sum_{k\in K_2}\beta_{i\ell k}^{(r)}m_j^{(n)} \\ &= \sum_{k\in K_1}\beta_{i\ell k}^{(r)}\{M_j^{(n)}-m_j^{(n)}\} \\ &\le (1-N\delta)\{M_j^{(n)}-m_j^{(n)}\}. \qquad (4.9)\end{aligned}$$

From (4.6) and (4.9) we find that

$$M_j^{(n+r)}-m_j^{(n+r)} \le (1-N\delta)\{M_j^{(n)}-m_j^{(n)}\}. \qquad (4.10)$$

Moreover, we have

$$\begin{aligned}M_j^{(r)}-m_j^{(r)} &= \max_{i,\ell}\{P_{ij}^{(r)}-P_{\ell j}^{(r)}\} \\ &\le \max_{i,\ell}\sum_{k\in K_1}\beta_{i\ell k}^{(r)} \le 1-N\delta. \qquad (4.11)\end{aligned}$$

Putting $n=r,2r,\ldots$ successively in (4.10) and (4.11) we obtain

$$M_j^{(tr)}-m_j^{(tr)} \le (1-N\delta)^t \to 0 \quad \text{as } t\to\infty. \qquad (4.12)$$

Thus $M_j^{(n)}$ and $m_j^{(n)}$ converge to the same limit u_j (say). Since

$$0<\delta\le m_j^{(n)}\le P_{ij}^{(n)}\le M_j^{(n)} \qquad (4.13)$$

it follows that $P_{ij}^{(n)}\to u_j$ as $n\to\infty$ and that $u_j>0$, as required. □

Corollary 4.1 *Under the conditions of Theorem* 4.1, *there exist constants* A *and* ρ *independent of* n *such that* $A>0, 0<\rho<1$, *and*

$$\left|P_{ij}^{(n)}-u_j\right| \le A\rho^n. \qquad (4.14)$$

Proof. From (4.4) and (4.5) we find that

$$m_j^{(1)}\le m_j^{(2)}\le\cdots\le M_j^{(2)}\le M_j^{(1)}. \qquad (4.15)$$

From this and (4.12) it follows that

$$\left|P_{ij}^{(tr)} - u_j\right| \leq M_j^{(tr)} - M_j^{(tr)} \leq (1 - N\delta)^t.$$

Therefore

$$\left|P_{ij}^{(n)} - u_j\right| \leq (1 - N\delta)^{\left[\frac{n}{r}\right]} \leq (1 - N\delta)^{\frac{n}{r}-1} \tag{4.16}$$

which is of the form (4.14) with $A = (1 - N\delta)^{-1}$ and $\rho = (1 - N\delta)^{\frac{I}{r}}$. Since $0 < 1 - N\delta < 1$ we have $A > 0$ and $0 < \rho < 1$ as claimed. □

Theorem 4.2 *Under condition* (4.1) *we have the following*:

(a) $a_j^{(n)} = P\{X_n = j\} \to u_j \quad as\ n \to \infty \quad (j = 0, 1, \ldots, N).$ (4.17)

(b) *The limit distribution* $\{u_j\}$ *is the unique solution of the equations*

$$u_j = \sum_{i=0}^{N} u_i P_{ij} \quad (j = 0, 1, \ldots, N), \ u_1 + u_2 + \cdots + u_N = 1. \tag{4.18}$$

Proof. We have

$$a_j^{(n)} = \sum_{i=0}^{N} a_i^{(0)} P_{ij}^{(n)} \to \sum_{i=0}^{N} a_i^{(0)} u_j = u_j,$$

which proves (a). Let $u = (u_0, u_1, \ldots, u_N)$ and $e = (1, 1, \ldots, 1)'$. From $P^{n+1} = P^n \cdot P$ and $P^n e = e$ we obtain

$$U = UP, \quad Ue = e, \tag{4.19}$$

where U is the matrix with $N + 1$ identical rows u. This shows that u is a solution of (4.18). Suppose $x = (x_0, x_1, \ldots, x_N)$ is a second solution of (4.18). Then $x = xP = xP \cdot P = xP^2$ and by induction $x = xP^n$ $(n \geq 1)$. Letting $n \to \infty$ in this we obtain $x = xU = u$, which proves (b). □

In the examples of Sec. 10.3 we obtained the limit distribution directly from $P_{ij}^{(n)}$. If we are merely interested in $\{u_j\}$, we need to only solve Eqs. (4.18).

Example 4.1 (The Bernoulli–Laplace Diffusion Model, Continued) We now consider the general case $a \geq 2$ and derive the limit distribution of the chain $\{X_n\}$. Equations (4.18) now become

$$u_j = u_{j+1}\left(\frac{j+1}{a}\right)^2 + u_{j-1}\left(1-\frac{j-1}{a}\right)^2 + u_j 2\frac{j}{a}\left(1-\frac{j}{a}\right) \quad (1 \le j \le a-1),$$

$$u_0 = u_1\left(\frac{1}{a}\right)^2, u_a = u_{a-1}\left(\frac{1}{a}\right)^2$$

$$u_0 + u_1 + \cdots + u_a = 1.$$

From these we find that $u_1 = a^2 u_0$, $u_2 = \left(\frac{a}{2}\right)^2 u_0$ and by induction

$$u_j = \binom{a}{j}^2 u_0 \quad (0 \le j \le a).$$

Also

$$1 = u_0 \sum_{j=0}^{a} \binom{a}{j}^2 = u_0 \binom{2a}{a}.$$

We conclude that

$$u_j = \binom{a}{j}^2 \Big/ \binom{2a}{a} \quad (0 \le j \le a). \qquad \square$$

10.5 Classification of States. Limit Theorems

We now consider the case of the discrete time Markov chain $\{X_n, n \ge 0\}$ on the state space $\{0, 1, 2, \ldots\}$. For each state j we define

$$\tau_j = \min\{n > 0 \colon X_n = j\} \quad \text{on} \quad \{X_0 = i\}. \tag{5.1}$$

We call τ_j the epoch of the first visit of the process to j. The (conditional) distribution of τ_j is given by $\{f_{ij}^{(n)},\ n \ge 1\}$, where

$$\begin{aligned} f_{ij}^{(n)} &= P\{\tau_j = n \mid X_0 = i\} \\ &= P\{X_1 \ne j, X_2 \ne j, \ldots, X_{n-1} \ne j, X_n = j \mid X_0 = i\}. \end{aligned} \tag{5.2}$$

Also, let

$$f_{ij} = \sum_{n=1}^{\infty} f_{ij}^{(n)} = P\{\tau_j < \infty \mid X_0 = i\} \le 1. \tag{5.3}$$

Then f_{ij} is the probability that starting from i the process ever visits j. In particular, f_{jj} is the probability of an eventual return to j. If $f_{jj} = 1$ we denote the mean return time as

$$\mu_j = E[\tau_j \mid X_0 = j] = \sum_{n=i}^{\infty} n f_{jj}^{(n)} \leq \infty. \tag{5.4}$$

Definitions

5.1 State j is persistent or transient according as $f_{jj} = 1$ or < 1.

5.2 A persistent state j is non-null or null according as $\mu_j < \infty$ or $\mu_j = \infty$.

5.3 Denote $d_j = g.c.d.\{n\colon P_{jj}^{(n)} > 0\}$. If $d_j = 1$ we say that state j is aperiodic; if $d_j > 1$, j has period d_j.

To understand the significance of the above definitions, let $X_0 = i$, $\tau_{j0} = 0$ and for $r \geq 1$ denote

$$\tau_{j1} = \min\{n > 0\colon X_n = j\}, \tag{5.5}$$

$$\tau_{jr} = \min\{n > \tau_{jr-1}\colon X_n = j\} \quad (r \geq 2). \tag{5.6}$$

We note that

$$\{\tau_{j1} = n\} = \{X_1 \neq j,\ X_2 \neq j, \ldots, X_{n-1} \neq j, X_n = j\} \in F_n, \tag{5.7}$$

where $\{F_n, n \geq 0\}$ is the history of the Markov chain $\{X_n\}$. Repeating this argument and using the time-homogeneity of the process we see that the random variables $\{\tau_{jr} - \tau_{jr-1}\ (r \geq 1)\}$ are independent, τ_{j1} having the distribution $\{f_{ij}^{(n)},\ n \geq 1\}$ and $\tau_{jr} - \tau_{jr-1}\ (r \geq 2)$ having the distribution $\{f_{jj}^{(n)},\ n \geq 1\}$. Thus we may view the Markov chain $\{X_n\}$ as a family of renewal processes corresponding to the different states of the chain. The jth renewal process $\{\tau_{jr},\ r \geq 0\}$ corresponds to state j of the chain; its initial lifespan has the distribution $\{f_{ij}^{(n)}\}$ and mean lifespan of the process is $\mu_j \leq \infty$. (We assume $f_{ij} > 0$ for $i \neq j$; otherwise the situation is trivial.) If j is persistent, then $\{\tau_{jr}\}$ is a renewal process whose lifespan distribution has a finite or infinite mean according as j is non-null or null. If j is transient, then $\{\tau_{jr}\}$ has a defective lifespan. If j has period $d_j > 1$, then the lifespan distribution is arithmetic with span d_j. The results now follow immediately.

Theorem 5.1 (i) *With probability* 1 *the Markov chain returns to a persistent state infinitely often, while it returns to a transient state finitely often.*

(ii) *State j is transient or persistent according as*

$$\sum_{n=0}^{\infty} P_{ij}^{(n)} < \infty \quad or \ = \infty. \tag{5.8}$$

(iii) *If j is transient or persistent null, then*

$$P_{ij}^{(n)} \to 0 \quad as \quad n \to \infty. \tag{5.9}$$

(iv) *If j is aperiodic persistent non-null, then*

$$P_{ij}^{(n)} \to \frac{f_{ij}}{\mu_j} \quad as \quad n \to \infty. \tag{5.10}$$

(v) *If j has period d_j and is persistent non-null, then*

$$P_{jj}^{(nd)} \to \frac{d_j}{\mu_j} \quad as \quad n \to \infty. \tag{5.11}$$

Proof. Let $Z_j = \{n \geq 0 \colon X_n = j\}$ be the subset of the index set consisting of epochs at which the Markov chain visits j. Thus

$$Z_j = \{\tau_{j0}, \tau_{j1}, \tau_{j2}, \ldots\}, \tag{5.12}$$

where

$$u_n = P\{n \in Z_j \mid X_0 = i\} = P_{ij}^{(n)} \quad (n \geq 0). \tag{5.13}$$

Denote

$$N_n = |Z_j \cap [0, n]|\,. \tag{5.14}$$

Then N_n is the number of times the Markov chain visits j during $[0, n]$. We have

$$U_n = E(N_n) = u_0 + u_1 + \cdots + u_n = \sum_{m=0}^{n} P_{ij}^{(m)}. \tag{5.15}$$

If j is transient, then the corresponding renewal process is terminating so that with probability 1, $N_n \to N < \infty$, where N is a random variable such that $\tau_{jN} < \infty$, $\tau_{jN+1} = \infty$. Also

$$U_n \to u_0 + u_1 + \cdots + u_N < \infty \quad \text{and} \quad u_n = P_{ij}^{(n)} \to 0.$$

If j is persistent (null or non-null), then $N = \infty$ (Z_j has infinitely many points) and $U_n \to \infty$. From the discrete renewal theory in the aperiodic case it follows that $P_{ij}^{(n)} \to f_{ij}/\mu_j$ as $n \to \infty$, the limit being interpreted as zero if $\mu_j = \infty$ (corresponding to a null state j). The periodic case also leads to the desired results in a similar manner. □

10.6 Closed Sets. Irreducible Chains

Definitions

6.1 We say that state j can be *reached* from i and write $i \to j$ if $P_{ij}^{(n)} > 0$ for some $n \geq 0$. If $i \to j$ and $j \to i$ we write $i \leftrightarrow j$. Since $P_{jj}^{(0)} = 1$ we have $j \leftrightarrow j$.

6.2 Two states are said to be of the same *type* if they have the same characteristics:

(i) both are aperiodic or both have the same period, or
(ii) both are transient or both persistent non-null or both persistent null.

Theorem 6.1 *If $i \leftrightarrow j$, then i, j are of the same type.*

Proof. We have $P_{ij}^{(N)} = \alpha > 0$ and $P_{ji}^{(M)} = \beta > 0$ for some $N \geq 0$, $M \geq 0$. Now for any n we have

$$P_{ii}^{(n+N+M)} \geq P_{ij}^{(N)} P_{jj}^{(n)} P_{ji}^{(M)} = \alpha\beta P_{jj}^{(n)}, \tag{6.1}$$

$$P_{jj}^{(n+N+M)} \geq P_{ji}^{(M)} P_{ii}^{(n)} P_{ij}^{(N)} = \alpha\beta P_{ii}^{(n)}. \tag{6.2}$$

From (6.1) and (6.2) we see that the two series $\Sigma P_{ii}^{(n)}$ and $\Sigma P_{jj}^{(n)}$ converge or diverge together so that i and j are either both persistent or both transient. If i is persistent null, $P_{ii}^{(n)} \to 0$, and from (6.1), $P_{jj}^{(n)} \to 0$ so that j is also persistent null. Finally, if i has period d, then since $P_{ii}^{(N+M)} \geq P_{ij}^{(N)} P_{ji}^{(M)} = \alpha\beta > 0$, $N + M$ must be a multiple of d, and from (6.1) and (6.2) it follows that j also has period d. □

Theorem 6.2 *If i is persistent and $i \to j$, then j is also persistent. Moreover, $f_{ji} = f_{ij} = 1$.*

Proof. We have for $j \neq i$,

$$\begin{aligned}
&P\{\tau_i > m+n \mid X_0 = i\} \\
&\quad = \sum_{k\neq i} P\{\tau_i > m, X_m = k \mid X_0 = i\} \cdot P\{\tau_i > m+n \mid X_m = k, X_0 = i\} \\
&\quad = \sum_{k\neq i} P\{\tau_i > m, X_m = k \mid X_0 = i\} \cdot P\{\tau_i > n \mid X_0 = k\} \\
&\quad \geq P\{\tau_i > m, X_m = j \mid X_0 = i\} \cdot P\{\tau_i > n \mid X_0 = j\}.
\end{aligned}$$

Letting $n \to \infty$ in this we obtain

$$1 - f_{ii} \geq P\{\tau_i > m, X_m = j \mid X_0 = i\} \quad (1 - f_{ji}). \tag{6.3}$$

Now $i \to j$ implies that for some m

$$P\{\tau_i > m, X_m = j \mid X_0 = i\} > 0. \tag{6.4}$$

If i is persistent, $f_{ii} = 1$ and so $f_{ji} = 1$. Thus $j \to i$ and by Theorem 6.1, j is also persistent. By symmetry it follows that $f_{ij} = 1$ and the proof is completed. □

Definitions

6.3 A subset C of the state space is said to be *closed* if

$$P_{ij} = 0 \quad \text{whenever } i \in C \text{ and } j \notin C. \tag{6.5}$$

In this case we have

$$P_{ij}^{(2)} = \sum_{k\in C} P_{ik}P_{kj} + \sum_{k\notin C} P_{ik}P_{kj} = 0.$$

and by induction

$$P_{ij}^{(n)} = 0 \quad (n \geq 1, i \in C, j \notin C). \tag{6.6}$$

Thus no state outside C can be reached from any state inside C. Hence

$$\sum_{j\in C} P_{ij}^{(n)} = 1 \quad \text{for } i \in C, \tag{6.7}$$

which shows that the states of C constitute a Markov chain which can be studied independently of states outside C.

We adopt the convention that the set of all states is closed.

6.4 If a closed set consists of a single state i, we say i is an *absorbing* state. In this case $P_{ii}^{(n)} = 1$ for all $n \geq 0$. Clearly, i is a persistent state.

6.5 A Markov chain is *irreducible* if it contains no closed sets other than the set of all states.

Theorem 6.3 *For any state i, let $C(i) = \{j\colon i \to j\}$. Then $C(i)$ is the smallest closed set containing i (closure of i). If i is persistent, then $C(i)$ is irreducible.*

Proof. Suppose $k \notin C(i)$. Then $i \nrightarrow k$. If for some $j \in C(i)$ we have $j \to k$, then $i \to j \to k$, which implies $i \to k$, *a* contradiction. Therefore $C(i)$ is closed. Clearly, $C(i)$ is the smallest closed set containing i, since if C is any closed containing i and $i \to j$, then $j \in C$.

Suppose now i is persistent. Since for every $j \in C(i)$ we have $i \to j$, we must have $j \to i$ by Theorem 6.2. Therefore $C(i)$ does not contain any closed set other than itself and $C(i)$ is thus irreducible. □

Theorem 6.4 *A Markov chain is irreducible iff every state can be reached from every other state. In this case all states are of the same type.*

Proof. If $i \leftrightarrow j$ for every i, j, then there are no closed sets other than the set of all states, so the chain is irreducible. Conversely, suppose that the chain is irreducible. Then for every i, the set $C(i)$ should coincide with the set of all states. This means that $i \leftrightarrow j$ for every i, j, as required. The remaining statement is a consequence of Theorem 6.1. □

Theorem 6.5 *The states of a Markov chain can be divided, in a unique manner, into non-overlapping sets T, C_1, $C_2, \ldots$ such that T consists of all transient states, and C_1, $C_2, \ldots$ are irreducible closed sets containing only persistent states of same type.*

Proof. If there are no persistent states, the theorem is obviously true. Suppose that i_1 is a persistent state and let $C_1 = C(i_1)$ be the set of all states that can be reached from i_1. By Theorem 6.3, C_1 is an irreducible closed set, by Theorem 6.2 all states in C_1 are persistent, and finally by Theorem 6.1 they are of the same type. If there are persistent states outside C_1 we proceed as before, and in this manner we obtain the sets C_1, $C_2, \ldots$. The theorem is completely proved. □

The following theorem gives a criterion to distinguish between persistent and transient states of an irreducible chain.

Theorem 6.6 *The states $\{0, 1, 2, \ldots\}$ of an irreducible chain are transient iff the equations*

$$y_i = \sum_{j=1}^{\infty} P_{ij} y_j \quad (i = 1, 2, \ldots) \tag{6.8}$$

admit of a nonzero bounded solution.

Proof. Suppose all states are transient then $f_{i0} < 1$ for at least one $i \geq 1$, since otherwise from

$$f_{00} = P_{00} + \sum_{i=1}^{\infty} P_{0i} f_{i0} \tag{6.9}$$

we will have $f_{00} = 1$, which is a contradiction. Now let

$$v_i^{(n)} = P\{\tau_0 > n \mid X_0 = i\} \quad (n \geq 1). \tag{6.10}$$

We have

$$v_i^{(1)} = \sum_{j=1}^{\infty} P_{ij}, \quad v_i^{(n)} = \sum_{j=1}^{\infty} P_{ij} v_j^{(n-1)} \quad (n \geq 2). \tag{6.11}$$

Since $v_i = \lim_{n\to\infty} v_i^{(n)} = 1 - f_{i0}$ we find from (6.11) that

$$v_i = \sum_{j=1}^{\infty} P_{ij} v_j \quad (i \geq 1). \tag{6.12}$$

Thus $\{v_i,\ i \geq 1\}$ is a nonzero bounded solution (6.8). Conversely, let $\{y_i,\ i \geq 1\}$ be an arbitrary bounded solution of (6.8). Thus $|y_i| < M$, where we can take $M = 1$ without loss of generality. We then have

$$|y_i| \leq \sum_{j=1}^{\infty} P_{ij} |y_j| \leq \sum_{j=1}^{\infty} P_{ij} = v_i^{(1)},$$

$$|y_i| \leq \sum_{j=1}^{\infty} P_{ij} v_j^{(1)} = v_j^{(2)}.$$

By induction $|y_i| \leq v_i^{(n)}$ $(n \geq 1)$. In the limit this gives $|y_i| \leq v_i$. If all states are persistent we have $v_i = 1 - f_{i0} = 0$ by Theorem 6.2, and therefore $y_i = 0$ $(i \geq 1)$. This means that Eqs. (6.8) have no nonzero bounded solution. □

10.7 Stationary Distributions

We now consider an irreducible Markov chain. By Theorem 6.5 its states are either all persistent or else all transient. For convenience we restrict our attention to the aperiodic case. By Theorem 5.1 we have for all i

$$\lim_{n\to\infty} P_{ij}^{(n)} = u_j \geq 0 \tag{7.1}$$

since $f_{ij} = 1$ by Theorem 6.2 in the persistent case. For the absolute (unconditional) probabilities we then have

$$P\{X_n = j\} = a_j^{(n)} = \sum_{i=0}^{\infty} a_i^{(0)} P_{ij}^{(n)} \to u_j. \tag{7.2}$$

Thus $\{u_j,\ j \geq 0\}$ is the limit distribution of $\{X_n\}$.

From Sec. 10.2 we recall the definition of a stationary distribution $\{v_j\}$. For this we have

$$v_j \geq 0, \quad v_j = \sum_{i=0}^{\infty} v_i P_{ij} \quad (j \geq 0). \tag{7.3}$$

As a consequence of (7.3) we also have for all $n \geq 0$

$$v_j = \sum_{i=0}^{\infty} v_i P_{ij}^{(n)} \quad (j \geq 0). \tag{7.4}$$

The following theorem establishes the connection between the stationary distribution and the limit distribution.

Theorem 7.1 *An irreducible aperiodic Markov chain has a stationary distribution iff its states are persistent non-null. In this case the stationary distribution is unique and is identical with the limit distribution $\{u_j\}$, so that*

$$u_j = \sum_{i=0}^{\infty} u_i P_{ij} \quad (j \geq 0) \tag{7.5}$$

$$u_j > 0, \quad \sum_{0}^{\infty} u_j = 1. \tag{7.6}$$

Proof. (i) If a stationary distribution exists, then (7.4) holds. If the states are all transient or persistent null, then $P_{ij}^{(n)} \to 0$ by Theorem 5.1 and

(7.4) gives $v_j = 0$ $(j \geq 0)$, which is not true since $\{v_j\}$ is a probability distribution. Therefore the states are persistent non-null.

(ii) Conversely, if all states are persistent non-null, then by Theorem 7.4, $u_j = \lim P_{ij}^{(n)}$ exists, where

$$u_j = \frac{f_{ij}}{\mu_j} = \frac{1}{\mu_j} > 0. \tag{7.7}$$

Now from $\sum_{j=0}^{J} P_{ij}^{(n)} \leq 1$ we obtain $u_0 + u_1 + \cdots + u_J \leq 1$. Since J is arbitrary, this gives

$$\sum_0^\infty u_j \leq 1. \tag{7.8}$$

Also, from

$$P_{ij}^{(m+n)} \geq \sum_{k=0}^{K} P_{ik}^{(n)} P_{kj}^{(m)}$$

we obtain, letting $n \to \infty$

$$u_j \geq \sum_{k=0}^{K} u_k P_{kj}^{(m)}.$$

Since K is arbitrary we find that

$$u_j \geq \sum_{k=0}^{\infty} u_k P_{kj}^{(m)}. \tag{7.9}$$

We shall show that equalities hold in (7.8) and (7.9). If a strict inequality holds for some j in (7.9), then

$$1 \geq \sum_0^\infty u_j > \sum_{j=0}^{\infty} \sum_{k=0}^{\infty} u_k P_{kj}^{(m)} = \sum_0^\infty u_k,$$

which is a contradiction. Hence for all j we must have

$$u_i = \sum_{k=0}^{\infty} u_k P_{kj}^{(m)} \quad (m \geq 0).$$

In particular, when $m = 1$ this gives $u_j = \sum_{k=0}^{\infty} u_k P_{kj}$. When $m \to \infty$ this gives $u_j = (\sum_0^\infty u_k) u_j$ so that $\sum_0^\infty u_k = 1$. Thus the limit distribution $\{u_j\}$ is stationary. If possible, let there be a second stationary distribution $\{v_j\}$. Then letting $n \to \infty$ in (7.4) we obtain

$$v_j = \left(\sum_{i=0}^{\infty} v_i\right) u_j = u_j$$

so that $\{u_j\}$ is unique. □

10.8 Examples of Infinite Markov Chains

10.8.1 *The Branching Process as a Markov Chain*

The notations and references are as in Chap. 8. We start with the representation

$$X_{n+1} = Y_1^{(n+1)} + Y_2^{(n+1)} + \cdots + Y_{X_n}^{(n+1)} \quad (n \geq 0), \tag{8.1}$$

where $\{Y_j^{(n+1)}, j \geq 1\}$ is a sequence of IID random variables whose distribution does not depend on n. This shows that $\{X_n,\ n \geq 0\}$ is a time-homogeneous Markov chain on the state space $\{0, 1, 2, \ldots\}$, with $P_{1j} = k_j \quad (j \geq 0\}$, where $\{k_j, j \geq 0\}$ is the offspring distribution. Let

$$K(s) = \sum_{j=0}^{\infty} k_j s^j \quad (0 < s < 1). \tag{8.2}$$

More generally, denote

$$\sum_{j=0}^{\infty} P_{1j}^{(n)} s^j = F_n(s) \quad (n \geq 0), \tag{8.3}$$

where $F_0(s) = s$ and $F_1(s) = K(s)$. For $i \geq 1$ we have

$$\sum_{j=0}^{\infty} P_{ij}^{(n)} s^j = [F_n(s)]^i \quad (i \geq 1). \tag{8.4}$$

It follows that in order to obtain $P_{ij}^{(n)}$ it suffices to calculate $F_n(s)$. From the Chapman–Kolmogorov equations

$$P_{1j}^{(m+n)} = \sum_{k=0}^{\infty} P_{1k}^{(m)} P_{kj}^{(n)} \tag{8.5}$$

we obtain, using (8.4),

$$F_{m+n}(s) = F_m \circ F_n(s) \quad (m \geq 0, n \geq 0). \tag{8.6}$$

This relation can be used to derive $F_n(s)$ $(n \geq 2)$ by induction.

In this chain the state 0 is absorbing, and all other states are therefore transient. Absorption at 0 means the extinction of the population. The extinction time T is given by

$$T = \min\{n\colon X_n = 0\} \text{ on } \{X_0 = i\}. \tag{8.7}$$

We have

$$P\{T \leq n \mid X_0 = i\} = P_{i0}^{(n)} \quad (n \geq 1). \tag{8.8}$$

The probability of ultimate extinction is given by

$$P\{T < \infty \mid X_0 = i\} = \lim_{n\to\infty} P_{i0}^{(n)}. \tag{8.9}$$

From (8.4) it is seen that $P_{i0}^{(n)} = \zeta_n^i$, where $\zeta_n = F_n(0)$. From this and

$$F_{n+1}(s) = K \circ F_n(s)$$

it follows that

$$\zeta_1 = K(0), \quad \zeta_{n+1} = K(\zeta_n) \quad (n \geq 1). \tag{8.10}$$

The probability of ultimate extinction (with one ancestor) is given by

$$\zeta = \lim_{n\to\infty} \zeta_n, \tag{8.11}$$

where from (8.10) it is seen that ζ satisfies the equation $s = K(s)$. An obvious root of this equation is $\zeta = 1$ and so it might be concluded that ultimate extinction is a sure event. Actually, the desired root is the smallest positive root of the equation. To prove this, we assume $0 < k_0 < 1$ and let s be an arbitrary root of $s = K(s)$ with $0 < s \leq 1$. Then since $K(s)$ is an increasing function of s, $\zeta_1 = K(0) < K(s) = s$, so $\zeta_2 = K(\zeta_1) < K(s) = s$, and by induction $\zeta_n < s$. This yields in the limit as $n \to \infty$, $\zeta < s$. Thus ζ is the smallest positive root of the equation $s = K(s)$. It turns out that $0 < \zeta < 1$ iff $\mu = K'(1) > 1$.

Since the states $1, 2, \ldots,$ are all transient, $P_{ij}^{(n)} \to 0$ as $n \to \infty$ for $i \geq 1$, $j \geq 1$. Thus, irrespective of the value of μ, the ultimate population size cannot attain any positive finite value. Therefore the only two possibilities are that the population either becomes extinct or becomes infinitely large (explodes) at some stage.

10.8.2 *The Queueing System GI/M/1*

This is a single server system in which customers arrive at the epochs of a renewal process $\{t_n,\ n \geq 0\}$ with lifespan distribution F, and their service times have density $\mu e^{-\mu x}$ $(0 < \mu < \infty)$. Denote by $Q(t)$ the number of customers in the system at time $t \geq 0$. We are interested in the sequence $\{Q_n,\ n \geq 0\}$, where $Q_n = Q(t_n-)$, the number of customers present just before the nth arrival $(n \geq 0)$. It is easy to see that

$$Q_{n+1} = (Q_n + 1 - X_{n+1})^+ \quad (n \geq 0), \tag{8.12}$$

where X_{n+1} is the number of service completions during $[t_n,\ t_{n+1})$. Our assumptions imply that $\{X_n,\ n \geq 1\}$ is a sequence of IID random variables with the distribution $\{k_j,\ j \geq 0\}$, where

$$k_j = \int_0^\infty e^{-\mu t} \frac{(\mu t)^j}{j!} F\{dt\} \quad (j \geq 0). \tag{8.13}$$

The mean of this distribution is μa, where a $(0 < a < \infty)$ is the mean lifespan of $\{t_n\}$. The quantity $\rho = (\mu a)^{-1}$ is called the traffic intensity of the system. We denote

$$E\left(z^{X_n}\right) = \sum_{j=0}^{\infty} k_j z^j \quad (0 < z < 1). \tag{8.14}$$

The relations (8.12) show that $\{Q_n,\ n \geq 0\}$ is a time-homogeneous Markov chain on the state space $\{0, 1, 2, \ldots\}$. Its transition probabilities P_{ij} are given by

$$P_{ij} = k_{i+1-j} \quad (0 < j \leq i+1), \quad P_{i0} = k_{i+1} + k_{i+2} + \cdots. \tag{8.15}$$

Since $k_j > 0$ $(j \geq 0)$, this chain is irreducible, so that the states are either all persistent or all transient. To derive its stationary distribution $\{u_j,\ j \geq 0\}$ we proceed as follows. From (8.12) we find that

$$P\{Q_{n+1} \leq j \mid Q_0 = i\} = \sum_{\nu=0}^{\infty} k_\nu P\{Q_n \leq j + \nu - 1 \mid Q_0 = i\}. \tag{8.16}$$

Letting $n \to \infty$ in this we obtain the equations

$$\sum_{\nu=(1-j)^+} k_\nu v_{j+\nu-1} \quad (j \geq 0), \tag{8.17}$$

where $v_j = u_0 + u_1 + \cdots + u_j$ $(j \geq 0)$. These lead to the following results for $\{U_j\}$.

Theorem 8.1 *For the GI/M/1 system with traffic intensity ρ we have the following. It $\rho \geq 1$ then $u_j = 0$ $(j \geq 0)$. If $\rho < 1$, then*

$$u_j = (1-\zeta)\zeta^j \quad (j \geq 0), \tag{8.18}$$

where ζ is the unique root of the equation $K(\zeta) = \zeta$ and $0 < \zeta < 1$.

Proof. The substitution $v_j = 1 - \zeta^{\zeta+1}$ $(\zeta \geq 0)$ reduces (8.17) to $\zeta = K(\zeta)$. Since $K'(1) = \rho^{-1}$ we know from Lemma (2.1) of Chap. 8 that $0 < \zeta < 1$ if $\rho < 1$ and $\zeta = 1$ if $\rho \geq 1$. Therefore if $\rho \geq 1$, $u_j = 0$ $(j \geq 0)$ which means that $u_j = 0$ $(j \geq 0)$. If $\rho < 1$, then $u_0 = v_0 = 1 - \zeta$, and $u_j = v_j - v_{j-1} = (1-\zeta)\zeta^j$ $(j \geq 1)$, as in (8.18). Uniqueness is guaranteed by Theorem 7.1. □

Solving (8.12) we arrive at the expression

$$Q_n = \max\{Q_0 + n - S_n,\ r - S_n + S_{n-r} \quad (0 \leq r \leq n) \quad (n \geq 0), \tag{8.19}$$

where $S_0 = 0$, $S_n = X_1 + X_2 + \cdots + X_n$ $(n \geq 1)$. From this we find that the higher order transition probabilities are given by

$$P\{Q_n \leq j \mid Q_0 = i\} = P\{n - S_n \leq j - i, r - S_n + S_{n-r} \leq j \quad (0 \leq r \leq n)\} \tag{8.20}$$

for $n \geq 1$. However we can apply the results of random walks to derive the properties of $\{Q_n\}$. Specifically, $\{S_n - n\ (n \geq 0)\}$ is a random walk on the state space $\{\cdots -1, 0, 1, 2, \ldots\}$. Denote

$$M_n = \max_{0 \leq r \leq n} (S_r - r), \quad m_n = \min_{0 \leq r \leq n} (S_r - r). \tag{8.21}$$

For simplicity we assume $Q_0 = 0$. Then since

$$r - S_n + S_{n-r} = (n - S_n) - (n - r - S_{n-r}) \stackrel{d}{=} r - S_r \quad (r \geq 0)$$

we find that

$$Q_n \stackrel{d}{=} \max_{0 \leq r \leq n} (r - S_r) = -m_n \quad (n \geq 0). \tag{8.22}$$

The distribution of m_n and M_n are given by the following Lemma. The notations are as in Chap. 8.

Lemma 8.1 *For the random walk* $\{S_n \; - n \; (n \geq 0)\}$ *we have*

$$(1-s)\sum_0^\infty s^n E\left(z^{m_n}\right) = \frac{z(1-\xi)}{z-\xi} \tag{8.23}$$

$$\sum_0^\infty s^n E\left(z^{M_n}\right) = \frac{1}{z-sK(z)} \cdot \frac{z-\xi}{1-\xi} \tag{8.24}$$

for $0 < s < 1, \; \xi(s) < z < 1.$

Proof. For $\xi(s) < z < 1$ we have

$$\sum_0^\infty s^n E\left(z^{S_n - n}\right) = \sum_0^\infty s^n \left[\frac{K(z)}{z}\right]^n = \frac{z}{z-sK(z)} \tag{8.25}$$

since $K(z)z^{-1} < \; s^{-1}$ by Lemma 2.1 of Chap. 8. We can write

$$\frac{(1-s)z}{z-sK(z)} = \frac{z(1-\xi)}{z-\xi} \cdot \frac{1-s}{z-sK(z)} \cdot \frac{z-\xi}{1-\xi}, \tag{8.26}$$

where

$$\frac{z(1-\xi)}{z-\xi} = \sum_{n=0}^\infty (1-\xi)\xi^n z^{-n}, \tag{8.27}$$

and it can be verified that

$$\frac{1-s}{z-sK(z)} \cdot \frac{z-\xi}{1-\xi} \tag{8.28}$$

is the p.g.f. of a non-negative random variable. From the Wiener–Hopf factorization of the random walk $\{S_n - n\}$ (see Theorem 8.2 of Chap. 6) we find that

$$(1-s)\sum_0^\infty s^n E(z^{S_n - n}) = (1-s)\sum_0^\infty s^n E(z^{m_n}) \cdot (1-s)\sum_0^\infty s^n E(z^{M_n}). \tag{8.29}$$

In view of the remarks made about (8.27) and (8.28) it follows that the two factors on the right side of (8.29) are indeed identical with (8.23) and (8.24), respectively. □

The above lemma yields the results concerning Q_n for finite n on well for $n \to \infty$. In particular (8.31) agrees with (8.18).

Theorem 8.2 (i) *For* $0 < s < 1$ *and* $1 < z < \xi(s)^{-1}$ *we have*

$$(1-s)\sum_{n=0}^{\infty} s^n E(z^{Q_n}) = \frac{1-\xi}{1-z\xi}. \tag{8.30}$$

(ii) *As* $n \to \infty$, $Q_n \xrightarrow{d} Q$, *where* $Q < \infty$ *iff* $\rho < 1$, *in which case*

$$E\left(z^Q\right) = \frac{1-\zeta}{1-z\zeta}. \tag{8.31}$$

Proof. The result (i) is essentially the same as (8.23) since $Q_n \stackrel{d}{=} -m_n$. As $n \to \infty$, $m_n \to m \geq -\infty$ with probability 1. Thus $Q_n \xrightarrow{d} Q = -m$, where by a Tauberian theorem

$$\begin{aligned} E(z^Q) &= \lim_{s\to 1-}(1-s)\sum_{0}^{\infty} s^n E(z^{Q_n}) \\ &= \frac{1-\zeta}{1-z\zeta} \text{ if } \rho < 1, \text{ and } = 0 \text{ if } \rho \geq 1. \end{aligned}$$

□

10.9 Continuous Time Markov Chains

The transition probabilities of a continuous time Markov chain $\{X(t), t \geq 0\}$ are defined by

$$P_{ij}(t) = P\{X(s+t) = j \mid X(s) = i\} \quad (s \geq 0,\ t \geq 0). \tag{9.1}$$

We have

$$P_{ij}(0) = \delta_{ij}, \quad P_{ij}(t) \geq 0, \quad \sum_{j=0}^{\infty} P_{ij}(t) = 1. \tag{9.2}$$

The Chapman–Kolmogorov equations are given by

$$P_{ij}(t+s) = \sum_{k=0}^{\infty} P_{ik}(s)P_{kj}(t) \quad (s \geq 0,\ t \geq 0). \tag{9.3}$$

In order to derive expressions for $P_{ij}(t)$ from (9.2) and (9.3) we need to assume further that

$$\lim_{t\to 0} \frac{P_{ij}(t) - \delta_{ij}}{t} = q_{ij}, \tag{9.4}$$

where

$$0 \leq q_{ij} < \infty \quad (i \neq j) \text{ and } \sum_{j=0}^{\infty} q_{ij} = 0 \tag{9.5}$$

so that $-\infty < q_{ii} \leq 0$ for each i. An immediate consequence of (9.4) and (9.5) is the following theorem.

Theorem 9.1 (i) *The transition probabilities are uniformly continuous functions of t.*
(ii) *Either $P_{ij}(t) \equiv 0$ or else $P_{ij}(t) > 0$ for large t.*

Proof. (i) From the Chapman–Kolmogorov equations we find that

$$\sum_{k=0}^{\infty} P_{ik}(s)P_{kj}(t) \geq P_{ii}(s)P_{ij}(t) \tag{9.6}$$

and

$$\begin{aligned}\sum_{k=0}^{\infty} P_{ik}(s)P_{kj}(t) &= P_{ii}(s)P_{ij}(t) + \sum_{k\neq i} P_{ik}(s)P_{kj}(t) \\ &\leq P_{ij}(t) + [1 - P_{ii}(s)].\end{aligned} \tag{9.7}$$

Since $P_{ii}(s) \to 1$ as $s \to 0$ we have $1 - \varepsilon \leq P_{ii}(s) \leq 1$ for sufficiently small s, the inequalities (9.6) and (9.7) lead to

$$\varepsilon P_{ij}(t) \leq P_{ij}(s+t) - P_{ij}(t) \leq \varepsilon. \tag{9.8}$$

This shows that $P_{ij}(t)$ is uniformly continuous in t.
(ii) From (9.8) we conclude that if $P_{ij}(t) > 0$ for some t, then $P_{ij}(s+t) > 0$ in some s-internal of fixed length. Therefore either $P_{ij}(t) = 0$ for all t or else $P_{ij}(t) > 0$ for large t. □

In the terminology of discrete time Markov chains, Theorem 9.1(ii) states that except in the trivial case when the transition probabilities vanish identically, the Markov chain is essentially irreducible and aperiodic. This yields the limit distribution of $\{X(t)\}$ as $t \to \infty$ (see Theorem 9.4).

Assumptions (9.4) and (9.5) lead to two systems of differential equations for $P_{ij}(t)$, as stated below.

Theorem 9.2 (i) *If assumptions* (9.4) *and* (9.5) *hold, then the transition probabilities satisfy the equations*

$$P'_{ij}(t) = \sum_{k=0}^{\infty} q_{ik} P_{kj}(t) \quad (t \geq 0). \tag{9.9}$$

(ii) *If, in addition, for each fixed* j *the limit*

$$\lim_{t \to 0} \frac{P_{ij}(t)}{t} \tag{9.10}$$

is reached uniformly with respect to i, *then*

$$P'_{ij}(t) = \sum_{k=0}^{\infty} P_{ik}(t) q_{kj} \quad (t \geq 0). \tag{9.11}$$

Proof. (i) From (9.3) we find that for $h > 0$

$$P_{ij}(t+h) = P_{ii}(h)P_{ij}(t) + \sum_{k \neq i} P_{ik}(h)P_{kj}(t).$$

This can be written as

$$\frac{P_{ij}(t+h) - P_{ij}(t)}{h} = \frac{P_{ii}(h) - 1}{h} P_{ij}(t) + \sum_{k \neq i} \frac{P_{ik}(h)}{h} P_{kj}(t).$$

Letting $h \to 0$ in this we obtain (9.9).

(ii) Again, we have

$$P_{ij}(t+h) = P_{ij}(t)P_{jj}(h) + \sum_{k \neq j} P_{ik}(t)P_{kj}(h),$$

which can be written as

$$\frac{P_{ij}(t+h) - P_{ij}(t)}{h} = P_{ij}(t) \frac{P_{jj}(h) - 1}{h} + \sum_{k \neq i} P_{ik}(t) \frac{P_{kj}(h)}{h}.$$

Letting $h \to 0$ in this and using condition (9.10) in the series on the right side we arrive at Eqs. (9.11). □

Equations (9.9) and (9.11) are, respectively, called the Kolmogorov backward and forward differential equations of the Markov chain. We seek a common unique solution of these two sets of equations such that (9.2) and (9.3) are satisfied. We denote

$$P(t) = (P_{ij}(t)), \quad Q = (q_{ij}). \tag{9.12}$$

Here Q is the general matrix of the chain. Conditions (9.2)–(9.4) can be expressed as

$$P(0) = I, \quad P(t) \geq 0, \quad P(t)e = e, \tag{9.2a}$$

$$P(t+s) = P(t)P(s), \tag{9.3a}$$

$$\lim_{t \to 0} \frac{P(t) - I}{t} = Q, \tag{9.4a}$$

where e is the column vector $(1, 1, \ldots)'$. The backward and forward equations are, respectively,

$$P'(t) = QP(t) \quad \text{and} \quad P'(t) = P(t)Q. \tag{9.13}$$

Integrating these we obtain

$$P(t) = I + \int_0^t QP(s)ds \quad P(t) = I + \int_0^t P(s)Qds. \tag{9.14}$$

The situation is simple in the case of a finite state space $\{0, 1, \cdots, N\}$, with $0 < N < \infty$. Here $P(t)$ and Q are square matrices of order $N + 1$. We need the concept of the norm of a matrix. For a square matrix A of order $N + 1$ we define the norm $\|A\|$ as

$$\|A\| = \sum_{i=0}^{N} \sum_{j=0}^{\infty} |a_{ij}|.$$

This norm has the following properties.

(a) $\|A\| = 0$ iff A is null matrix.
(b) If C is a constant, then $\|CA\| = |C| \, \|A\|$.
(c) If A and B are matrices of order $N + 1$, then

$$\|A + B\| \leq \|A\| + \|B\|, \quad \|AB\| \leq \|A\| \, \|B\|.$$

(d) If the elements of A are continuous functions of s, then

$$\left\| \int_{c_1}^{c_2} A(s)ds \right\| \leq \int_{c_1}^{c_2} \|A(s)\|ds \quad (c_1 < c_2).$$

It is easy to see that $m_1 = \|Q\| < \infty$. We shall ignore the case $m_1 = 0$.

Theorem 9.3 *For a Markov chain $\{X(t)\}$ on a finite state space $\{0, 1, 2, \ldots, N\}$ with $0 < N < \infty$, the series*

$$P(t) = \sum_{n=0}^{\infty} Q^n \frac{t^n}{n!} \tag{9.15}$$

converges to e^{tQ} uniformly in every interval $[0, T]$, and $P(t)$ is the common unique solution of the backward and forward equations satisfying the conditions (9.2a)–(9.4a).

Proof. (i) There can be no more than one solution to the forward equations. Suppose that P_1 and P_2 are two solutions and denote

$$m = \max_{0 \leq s \leq T} \|P_1(s) - P_2(s)\| \quad (0 < m_1 < \infty).$$

Then from (9.14) we find that

$$\|P_1 - P_2\| \leq \int_0^t \|P_1(s) - P_2(s)\| \, \|Q\| ds \leq m_1 \, mt$$

and by induction

$$\|P_1 - P_2\| \leq m_1 \frac{(mt)^n}{n!} \to 0 \quad (0 \leq t \leq T).$$

It follows that $\|P_1 - P_2\| = 0$ and so $P_1 = P_2$, which proves our assertion. A similar result holds for the backward equations.

(ii) We have

$$\left\| Q^n \frac{t^n}{n!} \right\| \leq \frac{(m_1 t)^n}{n!} \to 0 \quad (0 \leq t \leq T).$$

This shows that the series in (9.15) converges uniformly in $[0, T]$. Moreover, from (9.14) we find that

$$P(t) = I + \sum_{n=1}^{\infty} \int_0^t \frac{s^{n-1}}{(n-1)!} Q^n ds = I + \int_0^T P(s) Q ds$$

and similarly

$$P(t) = I + \int_0^t QP(s) ds.$$

Thus $P(t)$ satisfies both the forward and backward equations and by (i) it is the unique solution to both.

(iii) It can be easily verified that $P(t)$ satisfies conditions (9.2a)–(9.4a). □

The results of matrix algebra can be used to derive $P_{ij}(t)$ for finite t and investigate its limit behavior as illustrated in Example 10.3.2.

The situation is much more difficult in the general case, partly because of the possibility that the number of transitions (jumps) in any finite interval $(0, t]$ may not be finite; that is,

$$\sum_{j=0}^{\infty} P_{ij}(t) < 1.$$

(This happens in the case of the Pure Birth process of Example 10.10.2.) We shall not treat the general case here, but instead consider some important special cases, our analysis being based on the forward equations.

Theorem 9.4 *If no $P_{ij}(t)$ vanishes identically, the limit probabilities*

$$u_j = \lim_{t \to \infty} P_{ij}(t) \tag{9.16}$$

exist independently of i, where either $u_j = 0$ $(j \geq 0)$, or else $u = (u_0, u_1, \ldots)$ is the unique solution of the equations

$$uP(t) = u \tag{9.17}$$

such that $u > 0$ and $ue = 1$. (Thus u is the unique stationary distribution of the chain.)

Proof. (i) For $\delta > 0$ denote $X_n = X(n\delta)$ $(n \geq 0)$, so that $\{X_n, \ n \geq 0\}$ is a discrete time Markov chain with the n-step transition probability matrix $P(n\delta)$. By Theorem 9.1 we need to consider only the case where $P(n\delta) > 0$ for large n. In this case $\{X_n\}$ is irreducible and aperiodic, so lim $P_{ij}(n\delta)$ exits as $n \to \infty$. The same is true for all rational δ. Using uniform continuity of $P_{ij}(t)$ we conclude that the limits (9.16) exist independently of i.

(ii) In $P(t+s) = P(s)P(t)$ letting $s \to \infty$ we obtain (9.17). Letting $t \to \infty$ in (9.17) we find that $(ue)u = u$, which states that either $u = 0$ or else $u > 0$ and $ue = 1$. □

From (9.17) it follows that

$$0 = u\frac{P(t) - I}{t} \to uQ \quad \text{as } t \to 0$$

so that u satisfies the equation $uQ = 0$ or

$$\sum_{i \neq j} u_j q_{ij} = u_j(-q_{jj}) \quad (j \geq 0). \tag{9.18}$$

These equations imply that for each state j

$$E \text{ (entrance rate into } j) = \text{E (exit rate from } j). \tag{9.19}$$

For this reason Eqs. (9.18) and called the *balance equations* of the Markov chain.

10.10 Examples of Continuous Time Markov Chains

10.10.1 *The Poisson Process as a Markov Chain*

For this we have

$$q_{ii+1} = \lambda, \quad q_{ii} = -\lambda \quad (i \geq 0). \tag{10.1}$$

The backward and forward equations are, respectively, given by

$$P'_{ij}(t) = -\lambda P_{ij}(t) + \lambda P_{i+1,j}(t) \quad (t \geq 0) \tag{10.2}$$

and

$$P'_{ij}(t) = -\lambda P_{ij}(t) + \lambda P_{ij-1}(t) \quad (t \geq 0) \tag{10.3}$$

where $P_{ij}(t) = 0$ for $j < i$.

Theorem 10.1 (i) *For the Poisson process $\{X(t),\ t \geq 0\}$ the common unique solution of Eqs.* (10.2) *and* (10.3) *is given by*

$$P_{ij}(t) = e^{-\lambda t}\frac{(\lambda t)^{j-i}}{(j-i)!} \quad (j \geq i). \tag{10.4}$$

(ii) *For $j \geq i$ let*

$$\tau_j = \inf\{t\colon X(t) = j+1\} \text{ on } \{X(0) = i\}. \tag{10.5}$$

Then the random variables $\tau_i,\ \tau_{j+1} - \tau_j$ $(j \geq i)$ are IID with density $\lambda e^{-\lambda t}$.

Proof. (i) Integrating (10.2) and (10.3) we obtain

$$P_{ij}(t) = \delta_{ij}e^{-\lambda t} + \lambda \int_0^{\tau} P_{i+1,j}(s)e^{-\lambda(t-s)}ds \tag{10.6}$$

and

$$P_{ij}(t) = \delta_{ij}e^{-\lambda t} + \lambda \int_0^{t} P_{ij-1}(s)e^{-\lambda(t-s)}ds \tag{10.7}$$

for $t \geq 0$. To solve (10.6) we fix j and start with $i = j$. Then $P_{ii}(t) = e^{-\lambda t}$ and for $i = j - 1$

$$P_{j-1,j}(t) = \lambda \int_0^t e^{-\lambda s}e^{-\lambda(t-s)}ds = e^{-\lambda t}\lambda t.$$

By induction on i we arrive at result (10.4) as the unique solution of the backward equations. A similar procedure with (10.7) yields the same unique solution.

(ii) Since

$$\{\tau_i > t\} = \{X(s) = i \quad (0 \leq s \leq t)\} \text{ on } \{X(0) = i\} \tag{10.8}$$

we find that

$$P\{\tau_i > t | X(0) = i\} = P_{ii}(t) = e^{-\lambda t}, \tag{10.9}$$

so τ_i has density $\lambda e^{-\lambda t}$ independently of $X(0)$. From (10.8) we also see that τ_i is a stopping time for the process. Using the strong Markov property we conclude that the random variable $\tau_{i+1} - \tau_i$ is independent of τ_i. Also, on account of the time homogeneity of the process.

$$\begin{aligned} \tau_{i+1} - \tau_i &= \inf\{t - \tau_i \colon X(t) = i + 2\} \text{ on } \{X(\tau_i) = i + 1\} \\ &\stackrel{d}{=} \inf\{t' \colon X(t') = i + 2\} \text{ on } \{X(0) = i + 1\}, \end{aligned} \tag{10.10}$$

and therefore $\tau_{i+1} - \tau_i$ has density $\lambda e^{-\lambda t}$. Extending these arguments we arrive at the results in (ii). □

10.10.2 *The Pure Birth Process*

For this

$$q_{ii+1} = \lambda_i, \quad q_{ii} = -\lambda_i, \tag{10.11}$$

where $0 < \lambda_i < \infty$ $(i \geq 0)$. The backward and forward equations are, respectively,

$$P'_{ij}(t) = -\lambda_i P_{ij}(t) + \lambda_i P_{i+1,j}(t) \quad (t \geq 0) \tag{10.12}$$

and

$$P'_{ij}(t) = -\lambda_j P_{ij}(t) + \lambda_{j-1} P_{ij-1}(t) \quad (t \geq 0) \tag{10.13}$$

with $P_{ij}(t) = 0$ for $j < i$.

Theorem 10.2 (i) *For $j \geq i$ denote*

$$\tau_j = \inf\{t\colon X(t) = j+1\} \text{ on } \{X(0) = i\}. \tag{10.14}$$

Then the random variables τ_i, $\tau_j - \tau_{j-1}$ $(j \geq i+1)$ are independent, with densities $\lambda_j e^{-\lambda_j t}$ $(j \geq i)$.

(ii) *As $j \to \infty$, $\tau_j \to L \leq \infty$ with probability 1, where L is the lifetime of the process in the sense that*

$$X(t) < \infty \quad \text{for } t < L \quad \text{and} \quad = \infty \quad \text{for } t \geq L \tag{10.15}$$

and $L < \infty$ with probability 1 iff the series

$$\frac{1}{\lambda_i} + \frac{1}{\lambda_{i+1}} + \cdots < \infty \tag{10.16}$$

in which case

$$E[e^{-\theta L} \mid X(0) = i] = \prod_{j=i}^{\infty} \left(\frac{\lambda_j}{\theta + \lambda_j} \right) \quad (\theta > 0). \tag{10.17}$$

(iii) *Denote by $f_{ij}(t)$ the conditional of τ_j given $X(0) = i$. Then for $j \geq i+1$*

$$P_{ij}(t) = \int_0^t f_{ij-1}(s) e^{-\lambda_j (t-s)}\, ds \quad (t \geq 0) \tag{10.18}$$

and

$$\sum_{j=0}^{\infty} P_{ij}(t) = 1 \quad (t \geq 0,\ L = \infty)$$
$$< 1 \quad (0 \leq t < L < \infty). \tag{10.19}$$

(iv) *The transition probabilities $P_{ij}(t)$ as expressed by (10.18) satisfy both the backward and forward equations.*

Proof. (i) Integrating the backward and forward equations we obtain

$$P_{ij}(t) = \delta_{ij}e^{-\lambda_i t} + \lambda_i \int_0^t P_{i+1,j}(s)e^{-\lambda_i(t-s)}ds \tag{10.20}$$

and

$$P_{ij}(t) = \delta_{ij}e^{-\lambda_j t} + \lambda_{j-1} \int_0^t P_{ij-1}(s)e^{-\lambda_j(t-s)}ds \tag{10.21}$$

for $t \geq 0$. Both of these equations yields the result $P_{ii}(t) = e^{-\lambda_i t}$ $(t \geq 0)$. The rest of the proof is along the same lines as in Theorem 10.1(ii).

(ii) Since $\tau_j = \tau_i + (\tau_{i+1} - \tau_i) + \cdots + (\tau_j - \tau_{j-1})$ $(j \geq i+1)$ we see that the sequence $\{\tau_j,\ j \geq 1\}$ in monotone increasing, so $\tau_j \to L \leq \infty$ with probability 1. Therefore $E(e^{-\theta\tau_j}) \to E(e^{-\theta L})$ for $\theta > 0$. Now

$$\begin{aligned} E[e^{-\theta\tau_j} \mid X(0) = i] &= \prod_{k=i}^{j}\left(\frac{\lambda_k}{\theta + \lambda_k}\right) \to \prod_{k=i}^{\infty}\left(\frac{\lambda_k}{\theta + \lambda_k}\right) \\ &= 1 \Big/ \prod_{k=1}^{\infty}\left(1 + \frac{\theta}{\lambda_k}\right), \end{aligned}$$

where the infinite product converges iff series (10.16) converges. Accordingly, $L < \infty$ in this case and $E[e^{-\theta L} \mid X(0) = i]$ is given by (10.17). It is clear that

$$X(t) = \max\{j\colon \tau_{j-1} \leq t\} \tag{10.22}$$

so that $\{X(t) \geq j\} = \{\tau_{j-1} \leq t\}$. Letting $j \to \infty$ in this we find that

$$\{X(t) = \infty\} = \{t \geq L\}. \tag{10.23}$$

(iii) For $j \geq i+1$ we have

$$\begin{aligned} P_{ij}(t) &= P\{\tau_{j-1} \leq t < \tau_j \mid X(0) = i\} \\ &= \int_0^t f_{ij-1}(s)P\{\tau_j - \tau_{j-1} > t - s \mid \tau_{j-1} = s\} \\ &= \int_0^t f_{ij-1}(s)e^{-\lambda_j(t-s)}ds \quad (t \geq 0), \end{aligned}$$

in agreement with (10.18). From (10.23) we obtain

$$\sum_{k=0}^{\infty} P_{ik}(t) = P\{X(t) < \infty \mid X(0) = i\} = P\{L > t \mid X(0) = i\}$$

which leads to (10.19).

(iv) The convolution relations

$$f_{ij}(t) = \int_0^t f_{ij-1}(s)\lambda_j e^{-\lambda_j(t-s)} ds \quad (t \geq 0), \tag{10.24}$$

$$f_{ij}(t) = \int_0^t \lambda_i e^{-\lambda_i(t-s)} f_{i+1,j}(s) ds \quad (t \geq 0) \tag{10.25}$$

follow from (i) and lead to $\lambda_j P_{ij}(t) = f_{ij}(t)$. Using these results it can be easily verified that the $P_{ij}(t)$ as expressed by (10.18) satisfy both the backward and forward equations. □

Theorem 10.2(ii) demonstrates the possibility of the event $X(t) \to \infty$ in finite time (explosion). This situation can realistically occur in population growth models (for example, in the nonlinear growth model of Sec. 10.11).

The theoretical point of view of Theorem 10.2(ii) is as follows. If we incorporate ∞ into the state space (thus compactifying it), and assign to it the probability $1 - \sum_0^\infty P_{ij}(t)$, then (10.19) can be expressed as

$$\sum_{j=0}^{\infty} P_{ij}(t) + P\{X(t) = \infty\} = 1. \tag{10.19a}$$

We can also construct a return process, which starts from ∞, returns to a finite state i with a pre-assigned probability, and then moves among $\{0, 1, 2, \ldots\}$ before reaching ∞ for the second time. The procedure can be repeated more than once.

10.10.3 *The Pure Death Process*

This has the state space $\{0, 1, 2, \ldots, N\}$ with $0 < N < \infty$ and transition rates

$$q_{ii-1} = \mu_i, \; q_{ii} = -\mu_i \quad (0 \leq i \leq N) \tag{10.26}$$

with $\mu_0 = 0$. It is clear that $P_{ij}(t) = 0$ for $j > i$. In view of Theorem 9.3 it suffices to consider only the forward equations, which are

$$P'_{ij}(t) = -\mu_j P_{ij}(t) + \mu_{j+1} P_{ij+1}(t) \quad (0 \leq j \leq i). \tag{10.27}$$

The assumption $\mu_0 = 0$ implies that state 0 is absorbing. Of interest is the absorption time, namely the random variable

$$T = \inf\{t\colon X(t) = 0\} \text{ on } \{X(0) = i\}, \tag{10.28}$$

where $i \geq 1$. We have the following.

Theorem 10.3 *For the absorption T in the Pure Death Process we have*

$$E[e^{-\theta T} \mid X(0) = i] = \prod_{j=1}^{i} \left(\frac{\mu_j}{\theta + \mu_j} \right) \quad (\theta > 0). \tag{10.29}$$

Here $T < \infty$ with probability 1, and

$$E(T) = \frac{1}{\mu_1} + \frac{1}{\mu_2} + \cdots + \frac{1}{\mu_i} \quad (i \geq 1). \tag{10.30}$$

Proof. For $0 \leq j \leq i-1$ let

$$\tau_j = \inf\{t\colon X(t) = j\} \text{ on } \{X(0) = i\}. \tag{10.31}$$

Then $T = \tau_0$ and

$$\tau_0 = \tau_{i-1} + (\tau_{i-2} - \tau_{i-1}) + \cdots + (\tau_0 - \tau_1). \tag{10.32}$$

Proceeding as in Theorem 10.2 we arrive at result (10.29). The remaining results are obvious. □

10.10.4 *The Birth and Death Process*

This is a Markov chain whose transition rates are given by

$$q_{ii+1} = \lambda_i, \quad q_{ii-1} = \mu_i, \quad q_{ii} = -(\lambda_i + \mu_i) \quad (i \geq 0), \tag{10.33}$$

where $\lambda_i \geq 0$ $(i \geq 0)$, $\mu_i \geq 0$ $(i \geq 1)$ and $\mu_0 = 0$. If $\lambda_0 = 0$ state 0 is absorbing. The forward equations are given by

$$\begin{aligned} P'_{ij}(t) &= -(\lambda_j + \mu_j)P_{ij}(t) + \mu_{j+1}P_{ij+1}(t) + \lambda_{j-1}P_{ij-1}(t)(j \geq 1), \\ P'_{i0}(t) &= -\lambda_0 P_{i0}(t) + \mu_i P_{i1}(t) \end{aligned} \tag{10.34}$$

with the initial conditions $P_{ij}(0) = \delta_{ij}$. In special cases these equations can be solved to obtain expressions for $P_{ij}(t)$ for finite t as well as the stationary distribution of $X(t)$ under appropriate conditions. We do so for the population growth models of Sec. 10.11. Here we consider the general case and derive the stationary distribution.

Theorem 10.4 (i) *Suppose the Birth and Death Process has no absorbing states. Then the limit probabilities* $\{u_j, j \geq 0\}$ *exist.*

(ii) *Denote*

$$\pi_0 = 1, \; \pi_j = \frac{\lambda_0 \lambda_1 \cdots \lambda_{j-1}}{\mu_1 \mu_2 \cdots \mu_j} (j \geq 1). \tag{10.35}$$

If $\pi_0 + \pi_1 + \cdots = \infty$, *then* $u_j = 0$ $(j \geq 0)$. *Otherwise* $u_j = \pi_j u_0$ $(j \geq 0)$, *where* $u_0 = (\pi_0 + \pi_1 + \cdots)^{-1} > 0$.

Proof. The balance equations in this case reduce to

$$-(\lambda_j + \mu_j)u_j + \lambda_{j-1}u_{j-1} + \mu_{j+1}u_{j+1} = 0 \quad (j \geq 0). \tag{10.36}$$

Adding these over $0, 1, 2, \ldots, j$ we obtain $\mu_{j+1}u_{j+1} - \lambda_j u_j = 0$ $(j \geq 0)$. These lead to $u_j = \pi_j u_0$ $(j \geq 0)$ and the desired results. □

10.11 Models for Population Growth

10.11.1 *Some Deterministic Models*

In this section we describe some deterministic and stochastic models for population growth. The populations considered here may be human populations, bacterial colonies, or cosmic ray showers. However, in describing human populations, we take in to account only the female component of the population, and moreover, in the models described below we ignore the age distribution of the population.

In order to discriminate between various models, we emphasize the qualitative characteristics in each case. We are thus interested in the possible explosion or extinction of the population. It is also possible for the population to reach a finite size in the long run. These aspects may be illustrated by first considering deterministic (nonstochastic) models.

We shall denote by $x(t)$ the population size at time t, with the initial size $x(0) = i \geq 1$. Each deterministic model described here is actually a statement (a law) concerning the rate of growth dx/dt of the population. This leads to a differential equation, which has to be solved using the initial condition $x(0) = i$.

Example 11.1 (The Law of Natural Growth) This law states that

$$\frac{dx}{dt} = \alpha x \quad (x \geq 0), \tag{11.1}$$

where α is the constant rate of net increase (birth minus death) in the population. Integrating both sides of this equation over $[0, t]$ we obtain $\log x(t) - \log x(0) = \alpha t$, or

$$x(t) = ie^{\alpha t} \quad (t \geq 0). \tag{11.2}$$

If $\alpha > 0$, we see from (11.2) that as $t \to \infty$, $x(t) \to \infty$, which means that the population explodes eventually. This is in agreement with the observation of T.R. Malthus that a population, when unchecked, grows exponentially (the Malthusian law).

If $\alpha < 0$ in (11.2), then the conclusion is that $x(t) \to 0$ as $t \to \infty$; that is, the population becomes extinct eventually. If $\alpha = 0$, then

$$x(t) = i \quad (t \geq 0). \tag{11.3}$$

Here the population remains at a constant level (the case of zero population growth).

Example 11.2 (Linear Growth) This law states that

$$\frac{dx}{dt} = \alpha x + \nu, \tag{11.4}$$

where α, ν are constants, $\alpha \neq 0$ and $\nu > 0$. Here ν represents the rate of immigration into the population. The solution of (11.4) is found to be

$$x(t) = \left(i + \frac{\nu}{\alpha}\right)e^{\alpha t} - \frac{\nu}{\alpha}. \tag{11.5}$$

As $t \to \infty$,

$$x(t) \to \infty \text{ if } \alpha > 0, \text{ and } x(t) \to \frac{\nu}{|\alpha|} \text{ if } \alpha < 0. \tag{11.6}$$

The eventual behavior of the population depends on parameter α, but not on ν; it explodes or else it reaches a finite size according as $\alpha > 0$ or $\alpha > 0$.

Example 11.3 (The Logistic Law of Growth) This law states that

$$\frac{dx}{dt} = \alpha x \left(1 - \frac{x}{\beta}\right), \tag{11.7}$$

where α and β are constants, $\alpha > 0, \beta > i$. This gives

$$\frac{d^2x}{dt^2} = \alpha\left(1 - \frac{2x}{\beta}\right)\frac{dx}{dt},$$

which shows that the rate of growth dx/dt of the population increases for $x < \beta/2$ and decreases for $x > \beta/2$ (it will be found that $x(t) < \beta$ for all $t \geq 0$). This represents the impact of limited resources (food supply) that are available to the population. We can write (11.7) as

$$\frac{dx}{x} + \frac{1}{\beta}\frac{dx}{1 - \frac{x}{\beta}} = \alpha dt.$$

Integrating this over $(0, t)$ we find that

$$x(t)\left(1 - \frac{i}{\beta}\right) = i\left(1 - \frac{x(t)}{\beta}\right)e^{\alpha t}$$

or

$$x(t) = \frac{ie^{\alpha t}}{1 + \frac{i}{\beta}(e^{\alpha t} - 1)} \quad (t \geq 0). \tag{11.8}$$

Here

$$x(t) \to \beta \quad \text{as } t \to \infty \tag{11.9}$$

so that the population eventually reaches a finite size β.

Example 11.4 (Nonlinear Growth Rate) Suppose that

$$\frac{da}{dt} = \alpha x^2 \quad (t \geq 0), \tag{11.10}$$

where α is a positive constant. This gives

$$x(t) = \frac{i}{1 - i\alpha t} \quad (t \geq 0) \tag{11.11}$$

and

$$x(t) \to \infty \quad \text{as } t \to (i\alpha)^{-1}. \tag{11.12}$$

This shows that the population explosion takes place at the finite epoch of time $t_0 = (i\alpha)^{-1}$ This is in contrast with the models of Examples 11.1 and 11.2 with $\alpha > 0$, where the population size remains finite for all finite epochs $t \geq 0$ and explodes only eventually.

10.11.2 *Stochastic Models*

We shall denote by $X(t)$ the population size at time t. For the stochastic models described here it will turn out that $\{X(t), t \geq 0\}$ is a birth and death process. Let us denote its transition probabilities by

$$P_{ij}(t) = P\{X(\tau + t) = j \mid X(\tau) = i\}. \tag{11.13}$$

The analysis is based on the forward Kolmogorov equations of the process, which are

$$P'_{ij}(t) = -(\lambda_j + \mu_j)P_{ij}(t) + \lambda_{j-1}P_{ij} - 1(t) + \mu_{j+1}P_{ij+1}(t) \quad (j \geq 0) \tag{11.14}$$

λ_j and μ_j being, respectively, the birth and death rates (with $\lambda_0 \geq 0$, $\mu_0 = 0$) and $P_{ij}(t) = 0$ for $j < 0$. To solve these equations we introduce the (conditional) probability generating function (p.g.f.)

$$G(z,t) = \sum_{j=0}^{\infty} P_{ij}(t)z^j \quad (0 < z < 1) \tag{11.15}$$

with $G(z,0) = z^i$ $(i \geq 1)$. Equations (11.14) then yield a partial differential equation for $G(z,t)$, which needs to be solved. In many cases the solution will turn out to be a familiar p.g.f., such as that of the geometric or the negative binomial. In the general case the required (conditional) probability $P_{ij}(t)$ is the cofficient of z^j in the power series expansion of the solution $G(z,t)$, which can be computed numerically.

The (conditional) mean and the variance of $X(t)$ can be calculated from (11.15) as follows:

$$m(t) = E[X(t) \mid X(0) = i] = \sum_{j=0}^{\infty} jP_{ij}(t) = \left[\frac{\partial G}{\partial z}\right]_{z=1}, \tag{11.16}$$

$$\begin{aligned}\sigma^2(t) = \mathrm{Var}[X(t) \mid X(0) = i] &= \sum_{j=0}^{\infty} j^2 P_{ij}(t) - m(t)^2 \\ &= \left[\frac{\partial^2 G}{\partial z^2}\right]_{z=1} + m(t) - m(t)^2.\end{aligned} \tag{11.17}$$

If $P_{ij}(t)$ are known explicitly, we can obtain the limit probabilities

$$\lim_{t\to\infty} P_{ij}(t) = u_j \tag{11.18}$$

(which exist under appropriate conditions and are independent of i). In other cases we use the result

$$\sum_{j=0}^{\infty} u_j z^j = \lim_{t\to\infty} G(z,t) \tag{11.19}$$

to obtain the generating function of the limit distribution of $X(t)$.

10.11.3 *The Yule–Furry Model*

Consider a population in which (a) there is no interaction among individuals, and (b) during a time-interval of length h, each individual has a probability $\lambda h + o(h)$ of creating a new one. Let $X(t)$ be the number of individuals in the population (briefly, the population size) at time t; then our assumptions imply that $X(t)$ is a pure birth process with $\lambda_j = j\lambda (j \geq 1)$. The forward equations are given by

$$P'_{ij}(t) = -\lambda j P_{ij}(t) + \lambda(j-1)P_{ij-1}(t) \quad (j \geq 1), \tag{11.20}$$

where $P_{ij}(t) = 0$ for $j < i$. To solve these, let us define p.g.f. (11.15). We have

$$\frac{\partial G}{\partial t} = \sum_0^\infty P'_{ij}(t)z^j, \quad \frac{\partial G}{\partial z} = \sum_0^\infty jP_{ij}(t)z^{j-1}. \tag{11.21}$$

From (11.20) we then obtain the partial differential equation

$$\frac{\partial G}{\partial t} + \lambda z(1-z)\frac{\partial G}{\partial z} = 0. \tag{11.22}$$

This is a special case of Eq. (11.59) below. Its solution is found to be

$$G(z,t) = \left(\frac{pz}{1-qz}\right)^i, \tag{11.23}$$

where $p = e^{-\lambda t}$ and $q = 1 - e^{-\lambda t}$ This shows that the (conditional) distribution of $X(t)$ is a negative binomial, and

$$P_{ij}(t) = \binom{j-1}{j-i} p^i q^{j-i} (j \geq i). \tag{11.24}$$

The (conditional) mean and the variance are given by

$$EX(t) = ie^{\lambda t}, \quad \text{Var}X(t) = ie^{\lambda t}(e^{\lambda t}-1). \tag{11.25}$$

Since $p \to 0$ and $q \to 1$ as $t \to \infty$, we see from (11.24) that for $j \geq i$

$$P_{ij}(t) \to 0 \quad \text{as } t \to \infty. \tag{11.26}$$

It is therefore impossible for the population to reach any finite size eventually. We describe this phenomenon by stating that $X(t) \to \infty$ in distribution as $t \to \infty$ (population explosion).

The result $EX(t) = ie^{\lambda t}$ shows that *on the average* the population obeys the law of exponential growth (this is analogous to the deterministic model of Example 11.1 with $\alpha > 0$). We have here the additional information that the fluctuations around the mean $ie^{\lambda t}$ are also very large.

Since the series

$$\sum_{j=i}^{\infty} \frac{1}{\lambda_j} = \frac{1}{\lambda}\left(\frac{1}{i} + \frac{1}{i+1} + \cdots\right) = \infty, \tag{11.27}$$

the population size remains finite for all $i \geq 0$ (no explosion at finite time) by Theorem 10.2. This would not be the case if, for example, we had $\lambda_i = i^2\lambda$ (a nonlinear model). In this regard this model should be compared with the deterministic model of Example 11.4.

It turns out that, roughly speaking, $X(t) \to \infty$ exponentially. The following theorem is a more precise statement of this fact.

Theorem 11.1 *For the Yule–Furry model with $X(0) = 1$ we have*

$$\lim_{t\to\infty} P\{e^{-\lambda t}X(t) \leq x\} = 1 - e^{-x}. \tag{11.28}$$

Proof. We have

$$E[e^{-\theta e^{-\lambda t}X(t)}] = G(e^{-\theta e^{-\lambda t}}, t) \quad (\theta > 0),$$

where $G(z,t)$ is given by (11.23) with $i = 1$. Therefore

$$E[e^{-\theta e^{-\lambda t}X(t)}] = \frac{e^{-\lambda t}e^{-\theta e^{-\lambda t}}}{1 - (1 - e^{-\lambda t})e^{-\theta e^{-\lambda t}}}.$$

Expanding the numerator and the denominator of this last expression in terms of $e^{-\lambda t}$ we find that

$$\begin{aligned} E[e^{-\theta e^{-\lambda t}X(t)}] &= \frac{1 - \theta e^{-\lambda t} + o(e^{-\lambda t})}{1 + \theta - \theta e^{-\lambda t} + o(e^{e^{-\lambda t}})} = \frac{1 - \theta e^{-\lambda t}[1 + o(1)]}{1 + \theta - \theta e^{-\lambda t}[1 + o(1)]} \\ &\to \frac{1}{1+\theta}, \end{aligned} \tag{11.29}$$

where the limit is the Laplace transform of the density e^{-x}. □

10.11.4 *The Feller–Arley Model*

Consider the population model described in the previous section: but now let us introduce a third assumption, namely (c) during a time interval of length h, an individual has a probability $\mu h + o(h)$ of dying. We then obtain a birth and death process, in which $\lambda_j = j\lambda, \mu_j = j\mu (j \geq 0)$. The forward equations are found to be

$$P_{ij}(t) = -(\lambda+\mu)jP_{ij}(t) + \lambda(j-1)P_{ij-1}(t) + \mu(j+1)P_{ij+1}(t) \quad (j \geq 0), \tag{11.30}$$

where $P_{ij}(t) = 0$ for $j < 0$. Proceeding as in the previous case we obtain the differential equation

$$\frac{\partial G}{\partial t} + (1-z)(\lambda z - \mu)\frac{\partial G}{\partial z} = 0 \tag{11.31}$$

for the p.g.f. defined by (11.15). This is again a special case of Eq. (11.59) below. Its solution is given by

$$G(z,t) = \left[\frac{\mu}{\lambda}q + \left(1 - \frac{\mu}{\lambda}q\right)\frac{pz}{1-qz}\right]^i, \tag{11.32}$$

where

$$\begin{aligned} p \equiv p(t) &= (\lambda-\mu)/(\lambda e^{(\lambda-\mu)t} - \mu) \quad \text{if } \lambda \neq \mu \\ &= (1+\lambda t)^{-1} \quad \text{if } \lambda = \mu. \end{aligned} \tag{11.33}$$

and $q = 1 - p$. Result (11.32) confirms the fact that the population at time t comprises of i independent and identically distributed subpopulations generated by the i initial individuals (ancestors). For this reason it suffices to consider the case $i = 1$. Accordingly we find from (11.32) that for $i = 1$,

$$G(z,t) = \frac{\mu}{\lambda}q + \left(1 - \frac{\mu}{\lambda}q\right)\frac{pz}{1-qz}, \tag{11.34}$$

which is the p.g.f. of the modified geometric distribution. This gives

$$P_{10}(t) = \frac{\mu}{\lambda}q, \quad P_{1j(t)} = \left(1 - \frac{\mu}{\lambda}q\right)pq^{j-1} \quad (j \geq 1). \tag{11.35}$$

It is useful to distinguish between the two cases $\lambda = \mu$ and $\lambda \neq \mu$.

(i) Let $\lambda = \mu$. Substituting for p and q from (11.33) we find that

$$P_{10}(t) = \frac{\lambda t}{1+\lambda t}, \quad P_{1j}(t) = \frac{(\lambda t)^{j-1}}{(1+\lambda t)^{j+1}} \quad (j \geq 1). \tag{11.36}$$

We have

$$EX(t) = 1, \quad \mathrm{Var}X(t) = 2\lambda t. \tag{11.37}$$

Also, as $t \to \infty$

$$P_{10}(t) \to 1, \quad P_{1j}(t) \to 0 \quad (j \geq 1), \tag{11.38}$$

so that the population eventually becomes extinct with probability 1. The deterministic law of zero population growth (the model of Example 11.1 with $\alpha = 0$) leads to a constant population size (equal to 1), while the stochastic law predicts ultimate extinction.

(ii) Let $\lambda \neq \mu$. From (11.35) the mean and the variance are obtained as

$$EX(t) = e^{(\lambda-\mu)t}, \quad \mathrm{Var}X(t) = \frac{\lambda+\mu}{\lambda-\mu} e^{(\lambda-\mu)t}[e^{(\lambda-\mu)t} - 1]. \tag{11.39}$$

Also, since

$$\lim_{t\to\infty} p(t) = 1 - \frac{\lambda}{\mu} \text{ if } \lambda < \mu, \quad \text{and} = 0 \text{ if } \lambda \geq \mu, \tag{11.40}$$

we find that as $t \to \infty$,

$$P_{1j}(t) \to 0 \quad (j \geq 1) \tag{11.41}$$

and

$$P_{10}(t) \to 1 \text{ if } \lambda < \mu, \quad \text{and} \to \frac{\mu}{\lambda} \text{ if } \lambda > \mu. \tag{11.42}$$

This shows that regardless of $\lambda < \mu$ or $\lambda > \mu$, the population cannot reach a finite nonzero size eventually. If $\lambda < \mu$ it becomes extinct with probability 1, and if $\lambda > \mu$, it either becomes extinct with probability μ/λ, or explodes with probability $1 - \mu/\lambda(>0)$. A more complete statement is the following, where we denote

$$\rho = \frac{\lambda}{\mu} = \frac{\text{birth rate}}{\text{death rate}}. \tag{11.43}$$

Theorem 11.2 *If $\rho \leq 1$, the population described by the Feller–Arley model becomes extinct with probability* 1, *while if $\rho > 1$, the population either* *becomes extinct with probability* ρ^{-i}, *or explodes with probability* $1-\rho^{-i}$, *where* $i \geq 1$ *is the initial size of the population.*

The extinction time of the population is defined by the random variable

$$T = \min\{t\colon X(t) = 0\}. \tag{11.44}$$

Since $X(t) = 0$ implies that extinction must have occurred at or before epoch t, the distribution function of T is given by

$$F(t) = P\{T \leq t\} = P_{10}(t). \tag{11.45}$$

As we have already seen, eventual extinction is a sure event if $\rho \leq 1$. In this case we have the following.

Theorem 11.3 *In the Feller–Arley model with* $\rho \leq 1$, *the extinction time has the following properties*: (i) *If* $\rho = 1$, *then* T *has the density*

$$f(t) = \frac{\lambda}{(1+\lambda t)^2} \quad (t \geq 0) \tag{11.46}$$

and $E(T) = \infty$.

(ii) *If* $\rho < 1$, *then* T *has the exponential density* $\alpha e^{-\alpha t}$, *where* $\alpha(\mu-\lambda)^{-1}$ *is a random variable with the geometric distribution* $(1-\rho)\rho^{j-1} (j \geq 1)$. *Moreover,*

$$E(T) = \frac{1}{\lambda}\log(1-\rho)^{-1} \text{ and } E(T^2) = \frac{2}{\lambda\mu(1-\rho)}\sum_{1}^{\infty}\frac{\rho^j}{j^2} < \infty. \tag{11.47}$$

Proof. (i) Let $\rho = 1$. From (11.45) and (11.35) we obtain $F(t) = \lambda t(1+\lambda t)^{-1}$, which leads to (11.46). Also

$$E(T) = \int_0^\infty \frac{\lambda t}{(1+\lambda t)^2}dt = \infty.$$

(ii) Let $\rho < 1$. Then from (11.45) and (11.35) we obtain

$$1 - F(t) = \frac{(1-\rho)e^{-(\mu-\lambda)t}}{1-\rho e^{-(\mu-\lambda)t}} = Ee^{-(\mu-\lambda)tY}, \tag{11.48}$$

where the random variable Y has the geometric distribution $(1-\rho)\rho^{j-1}$ $(j \geq 1)$. This leads to the result concerning the density of T. We have

$$E(T) = E\left[\frac{1}{(\mu-\lambda)Y}\right] = \frac{1}{\mu(1-\rho)}\sum_{j=1}^{\infty}\frac{1}{j}(1-\rho)\rho^{j-1}$$
$$= \frac{1}{\lambda}\sum_{1}^{\infty}\frac{\rho^j}{j} = \frac{1}{\lambda}\log(1-\rho)^{-1}$$

and

$$E(T^2) = E\left[\frac{2}{(\mu-\lambda)^2Y^2}\right] = \frac{2}{\mu^2(1-\rho)^2}\sum_{1}^{\infty}\frac{1}{j^2}(1-\rho)\rho^{j-1}$$
$$= \frac{2}{\lambda\mu(1-\rho)}\sum_{1}^{\infty}\frac{\rho^j}{j^2}.$$

□

In the case $\rho > 1$, Theorem 11.2 tells us that the population has a positive probability $(1-\rho^{-1})$ of escaping extinction (survival). Now consider N identical populations developing independently under the assumptions of this section, all of them with initial population size unity. We will then have N independent copies of the Feller–Arley process, among which $N(1-\rho^{-1})$ will eventually survive. We now ask about the limit behavior of the population sizes among these surviving populations. It turns out that, roughly speaking, they grow at an exponential rate. A more precise statement is the following.

Theorem 11.4 *In the Feller–Arley model with $\rho > 1$ we have*

$$\lim_{t\to\infty} P\{e^{-(\lambda-\mu)t}X(t) \leq x \mid X(t) > 0\} = 1 - e^{-\left(1-\frac{1}{\rho}\right)x}. \tag{11.49}$$

Proof. We have

$$P\{X(t) = j \mid X(t) > 0\} = \frac{P_{1j}(t)}{1-P_{10}(t)} \quad (j = 1, 2, \cdots)$$
$$= 0 \quad (j = 0),$$

so that

$$E[z^{X(t)} \mid X(t) > 0] = \frac{1}{1-P_{10}(t)}\sum_{1}^{\infty}P_{ij}(t)z^j = \frac{G(z,t)-G(0,t)}{1-G(0,t)}$$
$$= \frac{pz}{1-qz} \tag{11.50}$$

in view of (11.32), with p given by (11.33). This result is true for all ρ. In our case this gives

$$E[e^{-\theta e^{-(\lambda-\mu)t}X(t)} \mid X(t) > 0] = \frac{pe^{-\theta e^{-(\lambda-\mu)t}}}{1 - qe^{-\theta e^{-(\lambda-\mu)t}}}.$$

Proceeding as in the proof of Theorem 11.1 we arrive at the desired result. □

10.11.5 *The Kendall Model*

The last two models dealt with closed populations, in which there was no possibility of emigration and immigration. It is clear that introducing an emigration factor would only increase the value of constant μ. To take into account immigration we introduce the new assumption that (d) during a time-interval of length h there is a probability $\nu h + o(h)$ of a new individual being added to the population from the outside world. We then have a birth and death process for which $\lambda_i = i\lambda + \nu$ and $\mu_i = i\mu (i \geq 0)$, the forward equations being

$$\begin{aligned} P_{ij}(t) = -(\lambda j + \mu j + v)P_{ij}(t) + [\lambda(j-1) + \nu] \\ \times P_{ij-1}(t) + \mu(j+1)P_{ij+1}(t) \quad (j \geq 0). \end{aligned} \tag{11.51}$$

Proceeding as before we obtain the differential equation

$$\frac{\partial G}{\partial t} + (1-z)(\lambda z - \mu)\frac{\partial G}{\partial z} = -\nu(1-z)G. \tag{11.52}$$

This will be solved in the next section and it will be found that

$$G(z,t) = \left[\frac{\mu}{\lambda}q + \left(1 - \frac{\mu}{\lambda}q\right)\frac{pz}{1-qz}\right]^i \left(\frac{p}{1-qz}\right)^{\nu/\lambda}, \tag{11.53}$$

where p is given by (11.33). The (conditional) distribution of $X(t)$ is thus the convolution of a negative binomial with the distribution obtained in the Feller–Arley model. For the mean and the variance we have

$$EX(t) = ie^{(\lambda-\mu)t} + \frac{\nu}{\lambda-\mu}[e^{(\lambda-\mu)t} - 1], \tag{11.54}$$

$$\begin{aligned} \text{Var}X(t) &= i\frac{\lambda+\mu}{\lambda-\mu}e^{(\lambda-\mu)t}[e^{(\lambda-\mu)t} - 1] \\ &\quad + \frac{\nu}{(\lambda-\mu)^2}[e^{(\lambda-\mu)t} - 1][\lambda e^{(\lambda-\mu)t} - \mu] \end{aligned} \tag{11.55}$$

if $\lambda \neq \mu$, and

$$EX(t) = i + \nu t, \quad \mathrm{Var}X(t) = i(2\lambda t) + \nu t(1 + \lambda t) \tag{11.56}$$

if $\lambda = \mu$. The following theorem describes the behavior of the population as $t \to \infty$. As before we let $\rho = \lambda/\mu$.

Theorem 11.5 *If $\rho \geq 1$ the population described by the Kendall model explodes (in distribution), while if $\rho < 1$ it attains a finite limit size with the negative binomial distribution*

$$\binom{-\nu/\lambda}{j}(1-\rho)(-\rho)^j \quad (j = 0, 1, 2, \ldots). \tag{11.57}$$

Proof. Using (11.40) we find that

$$\begin{aligned} \lim_{t\to\infty} G(z,t) &= 0 && \text{if } \rho \geq 1 \\ &= \left(\frac{1-\rho}{1-\rho z}\right)^{\nu/\lambda} && \text{if } \rho < 1, \end{aligned} \tag{11.58}$$

the limit in the case $\rho < 1$ being recognized as the p.g.f. of (11.57). □

10.11.6 *The Differential Equation*

In this section we solve the differential equation that arose from the Kendall model, namely

$$\frac{\partial G}{\partial t} + (1-z)(\lambda z - \mu)\frac{\partial G}{\partial z} = -\nu(1-z)G \tag{11.59}$$

with the initial condition $G(z, 0) = z^i$ $(i \geq 1)$. When $\mu = 0, \nu = 0$, this reduces to (11.22), the equation for the Yule-Furry model, and when $\nu = 0$, it reduces to (11.31) for the Feller–Arley model. To solve (11.59) we have to consider the auxiliary equations

$$\frac{dt}{1} = \frac{dz}{(1-z)(\lambda z - \mu)} = \frac{dG}{-\nu(1-z)G}. \tag{11.60}$$

From the first two terms in (11.60) we obtain

$$(\lambda - \mu)t = \log\frac{\lambda z - \mu}{1-z} + \text{const.}, \text{ or } \frac{\lambda z - \mu}{z-1}e^{(\mu-\lambda)t} = c_1, \tag{11.61}$$

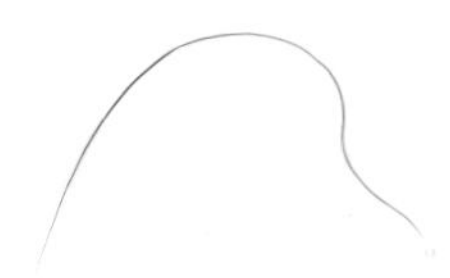

where c_1 is a constant. From the last two terms in (11.60) we obtain

$$\frac{\nu dz}{\lambda z - \mu} + \frac{dG}{G} = 0.$$

This gives $\nu\lambda^{-1}\log(\lambda z - \mu) + \log G = \text{constant}$, or

$$G(z,t)(\lambda z - \mu)^{\nu/\lambda} = c_2, \tag{11.62}$$

where c_2 is a constant. Eliminating one of the constants c_1, c_2 from (11.61) and (11.62) we obtain

$$G(z,t)(\lambda z - \mu)^{\nu/\lambda} = \phi\left(\frac{\lambda z - \mu}{z - 1}e^{(\mu-\lambda)t}\right), \tag{11.63}$$

with an unknown function ϕ. For $t = 0$ this gives

$$\phi\left(\frac{\lambda z - \mu}{z - 1}\right) = z^i(\lambda x - \mu)^{\nu/\lambda}.$$

Denoting $(\lambda z - \mu)/(z - 1) = x$, we have $z = (\mu - x)/(\lambda - x)$ and so

$$\phi(x) = \left(\frac{\mu - x}{\lambda - x}\right)^i \left(\frac{\mu - \lambda}{\lambda - x}x\right)^{\nu/\lambda}. \tag{11.64}$$

From (11.63) and (11.64) we finally obtain

$$G(z,t) = \left[\frac{\frac{\mu}{\lambda}q + (p - \frac{\mu}{\lambda}q)z}{1 - qz}\right]^i \left(\frac{p}{1 - qz}\right)^{\nu/\lambda}, \tag{11.65}$$

where p is given by (11.33). We have thus solved (11.59).

10.12 Problems for Solution

10.12.1 *Finite Markov Chains*

1. Consider a game played as follows: a group of $n+1$ persons is arranged in a line. The first person starts a rumor by telling his neighbor that the last person in line is a nonconformist. Each person in line then repeats this rumor to his neighbor, however, with probability $p > 0$, he reverses the sense of the rumor as it is told to him. What is the probability that the last person in line will be told he is a nonconformist if (i) $n = 6$, (ii) n is very large?

2. In a study of rainfall in Tel Aviv, Gabriel and Neumann [*Quart. J. R. Met. Soc.* **88** (1962)] formulated a Markov chain model for the occurrence of wet and dry days during the rainy period December, January, and February. The data comprising a total of 2437 days (in 27 years) were

	Actual day		
	Dry	Wet	Total
Preceding day	1049	350	1399
	351	687	1038

Using relative frequencies the transition probabilities were estimated as

$$P = \begin{Vmatrix} 0.750 & 0.250 \\ 0.338 & 0.662 \end{Vmatrix}.$$

(a) Show that the equilibrium probabilities are 0.575 (for a dry day) and 0.425 (for a wet day).

(b) A weather cycle may be defined as a wet spell followed by a dry spell or vice versa. Show that the expected length of a weather cycle is approximately a week.

3. *The Ehrenfest Model of Heat Exchange.* A certain number a of molecules (numbered $1, 2, \ldots, a$) are distributed in two containers. Each time a number is selected at random from $(1, 2, \ldots, a)$ and the molecule bearing that number is transferred from its container to the other. Let X_0 be the initial number of molecules in the first container and X_n the number of molecules in that container after the nth transfer ($n = 1, 2, \ldots$). (i) Show that $\{X_n, n \geq 0\}$ is a time-homogeneous Markov chain on the state space $(0, 1, 2, \ldots, a)$ and (ii) find its transition probability matrix. (For the historical background of this model, see Mark Kac, Probability, *Sci. Am.* **211**(3) (1964).)

4. *A Model for Social Mobility.* In a study of the relation between the social statuses of fathers and their sons, Prais [*J. Roy. Stat. Soc. A* **118** (1955)] used a Markov chain model to analyze the data obtained in the 1949 Social Survey from a series of interviews of about 3500 males, resident in England and Wales and aged 18 years and over. Social class is treated as if it is related to the male side of the family, largely because it is measured by occupation. Moreover, it is assumed that the influence

of one's ancestor in determining one's class is transmitted entirely through one's father; additional effects of earlier generations and the other side of the family are ignored. The social classes considered and the proportion of the 1949 population in them are given below:

Social class	Percentage of population in 1949
u: upper class (professional, high administrative, managerial, and executive)	7.5
m: middle class (supervisory, non-manual, and skilled)	63.4
ℓ: lower class (semi-skilled and unskilled manual)	29.1
Total	100.0

The recruitment probabilities are given by

	Sons		
Fathers	u	m	ℓ
u	0.45	0.48	0.07
m	0.05	0.70	0.25
ℓ	0.01	0.50	0.49

(i) Find the equilibrium distribution $\{\pi_i\}$ of the population among the social classes.

(ii) The index of immobility of the ith social class is defined as

$$a_i = \frac{1 - \pi_i}{1 - P_{ii}},$$

where P_{ii} is the diagonal element of the recruitment matrix. Show that

$$a_u = 1.66, \quad a_m = 1.23, \quad a_\ell = 1.23.$$

5. *The (s, S) Inventory Model.* An inventory is an amount of material stored for the purpose of future sale. Demands for this material arise

from time to time, and the inventory is subject to periodic review in order to determine whether or not new supplies have to be ordered. The (s, S) inventory model is defined in discrete time $n = 0, 1, 2, \ldots$, with two given numbers (s, S) such that $0 \leq s < S < \infty$. Let X_n be the inventory level at time $n, X_n < 0$ indicating a backlog. Let ξ_n be the demand during the time interval $(n-1, n)$; this demand is met or backlogged at time n. The following assumptions are made. (i) The demands $\xi_1, \xi_2, \ldots$ are independent and identically distributed random variables, which are also independent of X_n. (ii) The ordering policy is such that if $s < X_n \leq S$, no order is placed, while if $X_n \leq s$, an order is placed for an amount equal to $S - X_n$. The delivery of the amount ordered is immediate (no delivery lag).

(a) In the case $s = S - 1$ the policy is to order the amount equal to the demand every time a demand occurs (the so-called just-in-time policy), which may be appropriate when ordering costs are low compared with holding costs. Find the distribution of X_n in this case.

(b) In the case $s < S - 1$, show that $\{X_n, n \geq 0\}$ is a time-homogeneous Markov chain.

(c) Suppose that X_n and ξ_n are measured in discrete units and $P\{\xi_n = k\} = a_k (k = 0, 1, \ldots)$. For the model with $s = 1, S = 4$, and $a_k = 0$ for $k > 4$, obtain the transition probability matrix of the chain $\{X_n\}$.

6. *The Generalized Bernoulli-Laplace Model.* Here we have α white and β black balls, where $\alpha + \beta = 2a$ and $\alpha \leq a$. These $2a$ balls are distributed equally in two urns. At each trial one ball is selected at random from each urn, and these two balls are interchanged. Show that the distribution of the number of white balls in the first urn after an indefinitely large number of trials is given by

$$\binom{\alpha}{k}\binom{\beta}{a-k}\Big/\binom{2a}{a} \quad (k = 0, 1, \ldots, \alpha).$$

7. In the Ehrenfest model assume that initially all molecules are in the first container.

(a) Let $m_n = E(X_n)$. Show that m_n satisfies the recurrence relation $m_{n+1} = (1 - \frac{2}{a})m_n + 1$ $(n \geq 0)$ and hence that

$$m_n = \frac{a}{2} + \frac{a}{2}\left(1 - \frac{2}{a}\right)^n.$$

(b) Similarly show that

$$\sigma_n^2 = \mathrm{Var}(X_n) = \frac{a}{4} + \frac{1}{4}a(a-1)\left(1-\frac{4}{a}\right)^n - \frac{a^2}{4}\left(1-\frac{2}{a}\right)^{2n}.$$

10.12.2 *Infinite Markov Chains*

8. Let $P_{0j} = a_j$ $(j \geq 0), P_{i,i-1} = 1$ $(i \geq 1)$, where $a_j > 0$ $(j \geq 0)$. Show that all states are persistent, null or non-null according as $a = \sum_0^\infty ja_j = \infty$ or $< \infty$. In the latter case show that the stationary distribution is given by $u_j = (a+1)^{-1}(a_j + a_{j+1} + \cdots)$ $(j \geq 0)$.
9. Let $P_{0j} = a_j$ $(j \geq 0), P_{i,i_1} = q, P_{ii} = p$ $(i \geq 1)$.

   (i) Show that the first visit time to state 0 has the distribution

   $$f_{i0}^{(n)} = \binom{n-1}{n-i} p^{n-i} q^i \quad (n \geq i \geq 1).$$

   (ii) Use the result in (i) to obtain the distribution $\{f_{00}^{(n)}, n \geq 1\}$ of the return time of state 0.
   (iii) Show that all states are persistent, null or non-null according as $a = \sum_0^\infty ja_j = \infty$ or $< \infty$. In the latter case show that the stationary distribution is given by

   $$u_0 = q(a+q)^{-1}, \quad u_j = (a+q)^{-1}(a_j + a_{j+1} + \cdots) \quad (j \geq 1).$$

10. For $i \geq 0$, let $P_{i0} = q_i, P_{i,i+1} = p_i$. Show that $f_{00}^{(n)} = Q_{n-1} - Q_n$, where

    $$Q_0 = 1, \quad Q_n = (1-q_0)(1-q_1)\cdots(1-q_{n-1}) \quad (n \geq 1).$$

    Hence show that the states are

    (i) transient if $\sum q_i < \infty$,
    (ii) persistent null if $\sum q_i = \infty, \sum Q_n = \infty$, and
    (iii) persistent non-null if $\sum q_i = \infty,\ \mu = \sum Q_n < \infty$.
    In the last case show that the stationary distribution is given by $u_j = Q_j\mu^{-1}$ $(j \geq 0)$.

11. Let

$$P_{00} = q_0, \quad P_{01} = p_0,$$
$$P_{i,i-1} = q_i, \quad P_{i,i+1} = p_i \quad (i \geq 1).$$

Further, let $L_0 = 1, L_i = (q_1 q_2 \ldots q_i)(p_1 p_2 \ldots p_i)^{-1} \quad (i \geq 1)$. Show that the states are

(i) transient if $\Sigma L_i < \infty$,
(ii) persistent null if $\Sigma L_i = \infty, \Sigma(p_i L_i)^{-1} = \infty$, and
(iii) persistent non-null if $\Sigma L_i = \infty, \Sigma(p_i L_i)^{-1} < \infty$.

Find the stationary distribution in the last case.

12. For $i \geq 0$, let $P_{i0} = a_i, P_{i,i+2} = 1 - a_i$. Also, let

$$A_n = (1 - a_0)(1 - a_2) \cdots (1 - a_{2n}) \quad (n \geq 1).$$

Show that the states 1,3,5,... are all transient, while the states 0,2,4,$\cdots$ form a closed set, with all of its states

(i) transient if $\sum a_{2i} < \infty$,
(ii) persistent null if $\sum a_{2i} = \infty, \ \sum A_n = \infty$, and
(iii) persistent non-null if $\sum a_{2i} = \infty, \ \sum A_n < \infty$.

10.12.3 *Continuous Time Markov Chains*

13. Suppose that $\{X(t), t \geq 0\}$ is a time-homogeneous Markov chain with the transition probabilities

$$P_{ij}(t) = \binom{j-1}{j-i} e^{-it}(1 - e^{-t})^{j-i} \quad \text{for } j \geq i$$
$$= 0 \qquad \text{for } j < i.$$

We are given $X(0) = 2$, and some action is being planned as soon as $X(t)$ reaches level 5. Find the expected time for this action to take place.

14. A population comprised of N individuals (N fixed) is subject to a certain infectious disease. Suppose that in any time-interval $(t, t + h)$, an infected person will cause any non-infected person to become infected with probability $\lambda h + o(h)$, where $\lambda > 0$. Once infected, an individual remains in that state forever. Initially, there is exactly one

infected individual. Let $X(t)$ denote the number of infected persons at time t.

(a) Model $X(t)$ as a pure birth process, and calculate its parameters λ_n.

(b) Let T denote the time at which the population becomes totally infected. Find the mean and the variance of T.

15. A system consists of N components hooked up in parallel (so that it needs exactly one of these components to function). Each component has a constant failure rate α. Initially all N components are functioning. Let $X(t)$ denote the number of components that are functioning at time t.

(a) Model $X(t)$ as a pure death process, and find its parameters μ_n.

(b) Let T denote the lifetime of the system. Find the mean and the variance of T.

(c) Prove that the transition probabilities of $X(t)$ are given by

$$P_{ij}(t) = \binom{i}{j} e^{-j\lambda t}(1 - e^{-\lambda t})^{j-i}(j \leq i).$$

16. People are immigrating to a certain country at an exponential rate λ. Each arrival stays in this country for an exponential amount of time having mean μ^{-1}. Let $X(t)$ denote the number of people living there at time t.

(a) Model $\{X(t), t \geq 0\}$ as a birth and death process, and find its parameters.

(b) Find the stationary distribution of the process, and its mean and variance.

17. A collection of particles act independently in giving rise to succeeding generations of particles in the following manner. Each particle, from the time it appears, waits for a random length of time having an exponential density with parameter λ and then either splits into two identical parts with probability p or else disappears with probability $1-p$. Also, new particles immigrate into the system in a Poisson process with parameter ν and then gives rise to succeeding generations as the others do. Denote by $X(t)$ the number of particles present at time t.

(a) Show that $\{X(t), t \geq 0\}$ is a birth and death process, and find its parameters.

(b) Find the stationary distribution of $X(t)$.

18. A bottling plant has three machines and one repairman. Each machine randomly breaks down at an exponential rate of two times per hour, and the repairman fixes it at an exponential rate of three machines per hour. Whenever all three machines are out of order, the plant manager assists the repairman, and their total service rate becomes four machines per hour. Denote by $X(t)$ the number of machines that are broken at time t.
    (a) Model $\{X(t), t \geq 0\}$ as a birth and death process, and find its state space and parameters.
    (b) Find the stationary distribution and its mean.

19. An electric circuit supplies N welders who use the current only intermittently. If at epoch t a welder uses current, the probability that he ceases using it sometime during $(t, t+h)$ is $\mu h + o(h)$. If at epoch t he requires no current, the probability that he calls for current sometime during $(t, t+h]$ is $\lambda h + o(h)$. The welders work independently of each other. Let $X(t)$ be the number of welders using current at time t.
    (a) Model $\{X(t), t \geq 0\}$ as a birth and death process, and find its parameters.
    (b) Find the stationary distribution of the process.

10.12.4 *Models for Population Growth*

20. Consider the Yule–Furry process with one initial individual. Suppose this individual's lifetime is a random variable with the exponential density $\alpha e^{-\alpha t}$. Prove that the distribution of the number of offspring due to him and his descendents at the time of his death is given by

$$\frac{\alpha}{\lambda} \cdot \frac{\Gamma(j+1)\Gamma\left(\frac{\alpha}{\lambda}+1\right)}{\Gamma\left(j+\frac{\alpha}{\lambda}+2\right)} \quad (j \geq 0).$$

21. In the Feller–Arley process show that the distribution of the total number of direct descendents of an individual during his lifetime is given by

$$\frac{\mu}{\lambda+\mu}\left(\frac{\lambda}{\lambda+\mu}\right)^j \quad (j \geq 0).$$

22. For the Feller–Arley process prove the following:
    (i) If $\rho = 1$, then $\lim_{t\to\infty} E[e^{-\theta X(t)/t} \mid X(t) > 0] = (1+\lambda\theta)^{-1}$.
    (ii) If $\rho < 1$, then $\lim_{t\to\infty} E[z^{X(t)} \mid X(t) > 0] = \frac{(1-\rho)z}{1-\rho z}$.

23. For the Kendall process prove the following: if $\rho = 1$, then $\lim_{t\to\infty} E[e^{-\theta X(t)/t}] = (1+\lambda\theta)^{-\nu/\lambda}$.
24. **A Continuous Time Branching Process**. Assume that during a time interval of length h, an individual in the population gives way to a new distribution of individuals with the p.g.f.

$$z + g(z)h + o(h).$$

Show that the p.g.f. $G(z,t)$ satisfies the relation

$$G(z,t+s) = G(G(z,s),t).$$

Use this result to show that $G(z,t)$ satisfies the two differential equations

$$\frac{\partial G}{\partial t} = g(z)\frac{\partial G}{\partial z}, \quad \frac{\partial G}{\partial \tau} = g(G(z,t)).$$

The solution of the second equation here can be written in the form

$$t = \int_z^{G(z,t)} \frac{du}{g(u)}.$$

25. **A Contagion Model**. A population growth is subject to the assumptions (a) and (b) of Sec. 11.3. In addition (c) during a time interval of length h, there is a probability $\nu h + o(h)$ of a new individual being added to the population. Initially there are i (≥ 0) individuals. Prove the following:

    (i) The (conditional) distribution of the population size $X(t)$ is the negative binomial

$$P_{ij}(t) = e^{-(i\lambda+\nu)t}\binom{-i-\nu/\lambda}{j-i}(e^{-\lambda t}-1)^{j-i} \quad (j \geq i).$$

    (ii) Eventual population explosion is a sure event.

26. A population growth is subject to the following assumptions. During a time interval of lenght h, (a) there is a probability $\nu h + o(h)$ of a new individual being added to the population and (b) an individual has a probability $\mu h + o(h)$ of dying. Let $X(t)$ be the population size

at time t, with $X(0) = i \geq 0$. Prove the following:

(i) The (conditional) distribution of $X(t)$ is given by

$$P_{ij}(t) = \sum_{k=0}^{\min(i,j)} \binom{i}{k} p^k q^{i-k} e^{-\frac{\nu q}{\mu}} \left(\frac{\nu q}{\mu}\right)^{j-k} \Big/ (j-k)!$$

with $p = e^{-\mu t}$ and $q = 1 - e^{-\mu t}$.

(ii) The (conditional) mean and the variance of $X(t)$ are given by

$$EX(t) = \left(i - \frac{\nu}{\mu}\right) e^{-\mu t} + \frac{\nu}{\mu}, \quad \mathrm{Var} X(t) = \left(i e^{-\mu t} + \frac{\nu}{\mu}\right)(1 - e^{-\mu t}).$$

(iii) The limit distribution of $X(t)$ as $t \to \infty$ is Poisson with mean ν/μ.

27. **A Problem in Consumer Economics.** Suppose that the market for a product such as cellphone consists of a finite number N of potential consumers. Initially a certain number a (>0) of the consumers have adopted the product, the others being nonconsumers.

(i) *Deterministic model*: Denote by $x(t)$ the number of consumers who have adopted the product at time t. Assume that

$$\frac{dx}{dt} = \beta x \left(1 - \frac{x}{N}\right).$$

Note that this leads to the logistic law for $x(t)$, as in Example 11.3.

(ii) *Stochastic model*: Denote by $X(t)$ the number of consumers who have adopted the product at time t. Assume that if $X(t) = j$, then during the time interval $(t, t+h]$, each of the nonconsumers will adopt the product with probability $\frac{\lambda}{N} jh + o(h)$. Show that $\{X(t), t \geq 0\}$ is a pure birth process, with $X(0) = a$. For $b > a$, let T_b be the time at which the bth consumer will adopt the product. As $b \to \infty, N - b \to \infty$, show that

$$P\left\{\lambda T_b - \log N - \log N \frac{b}{N-b} \leq x\right\} \to \sum_{j=0}^{a-1} e^{-e^{-x}} \frac{(e^{-x})^j}{j!}.$$

Appendix A: Tauberian Theorems

A.1 Introduction

We start with the statement of Abel's (1826) theorem on power series. Let

$$\sum_{0}^{\infty} a_n z^n = A(z) \quad \text{for } |z| < 1. \tag{1.1}$$

Abel's Theorem. *If* $\sum_0^\infty a_n$ *converges, then*

$$\lim_{z \to 1-} A(z) = \sum_{0}^{\infty} a_n. \tag{1.2}$$

A generalization of Abel's theorem was given by Appell (1878). *It states the following:*

Theorem (Appell) *For* $\rho \geq 0$ *suppose*

$$a_0 + a_1 + \cdots + a_n \sim \frac{cn^\rho}{\Gamma(\rho + 1)} \quad (n \to \infty) \tag{1.3}$$

for some c. Then

$$A(z) \sim \frac{c}{(1 - z)^\rho} \quad (z \to 1-). \tag{1.4}$$

The converse of Abel's theorem is false. As an example, consider the series

$$\sum_{n=0}^{\infty} (-1)^n z^n = \frac{1}{1 + z} \quad \text{for } |z| < 1.$$

Here $\lim_{z\to 1} A(z) = 1/2$, but the series $\sum_0^\infty (-1)^n$ does not converge. The following partial converse of Abel's theorem was given by Tauber (1897):

Theorem (Tauber) *Let* $\lim_{z\to 1-} A(z) = a$. *If* $a_n = o(1/n)$, *then*

$$\sum_0^\infty a_n = a. \tag{1.5}$$

Tauber's conclusion holds under less restrictive conditions concerning a_n. Littlewood (1910) replaced the condition $a_n = o(1/n)$ by $a_n = 0(1/n)$. A converse of Appell's theorem was proved by Hardy and Littlewood (1913) as follows:

Theorem (Hardy and Littlewood) *Let the power series* (1.1) *with non-negative coefficients be such that*

$$A(z) \sim \frac{c}{(1-z)^\rho} \quad (z \to 1-) \tag{1.6}$$

for some $\rho > 0$. *Then*

$$a_0 + a_1 + \cdots + a_n \sim \frac{cn^\rho}{\Gamma(\rho+1)} \quad (n \to \infty). \tag{1.7}$$

We call a theorem *Abelian* if properties of the power series (generating function) are found from those of its coefficients. We call a theorem *Tauberian* if properties of the coefficients are found from those of the generating function.

Let us consider the case where the coefficients a_n are non-negative. Putting $z = e^{-\theta}$ $(\theta > 0)$ we can write (1.1) as

$$\phi(\theta) = \sum_{n=0}^\infty a_n e^{-n\theta}. \tag{1.8}$$

Let us now define the measure U by setting

$$U(t) = U\{(0,t]\} = \sum_{n\le t} a_n. \tag{1.9}$$

We can then write

$$\phi(\theta) = \int_0^\infty e^{-\theta t} U\{dt\}, \tag{1.8$'$}$$

that is, ϕ is the Laplace transform of the measure U. We can consider more general measures U and their transforms ϕ, and it turns out that Abelian and Tauberian theorems hold for U and ϕ. (However, theorems for power series are usually obtained as corollaries of theorems for Laplace transforms.) The simplest Abelian theorem for measures is the following: if U is a finite measure (that is, if $U(\infty) = a$), then

$$\lim_{\theta \to 0+} \phi(\theta) = a. \tag{1.10}$$

A generalization of Appell's theorem is the following:

$$U(t) \sim ct^\rho/\Gamma(\rho+1) \quad (t \to \infty) \ \Rightarrow \ \phi(\theta) \sim c\theta^{-\rho} \quad (\theta \to 0+). \tag{1.11}$$

The equivalent of the Hardy–Littlewood theorem for Laplace transforms is the following:

$$\phi(\theta) \sim c\theta^{-\rho} \quad (\theta \to 0+) \ \Rightarrow \ U(t) \sim ct^\rho/\Gamma(\rho+1) \quad (t \to \infty). \tag{1.12}$$

Considerable simplification was made possible by the notion of *regular variation* introduced by Karamata (1930) and the notion of *slow variation* introduced by Schmidt (1925). Karamata proved the following general result extending (1.12).

Karamata's Tauberian Theorem *Let* $0 \le \rho < \infty$ *and* L *a slowly varying function. Then*

$$\phi(\theta) \sim \theta^{-\rho} L(1/\theta) \ \Rightarrow \ U(t) \sim t^\rho L(t)/\Gamma(\rho+1) \quad (t \to \infty). \tag{1.13}$$

In a series of papers published during 1963–1969 Feller gave a simplified treatment of this area, using probability arguments based on convergence of measures. (It must be noted that while the monotonicity of U and hence the interpretation of U as a measure is an essential condition for Tauberian theorems, it is quite unnecessary for Abelian theorems, as the coefficients a_n in (1.1) need not be non-negative.) A substantial part of Feller's treatment is found in his book (1971) (Chap. VIII, Secs. 8 and 9, and Chap. XIII, Sec. 5). However, there are some gaps in his proofs which need to be filled. Supplementary material will be found in the monographs by De Haan (1970), Postnikov (1980), and Seneta (1976). In this appendix we present some of the important results in this area.

A.2 Slow and Regular Variation

A positive function L on $(0, \infty)$ *varies slowly* at ∞ if for each $x > 0$

$$\frac{L(tx)}{L(t)} \to 1 \quad \text{as } t \to \infty. \tag{2.1}$$

It varies slowly at the point a if $L(1/x - a)$ varies slowly at ∞.

Examples

2.1 The function $L(x) = c$ $(0 < c < \infty)$ clearly varies slowly at ∞. More generally, if $L(x) \to c$ as $x \to \infty$, then L varies slowly at ∞.
2.2 The function $L(x) = |\log x|$ varies slowly at ∞.
2.3 The functions e^x, $\sin x$ do not vary slowly.

A positive function U on $(0, \infty)$ *varies regularly* at ∞ if

$$U(x) = x^\rho L(x) \quad (-\infty < \rho < \infty) \tag{2.2}$$

with L slowly varying. We call ρ the *exponent* of U.

In what follows we shall omit the term "at ∞" if it is clear from the context.

Theorem 2.1 *A positive monotone function U varies regularly iff for each $x > 0$,*

$$\frac{U(tx)}{U(t)} \to \psi(x) \quad (t \to \infty) \tag{2.3}$$

on a dense set D of points, and ψ is finite and positive in some interval. In this case $\psi(x) = x^\rho$ with $-\infty < \rho < \infty$.

Proof. Suppose U is positive monotone and $U(x) = x^\rho L(x)$ with ρ finite and L slowly varying. Then for $x > 0$

$$\frac{U(tx)}{U(t)} = \frac{(tx)^\rho L(tx)}{t^\rho L(t)} = x^\rho \frac{L(tx)}{L(t)} \to x^\rho$$

as $t \to \infty$, as required. Conversely, let U be positive monotone and (2.3) hold. We shall prove that $\psi(x) = x^\rho$, with ρ finite. We have

$$\frac{U(tx_1x_2)}{U(t)} = \frac{U(tx_1x_2)}{U(tx_2)} \cdot \frac{U(tx_2)}{U(t)}.$$

Letting $t \to \infty$ in this we obtain

$$\psi(x_1 x_2) = \psi(x_1)\psi(x_2) \quad \text{for } x_1, x_2 \in D. \tag{2.4}$$

If $\psi(x_1) = \infty$ then $\psi(x_1^n) = \infty$ and $\psi(x_1^{-n}) = 0$ for all n. Since ψ is monotone, this implies that either $\psi(x) \equiv \infty$ or $\psi(x) \equiv 0$ for $x \in D$, which is not true. We shall therefore consider the case where ψ is finite and positive, and ψ is defined everywhere by right continuity. The only solution of (2.4) which is bounded in finite intervals is known to be $\psi(x) = x^\rho$, with finite ρ. Since

$$\frac{(tx)^{-\rho}U(tx)}{t^{-\rho}U(t)} \to 1 \quad \text{as } t \to \infty$$

it follows that $x^{-\rho}L(x)$ varies slowly, or U varies regularly with exponent $\rho, -\infty < \rho < \infty$. □

Theorem 2.2 *A positive monotone function U varies regularly iff there exist two sequences $\{a_n\}$ and $\{\lambda_n\}$ of positive numbers such that*

$$a_n \to \infty, \quad \lambda_{n+1}/\lambda_n \to 1 \quad \text{as } n \to \infty \tag{2.5}$$

and the limit

$$\lim_{n\to\infty} \lambda_n U(a_n x) = \chi(x) \le \infty \tag{2.6}$$

exists on a dense set and χ is finite and positive in some interval. In this case $\chi(x) = \chi(1)x^\rho, -\infty < \rho < \infty$.

Proof. It suffices to consider the case where U is non-increasing.

(i) Suppose there exist two sequences $\{a_n\}$ and $\{\lambda_n\}$ with the prescribed properties. For given $t > 0$, let $n = \min\{m\colon a_{m+1} > t\}$. Then $a_n \le t < a_{n+1}$, and

$$\frac{U(a_{n+1}x)}{U(a_n)} \le \frac{U(tx)}{U(t)} \le \frac{U(a_n x)}{U(a_{n+1})}. \tag{2.7}$$

Under the given conditions the extreme members in (2.7) converge to $\chi(x)/\chi(1)$ as $n \to \infty$. Therefore,

$$\frac{U(tx)}{U(t)} \to \frac{\chi(x)}{\chi(1)} \quad \text{as } t \to \infty.$$

By Theorem 2.1, U varies regularly, and $\chi(x) = \chi(1)x^\rho$, with ρ finite.

(ii) Suppose U varies regularly with exponent $\rho \le 0$. We assume that $U(\infty) = 0$. Let $a_n = \inf\{x : U(x) \le cn^{-1}\}$ $(c > 0)$, so that $a_n \to \infty$ as $n \to \infty$. We have $U(a_{n}+) \le cn^{-1} \le U(a_{n}-)$, and so

$$\frac{U(a_{n}+)}{U(a_n)} \le \frac{cn^{-1}}{U(a_n)} \le \frac{U(a_{n}-)}{U(a_n)}. \tag{2.8}$$

For given $\varepsilon > 0$, $x < 1$ there exists a $t_0 \equiv t_0(x, \varepsilon)$ such that for $t > t_0$

$$1 \le \frac{U(t-)}{U(t)} \le \frac{U(tx)}{U(t)} < x^{\rho} + \varepsilon.$$

This gives

$$\lim_{t\to\infty} \frac{U(t-)}{U(t)} = 1,$$

and similarly

$$\lim_{t\to\infty} \frac{U(t+)}{U(t)} = 1.$$

From (2.8) we therefore obtain $nU(a_n) \to c$ as $n \to \infty$. Choose $\lambda_n = n$.

$$\lim_{n\to\infty} \lambda_n U(a_n x) = \lim_{n\to\infty} \lambda_n U(a_n) \cdot \frac{U(a_n x)}{U(a_n)} \to cx^{\rho}$$

as required. We have thus proved (2.6), with the sequences $\{a_n\}$ and $\{\lambda_n\}$ satisfying (2.5). □

A.3 Measures and Their Transforms

Let U be a measure on $[0, \infty)$ and

$$\phi(\theta) = \int_0^{\infty} e^{-\theta x} U\{dx\} \tag{3.1}$$

converge for $\theta > a$. We call ϕ the Laplace transform of U. We shall sometimes write $U(x) = U\{(0, x]\}$.

Examples

3.1 Let $U = F$, where F is a proper distribution on $[0, \infty)$. Here $U(\infty) = 1$.

3.2 Let $U(x) = \int_0^x [1 - F(y)]dy$, so that U is absolutely continuous, with a monotone density $u(x) = 1 - F(x)$. We have $U(\infty) = \mu \leq \infty$, where μ is the mean of the distribution F. Moreover,

$$\phi(\theta) = \frac{1 - F^*(\theta)}{\theta} \quad (\theta > 0)$$

where F^* is the Laplace transform of F.

3.3 Let F_n be the n-fold convolution of F with itself ($n \geq 1$) and F_0 a distribution concentrated at zero. Let $U(x) = \sum_0^\infty F_n(x)$. Then

$$\phi(\theta) = \frac{1}{1 - F^*(\theta)} \quad (\theta > 0).$$

3.4. Let $U(x) = x^\rho (\rho \geq 0)$. Then $\phi(\theta) = \Gamma(\rho + 1)\theta^{-\rho}$ $(\theta > 0)$.

Given a measure U with the Laplace transform ϕ for $\theta \geq \theta_0$ we construct a distribution measure F by setting

$$F\{dx\} = \frac{e^{-\theta_0 x} U(dx)}{\phi(\theta_0)}. \tag{3.2}$$

The Laplace transform of F is given by

$$\psi(\theta) = \frac{\phi(\theta + \theta_0)}{\phi(\theta_0)} \quad (\theta \geq 0). \tag{3.3}$$

This construction leads to the following theorems.

Theorem 3.1 *A measure U is uniquely determined by its Laplace transform ϕ in some interval $a < \theta < \infty$.*

Theorem 3.2 (Extended Continuity Theorem) *Let $\{U_n\}$ be a sequence of measures and $\{\phi_n\}$ the sequence of the corresponding Laplace transforms.*

If $\phi_n(\theta) \to \phi(\theta)$ for $\theta > a$, then ϕ is the Laplace transform of a measure U and $U_n \to U$.

Conversely, if $U_n \to U$ and the sequence $\{\phi_n(a)\}$ is bounded, then $\phi_n(\theta) \to \phi(\theta)$ for $\theta > a$.

A.4 Tauberian Theorems

We now consider a measure U such that its transform ϕ converges for $\theta > 0$. Any relation between the behavior of $\phi(\theta)$ as $\theta \to 0+$ and the behavior of

$U(x)$ as $x \to \infty$ is called a Tauberian theorem. In deriving the following results it is useful to observe that with $t\tau = 1$,

$$\int_0^\infty e^{-\theta x} U\{tdx\} = \int_0^\infty e^{-\theta\tau y} U\{dy\} = \phi(\tau\theta). \tag{4.1}$$

As $t \to \infty, \tau \to 0+$.

Lemma 4.1 *If U varies regularly, then the function $\phi(\tau)/U(t)$ is bounded.*

Proof. We have

$$\begin{aligned}\phi(\tau) &= \int_0^\infty e^{-\tau x} U\{dx\} \\ &= \int_0^t e^{-\tau x} U\{dx\} + \sum_{n=1}^\infty \int_{2^{n-1}t}^{2^n t} e^{-\tau x} U\{dx\} \\ &\leq U(t) + \sum_{n=1}^\infty e^{-2^{n-1}} U(2^n t).\end{aligned}$$

Now since U varies regularly, $U(tx)/U(t) \to x^\rho$ (with ρ finite) as $t \to \infty$, by Theorem 2.1. So, for $t > t_0$ we have $U(2t)/U(t) < 2^{\rho+1}$. This gives $U(2^n t) < 2^{n(\rho+1)} U(t)$, and therefore

$$\frac{\phi(\tau)}{U(t)} \leq 1 + \sum_{n=1}^\infty e^{-2^{n-1}} 2^{n(\rho+1)} < \infty$$

for $t > t_0$, as required. □

Theorem 4.1 *The measure U varies regularly, with*

$$U(t) \sim \frac{t^\rho}{\Gamma(\rho+1)} L(t) \quad (t \to \infty) \tag{4.2}$$

$(0 \leq \rho < \infty)$ *iff its Laplace transform*

$$\phi(\theta) \sim \theta^{-\rho} L(1/\theta) \quad (\theta \to 0+). \tag{4.3}$$

Proof. (i) Suppose that (4.2) holds. Consider the family of measures $\{U_t, t \geq 0\}$, where

$$U_t(x) = \frac{U(tx)}{U(t)}.$$

The transform of U_t is given by

$$\phi_t(\theta) = \frac{\phi(\tau\theta)}{U(t)}.$$

By Lemma 4.1 we know $\phi_t(\theta)$ is bounded at $\theta = 1$. From (4.2) we find that

$$U_t(x) \to x^\rho \quad \text{as } t \to \infty.$$

Therefore, by Theorem 3.2,

$$\phi_t(\theta) \to \int_0^\infty e^{-\theta x} d(x^\rho) = \frac{\Gamma(\rho+1)}{\theta^\rho}$$

and thus

$$\frac{\phi(\tau)}{\tau^{-\rho}L(1/\tau)} = \frac{\phi(\tau)}{U(t)} \cdot \frac{U(t)}{t^\rho L(t)} \to 1$$

as $t \to \infty$. This gives $\phi(\tau) \sim \tau^{-\rho}L(1/\tau)$ as $\tau \to 0+$, which is the desired result (4.3).

(ii) Conversely, let 4.3 hold. Then

$$\frac{\phi(\tau\theta)}{\phi(\tau)} \to \theta^{-\rho} \quad \text{as } \tau \to 0+.$$

Since $\phi(\tau\theta)/\phi(\tau)$ is the transform of the measure $W_t(x) = U(tz)/\phi(\tau)$, we conclude from Theorem 3.2 that

$$\frac{U(tx)}{\phi(\tau)} \to \frac{x^\rho}{\Gamma(\rho+1)} \quad \text{as } t \to \infty.$$

Therefore,

$$\frac{U(t)}{t^\rho L(t)} = \frac{U(t)}{\phi(\tau)} \cdot \frac{\phi(\tau)}{\tau^{-\rho}L(1/\tau)} \to \frac{1}{\Gamma(\rho+1)} \quad \text{as } t \to \infty.$$

□

Theorem 4.1 is actually the main Tauberian theorem. However, we need to deal with the two special cases where (i) U is absolutely continuous, and

(ii) U is atomic. Considering first case (i), let u be a non-negative function on $[0, \infty)$ and

$$U(t) = \int_0^t u(s)ds. \tag{4.4}$$

The following theorem relates the regular variation of u to that of U and vice versa.

Theorem 4.2 *Let $0 < \rho < \infty$. If*

$$u(t) \sim t^{\rho-1} L(t)/\Gamma(\rho) \tag{4.5}$$

then

$$U(t) \sim t^{\rho} L(t)/\Gamma(\rho + 1). \tag{4.6}$$

Conversely, if (4.6) *holds and u is monotone, then* (4.5) *holds.*

The remarkable feature of the above result is that for the converse we need an additional condition, namely, the monotonicity of the density. For the proof we need some further properties of regular variation. These are described in the following two lemmas.

Lemma 4.2 *Let u be as in* (4.4). *If u varies regularly with exponent $\rho - 1$ we have the following:*

(i) *As $t \to \infty, U(t)$ converges if $\rho < 0$ and diverges if $\rho > 0$.*
(ii) *For $\rho \geq 0, U(t)$ varies regularly with exponent ρ.*

Proof. (i) If $U(\infty) = \infty$, then

$$\frac{U(tx)}{U(t)} = \frac{\int_0^{tx} u(s)ds}{\int_0^t u(s)ds} = \frac{\int_0^t xu(xy)dy}{\int_0^t u(y)dy}$$

and so

$$\lim_{t\to\infty} \frac{U(tx)}{U(t)} = \lim_{t\to\infty} \frac{xu(xt)}{u(t)} = x^{\rho}. \tag{4.7}$$

Since U is non-decreasing, we must have $\rho \geq 0$.

If $U(\infty) < \infty$, then consider the function

$$V(t) = \int_t^\infty u(x)dx$$

which converges to zero. Proceeding as before we find that

$$\lim_{t\to\infty} \frac{V(tx)}{V(t)} = x^\rho \tag{4.8}$$

Since V is non-increasing, we must have $\rho \le 0$.

(ii) If $\rho > 0$ we have shown in (4.7) that U varies regularly with exponent ρ. If $\rho = 0$ and $U(\infty) = \infty$ then (4.7) shows that U varies slowly. If $\rho = 0$ and $U(\infty) < \infty$ then it is trivial that U is slowly varying. Thus for $\rho \ge 0$ we see that U varies regularly with exponent ρ. □

Lemma 4.3 *Let u be as in* (4.4).

(i) *if u varies regularly with exponent $\rho - 1$ ($\rho \ge 0$), then*

$$\lim_{t\to\infty} = \frac{tu(t)}{U(t)} = \rho. \tag{4.9}$$

(ii) *Conversely, if* (4.9) *holds with $\rho > 0$, then u and U vary regularly with exponents $\rho - 1$ and ρ, respectively. If $\rho = 0$, then U varies slowly.*

Proof. (i) Let $u(t) \sim t^{\rho-1}L(t)$ with $\rho \ge 0$ and L slowly varying. Then from Lemma 4.2(ii) we find that $U(t) \sim t^p L_1(t)$ with L_1 slowly varying. We are required to prove that $L(t)/L_1(t) \to -\rho$ as $t \to \infty$. For this purpose let

$$\frac{tu(t)}{U(t)} = \eta(t). \tag{4.10}$$

Then η is slowly varying. For $x > 1$ we obtain

$$\int_t^{tx} \frac{u(s)}{U(s)} ds = \int_t^{tx} \frac{\eta(s)}{s} ds$$

or

$$\log \frac{U(tx)}{U(t)} = \int_1^x \eta(ty)\frac{dy}{y} = \eta(t) \int_1^x \frac{\eta(ty)}{\eta(t)} \cdot \frac{dy}{y}. \tag{4.11}$$

As $t \to \infty$ the integrand in the last integral in (4.11) tends to $1/y$, while the first term in (4.11) tends to $\rho \log x$, and Fatou's theorem implies that $\limsup \eta(t) \le \rho$. Thus η is bounded, so there is a subsequence $t_n \to \infty$ with $\eta(t_n) \to c' \le \rho$. Since η varies slowly, $\eta(t_n y) \to c'$ for every $y > 0$, and by the dominated convergence theorem (4.11) gives $\rho \log x = c' \log x$. Hence $c' = \rho$ for every subsequence $\{t_n\}$. Therefore $\eta(t) \to \rho$ as $t \to \infty$, as required.

(ii) Suppose (4.9) holds with $\rho \geq 0$ and define $\eta(t)$ by (4.10), so that $\eta(t) \to \rho$ as $t \to \infty$. This gives

$$\log \frac{U(tx)}{U(t)} = \int_1^x \eta(ty) \frac{dy}{y} \to \rho \log x$$

or

$$\frac{U(tx)}{U(t)} \to x^\rho \quad \text{as } t \to \infty. \tag{4.12}$$

If $\rho = 0$, U is slowly varying. If $\rho > 0$ then

$$U(t) \sim t^\rho L(t) \quad (t \to \infty)$$

with L slowly varying. From (4.9) it follows that

$$u(t) \sim \rho t^{-1} U(t) \sim \rho t^{\rho - 1} U(t)$$

as required. □

Proof of Theorem 4.2. The first part follows from Lemma 4.3(i). To prove the converse, consider the family of measures $\{V_t\}$, where

$$V_t(x) = \frac{U(tx)}{U(t)}.$$

By Theorem 2.1,

$$V_t(x) \to x^\rho \quad \text{as } t \to \infty.$$

Since $V_t(x)$ has density $tu(tx)/U(t)$ it follows that

$$\int_a^b \frac{tu(tx)}{U(t)} dx \to b^\rho - a^\rho. \tag{4.13}$$

Since U is monotone this implies that $tu(tx)/U(t)$ is bounded. By the selection theorem there exists a sequence $\{t_n\}$, $t_n \to \infty$, such that

$$\frac{t_n u(t_n y)}{U(t_n)} \to \psi(y) \quad \text{as } n \to \infty$$

at each point of continuity y of ψ. Here on account of (4.13)

$$\int_a^b \psi(y)dy = b^\rho - a^\rho,$$

which gives $\psi(y) = \rho y^{\rho-1}$. This limit is the same for all sequence $\{t_n\}$, and therefore

$$\frac{tu(ty)}{U(t)} \to \rho y^{\rho-1} \quad \text{as } t \to \infty \tag{4.14}$$

and

$$\frac{u(t)}{t^{\rho-1}L(t)} = \frac{tu(t)}{U(t)} \cdot \frac{U(t)}{t^\rho L(t)} \to \frac{\rho}{\Gamma(\rho+1)} = \frac{1}{\Gamma(\rho)},$$

which is the desired result. □

Theorem 4.3 *Let u be a non-negative function on $[0, \infty)$, and*

$$\phi(\theta) = \int_0^\infty e^{-\theta x} u(x)dx \tag{4.15}$$

converge for $\theta > 0$. Moreover, let $0 < \rho < \infty$.

(i) *If $u(t) \sim t^{\rho-1}L(t)/\Gamma(\rho)$, then $\phi(\theta) \sim \theta^{-\rho}L(1/\theta)$.*
(ii) *Conversely, if $\phi(\theta) \sim \theta^{-\rho}L(1/\theta)$ and u is monotone, then $u(t) \sim t^{\rho-1}L(t)/\Gamma(\rho)$.*

Proof. In each case, let $U(t) = \int_0^t u(s)ds$, so that u is the density of the measure U.

(i) If $u(t) \sim t^{\rho-1}L(t)/\Gamma(\rho)$, then Lemma 4.3(i) gives $U(t) \sim t^\rho L(t)/\Gamma(\rho+1)$, and Theorem 4.1 gives $\phi(\theta) \sim \theta^{-\rho}L(1/\theta)$.
(ii) If $\phi(\theta) \sim \theta^{-\rho}L(1/\theta)$, then Theorem 4.1 gives $U(t) \sim t^\rho L(t)/\Gamma(\rho+1)$. Since u is monotone, Theorem 4.2 gives $u(t) \sim t^{\rho-1}L(t)/\Gamma(\rho)$. □

Examples

4.1 If $u(t) \to c$ $(0 < c < \infty)$, then $\theta\phi(\theta) \to c$.
4.2 Let $u(x) = 1 - F(x)$ and $\alpha < 1$. Then $1 - F(t) \sim t^{-\alpha}L(t)/\Gamma(1-\alpha)$ iff $1 - F^*(\theta) \sim \theta^\alpha L(1/\theta)$.

Theorem 4.4 *Let $a_n \geq 0$ and $A(z) = \sum_0^\infty a_n z^n$ $(0 \leq z < 1)$. Then*

$$a_0 + a_1 + \cdots + a_n \sim \frac{n^\rho}{\Gamma(\rho+1)} L(n) \quad (n \to \infty) \tag{4.16}$$

with $0 \leq \rho < \infty$ iff

$$A(z) \sim (1-z)^{-\rho} \, L\left(\frac{1}{1-z}\right) \quad (z \to 1-). \tag{4.17}$$

If $a_n \sim n^{\rho-1}L(n)/\Gamma(\rho)$ with $0 < \rho < \infty$, then (4.17) holds. Conversely, if (4.17) holds with $0 < \rho < \infty$ and the sequence $\{a_n\}$ is monotone, then $a_n \sim n^{\rho-1}L(n)/\Gamma(\rho)$.

Proof. These results follow Theorems 4.1 and 4.3 if we consider a measure U with density u defined by

$$u(x) = a_n \quad \text{for } n \leq x < n+1 \quad (n \geq 0). \qquad \square$$

A.5 Problems

1. Show that if L_1 and L_2 are slowly varying functions, so are (i) L_1^α (α finite); (ii) $L_1 L_2$; (iii) $L_1 + L_2$.
2. Prove that a function L varies slowly iff it is of the form

$$L(x) = a(x) \exp\left\{ \int_1^x \frac{b(y)}{y} dy \right\},$$

   where $b(x) \to 0$ and $a(x) \to c < \infty$ as $x \to \infty$.
3. Let L be a slowly varying function. Prove the following:
   (i) The convergence $L(tx)/L(t) \to 1 (t \to \infty)$ is uniform in $a \leq x \leq b$, where $0 < a < b < \infty$.
   (ii) For any $\delta > 0, t^\delta L(t) \to \infty, t^{-\delta} L(t) \to 0$ as $t \to \infty$.
4. If F_1 and F_2 are distributions on $[0, \infty)$ such that as $t \to \infty$

$$1 - F_i(t) \sim t^{-\alpha} L_i(t) \quad (i = 1, 2),$$

   show that for $G = F_1 * F_2$,

$$1 - G(t) \sim t^{-\alpha}[L_1(t) + L_2(t)].$$

5. Let K be the compound Poisson distribution

$$K(x) = \sum_{n=0}^{\infty} e^{-\lambda} \frac{\lambda^n}{n!} F_n(x),$$

where for $n \geq 1, F_n$ is the n-fold convolution of a distribution F with itself, F is concentrated on $[0, \infty)$, and F_0 is degenerate at zero. If as $t \to \infty$,

$$1 - F(t) \sim t^{-\alpha} L(t)$$

with $0 \leq \alpha < \infty$, then show that

$$1 - K(t) \sim \lambda t^{-\alpha} L(t).$$

6. Let F be a distribution on $(-\infty, \infty)$ and $1 - F(t) + F(-t) \sim t^{-\alpha} L(t)$ $(t \to \infty)$. Prove that moments of order $\beta < \alpha$ exist, while moments of order $\beta > \alpha$ do not.

Additional Reading

Bingham, N, Goldie, C and Teugels, J (1987). *Regular Variation. Encyclopedia of Math, and Its Appl. 27.* Cambridge: Cambridge University Press.

Feller, W (1971). *An Introduction to Probability Theory and Its Applications, Volume 2, 2nd Edition.* New York: John Wiley.

Haan, L de (1970). *On Regular Variation and Its Application To the Weak Convergence of Sample Extremes.* Amsterdam: Math. Centrum.

Postnikov, AG (1980). Tauberian Theory and Its Applications. Proc. Steklov; Inst. of Math. Providence: American Math. Society.

Sereta, E (1976). *Regular Varying Functions. Lecture Notes in Math. 508.* Berlin: Springer Verlag.

Appendix B: Some Asymptotic Relations

1. For the standard normal density $n(x)$ and its d.f. $N(x)$ we have

$$1 - N(t) = t^{-1}n(t) + 0(t^{-1}) \quad (t \to \infty). \tag{1}$$

Proof. We have

$$1 - N(t) = \int_t^\infty n(x)dx = \int_t^\infty \frac{x}{t} n(x)dx - \int_t^\infty \left(\frac{x}{t} - 1\right) n(x)dx.$$

Here

$$\int_t^\infty \frac{x}{t} n(x)dx = \left[-\frac{1}{t} n(x)\right]_t^\infty = \frac{1}{t} n(t)$$

and

$$\begin{aligned}\int_t^\infty \left(\frac{x}{t} - 1\right) n(x)dx &< \int_t^\infty \left(\frac{x}{t} - 1 + \frac{3}{x^4}\right) n(x)dx \\ &= \left[\left(-\frac{1}{t} + \frac{1}{x} - \frac{1}{x^3}\right) n(x)\right]_t^\infty = \frac{1}{t^3} n(t).\end{aligned}$$

We can therefore write

$$1 - N(t) = t^{-1}n(t) + R(t),$$

where $|tR(t)| \to 0$ as $t \to \infty$. □

2. For the stable d.f. $G_{1/2}(x)$ with exponent $1/2$ we have

$$1 - G_{1/2}(t) = \sqrt{\frac{2}{\pi t}} + 0\left(t^{-1/2}\right) \quad (t \to \infty). \tag{2}$$

Proof. We have

$$G_{1/2}(t) = 2\left[1 - N\left(\frac{1}{\sqrt{t}}\right)\right]$$

so that

$$1 - G_{1/2}(t) = 2N\left(\frac{1}{\sqrt{t}}\right) - 1 = 2\int_0^{1/\sqrt{t}} n(x)dx.$$

Now for $h > 0$

$$\int_0^h n(x)dx = \int_0^h n(0)dx - \int_0^h [n(0) - n(x)]dx = \frac{h}{\sqrt{2\pi}} - R,$$

where

$$R = \int_0^h [n(0) - n(x)]dx \leq \frac{h}{\sqrt{2\pi}} e^{-1/2h^2}$$

so that $h^{-1}R \to 0$ as $h \to 0+$. This leads to (2). □

3. Let $F(x)$ be the Cauchy d.f. Then

$$1 - F(t) = \frac{t^{-1}}{\pi} + 0(t^{-1}) \quad (t \to \infty). \tag{3}$$

Proof. We have

$$1 - F(t) = \int_t^\infty \frac{1}{\pi} \cdot \frac{dx}{1 + x^2} = \frac{1}{\pi t} - \frac{1}{\pi}\int_t^\infty \left(\frac{1}{x^2} - \frac{1}{1 + x^2}\right) dx$$

$$= \frac{1}{\pi t} - \frac{1}{\pi}\int_t^\infty \frac{dx}{x^2(1 + x^2)}.$$

Here

$$\int_t^\infty \frac{dx}{x^2(1 + x^2)} < \int_t^\infty \frac{dx}{x^4} = \frac{1}{3t^3}.$$

Therefore $1 - F(t) = (\pi t)^{-1} + R(t)$, where $|tR(t)| \to 0$ as $t \to \infty$. □

INDEX